The Auditor's Guide to Blockchain Technology

The 21st century has been host to a number of information systems technologies in the areas of science, automotive, aviation and supply chain, among others. But perhaps one of its most disruptive is blockchain technology whose origin dates to only 2008, when an individual (or perhaps a group of individuals) using the pseudonym Satoshi Nakamoto published a white paper entitled *Bitcoin: A peer-to-peer electronic cash system in an attempt to address the threat of "double-spending" in digital currency.*

Today, many top-notch global organizations are already using or planning to use blockchain technology as a secure, robust and cutting-edge technology to better serve customers. The list includes such well-known corporate entities as JP Morgan, Royal Bank of Canada, Bank of America, IBM and Walmart.

The tamper-proof attributes of blockchain, leading to immutable sets of transaction records, represent a higher quality of evidence for internal and external auditors. Blockchain technology will impact the performance of the audit engagement due to its attributes, as the technology can seamlessly complement traditional auditing techniques. Furthermore, various fraud schemes related to financial reporting, such as the recording of fictitious revenues, could be avoided or at least greatly mitigated. Frauds related to missing, duplicated and identical invoices can also be greatly curtailed.

As a result, the advent of blockchain will enable auditors to reduce substantive testing as inherent and control audit risks will be reduced thereby greatly improving an audit's detection risk. As such, the continuing use and popularity of blockchain will mean that auditors and information systems security professionals will need to deepen their knowledge of this disruptive technology.

If you are looking for a comprehensive study and reference source on blockchain technology, look no further than *The Auditor's Guide to Blockchain Technology: Architecture, Use Cases, Security and Assurance*. This title is a must read for all security and assurance professionals and students looking to become more proficient at auditing this new and disruptive technology.

Shaun Aghili is a full professor of management and lead faculty in the Master of Information Systems Assurance Management (MISAM) program at Concordia University of Edmonton. MISAM is Canada's first and only graduate-level program in information systems auditing.

Shaun holds a total of 15 professional designations. These include certifications as a management accountant, a financial services and government sector internal auditor and risk assurance specialist, a fraud examiner, an information systems auditor and an information systems and cloud security professional. He also holds a number of blockchain-related certifications such as Certified Blockchain Expert, Certified Blockchain Solutions Architect, Certified Blockchain Security Professional and Certified Blockchain Project Manager.

Shaun is the author of over 90 published articles, book chapters and conference proceedings. His latest book is entitled *Fraud Auditing Using CAATT: A Manual for Auditors and Forensic Accountants to Detect Organizational Fraud.* Shaun has also been awarded a certificate of merit for 'outstanding character and excellence in contributing to the literature for the advancement of management accounting and financial management' by the Institute of Management Accountants. His current research interest revolves around the application of blockchain technology to various business uses.

Security, Audit and Leadership Series

Series Editor: Dan Swanson, Dan Swanson and Associates, Ltd., Winnipeg, Manitoba, Canada.

The ***Security, Audit and Leadership Series*** publishes leading-edge books on critical subjects facing security and audit executives as well as business leaders. Key topics addressed include Leadership, Cybersecurity, Security Leadership, Privacy, Strategic Risk Management, Auditing IT, Audit Management and Leadership

Agile Audit Transformation and Beyond
Toby DeRoche

Information System Audit
How to Control the Digital Disruption
Philippe Peret

The Security Hippie
Barak Engel

Finding Your Granite
My Four Cornerstones of Personal Leadership
Douglas P. Pflug

Strong Security Governance through Integration and Automation
A Practical Guide to Building an Integrated GRC Framework for Your Organization
Priti Sikdar

Say What!? Communicate with Tact and Impact
What to Say to Get Results at Any Point in an Audit
Ann M. Butera

Auditing Information and Cyber Security Governance
A Controls-Based Approach
Robert E. Davis

The Security Leader's Communication Playbook
Bridging the Gap between Security and the Business
Jeffrey W. Brown

Modern Management and Leadership
Best Practice Essentials with CISO/CSO Applications
Mark Tarallo

Rising from the Mailroom to the Boardroom
Unique Insights for Governance, Risk, Compliance and Audit Leaders
Bruce Turner

Operational Auditing
Principles and Techniques for a Changing World (Second Edition)
Hernan Murdock

CyRM^{SM}
Mastering the Management of Cybersecurity
David X Martin

The Auditor's Guide to Blockchain Technology

Architecture, Use Cases, Security and Assurance

Shaun Aghili

CRC Press is an imprint of the
Taylor & Francis Group, an informa business

First edition published 2023
by CRC Press
6000 Broken Sound Parkway NW, Suite 300, Boca Raton, FL 33487-2742

and by CRC Press
2 Park Square, Milton Park, Abingdon, Oxon, OX14 4RN

CRC Press is an imprint of Taylor & Francis Group, LLC

ISBN: 978-1-032-07824-3 (hbk)
ISBN: 978-1-032-07825-0 (pbk)
ISBN: 978-1-003-21172-3 (ebk)

DOI: 10.1201/9781003211723

Typeset in Times
by Newgen Publishing UK

Contents

Foreword xix
Alison Yacyshyn

Preface xxi

Chapter 1 Blockchain Technology: Creating Trust in a Trustless Environment 1

Harriet Tenge and Maureen Okello

Blockchain Defined 1
Blockchain History 2
Blockchain Technology versus Traditional Database Solutions 3
The Byzantine Generals Dilemma 3
Blockchain Types 5
Blockchain Components: A Primer 6
Centralized Networks 7
Decentralized Networks 7
Distributed Networks 8
The Block 9
Shared Ledger 9
Chaining Blocks 10
Blockchain Attributes and Benefits 10
Blockchain as a ZTA Tool 12
Conclusions and Recommendations 12
Core Concepts 12
Activity for Better Understanding 14
References 14

Chapter 2 Blockchain Architecture, Components and Considerations 17

Aafreen Fathima Altaf Hussain, Temitope Ipentan, Mahakpreet Singh and Grace Moyo Adeyemi

Blockchain Participants and Roles 17
Blockchain Types 18
Public Blockchains 18
Private Blockchains 18
Open versus Closed Blockchains 19
Consensus 20
Consensus Algorithms 20
A Cryptography Primer 23
Symmetric Cryptography Algorithms 24
Asymmetric Cryptography Algorithms 25
Hybrid Cryptography 25
Homomorphic Cryptography 28
Elliptic Curve Cryptography 28
Zero-Knowledge Proof 28
Hash Functions 30
How Does a Hash Work? 30

Where Are Hash Functions Used? 30
Security Properties of Hash Functions 31
Real-World Hash Applications 31
Common Cryptographic Hash Functions 31
Merkle Trees 33
The Working of Merkle Trees 33
Benefits of Merkle Trees 34
Digital Signatures 34
Digital Signature Generation 34
Digital Signature Verification and Validation 34
Access Control, Identity Management and Membership Service Providers 35
Identity Management 36
Membership Service Providers 37
Crypto Wallets 37
Types of Wallets 37
Cold Wallet or Cold Storage 38
How Are Keys Stored in a Wallet? 38
Inter-Planetary File Systems 39
Smart Contracts 40
Smart Contract Lifetimes 41
How Developers Ensure Users Cannot Exploit Bugs or Unintended Functionality 41
Turing Completeness 41
Untrusted Code 42
Conclusions and Recommendations 42
Core Concepts 42
Activity for Better Understanding 44
References 44

Chapter 3 Blockchain Tokens and Cryptocurrencies 49

Lavanya Vaddi, Jaskaran Singh Chana and Gurjot Singh Kocher

Blockchain Governance 49
Popular Blockchain Governance Strategies 50
Blockchain Forks 50
Why Forks Happen 51
Types of Forks 51
The DAO Hack 52
Tokens in Blockchains 52
Types of Tokens 52
Token Standards 53
Ethereum Gas and EIP 1559 54
Blockchain Token Use Cases 55
Cryptocurrencies 56
Popular Cryptocurrency Types 56
Cryptocurrency Exchanges 58
Summary and Conclusions 59
Core Concepts 59
Activity for Better Understanding 60
References 60

Chapter 4 The Advent of the Triple Entry Accounting: Implications for Accountants and Auditors ... 63

Ibukunoluwa Mabo, Oluwatobiloba Iroko and Oyegbenga Oyenekan

The History of Accounting Entries ... 63
Single Entry ... 63
Double Entry ... 64
Triple-Entry Accounting ... 64
Double-Entry versus Triple-Entry Accounting ... 64
The Effect of Triple-Entry Accounting for Accountants ... 65
The Implications of Blockchain for Auditors ... 65
Financial Fraud Mitigation Using Blockchain Technology ... 66
Financial Statement Fraud ... 66
Skimming ... 67
Cash Larceny ... 67
Register Disbursement Schemes ... 67
Check Tampering ... 67
Billing Schemes ... 68
Payroll Schemes ... 68
Expense Reimbursement Schemes ... 68
Misuse of Inventory ... 68
Other Non-ACFE-Related Blockchain Threats ... 68
Double-Spending Attacks ... 68
Exchange Hacks ... 69
Social Engineering ... 69
Malware ... 69
A Proposed Purchase Cycle Blockchain Audit Checklist ... 69
Fraud Schemes Associated with Purchase Cycle ... 69
Public and Private Sector Triple-Entry Accounting Challenges ... 72
Summary and Conclusions ... 72
Core Concepts ... 73
Activity for Better Understanding ... 74
References ... 74

Chapter 5 Blockchain Use in the Financial Services Sectors ... 77

Antonio Ramirez, Bhanu Theja Satyani, Jovid Ismailov and Lovepreet Singh

Blockchain Use in Financial Services ... 77
Blockchain in the Banking Sector ... 78
Banking Payments Systems ... 78
The Embedded Supervision Principles in Blockchains ... 79
Use of Blockchain on Exchanges ... 80
Blockchain-Based Transaction Clearance and Settlements ... 80
The Investment Management Sector (Mutual Funds) ... 80
Investment Management Use Cases ... 81
Know Your Customer Considerations ... 81
Streamline Compliance and Process Duplication Avoidance ... 82
Blockchain as a Money Laundering Prevention Tool ... 82
Blockchain and Consumer Credit ... 82
Cardholder Identity and Verification Processes ... 83

Microloans Management for Users in Developing Countries 83
A Corda Platform Use Case 84
Blockchain and Trade Financing 85
International Trade Use Cases 85
Blockchain in the Insurance Sector 86
Blockchain Solutions for Various Insurance Use Cases 86
Blockchain and Group Insurance Benefits 88
Summary and Conclusions 88
Core Concepts 89
Activity for Better Understanding 89
References 90

Chapter 6 Blockchain and Supply Chain Management 93

Divya Garg, Siddhanth Karunakar Poojary, Jasjeet Singh Raikhi and Purav Thakkar

Supply Chain Defined 93
Different Types of Supply Chain 94
Supply Chain Components 94
Supply Chain Fraud Risks 95
Counterfeit Goods 95
Fraudulent Billing 95
Misappropriation of Assets 96
Food Fraud 96
Mislabeling – Some Major Food Fraud Cases 97
Blockchain Benefits in Supply Chain Management 98
Walmart's Supply Chain Blockchain 99
Maersk 100
DeBeers Jewelers 100
Summary and Conclusions 100
Core Concepts 101
Activity for Better Understanding 102
References 102

Chapter 7 Ethereum, Hyperledger and Corda: A Side-by-Side Comparison of Capabilities and Constraints for Developing Various Business Case Uses 105

Manroop Kaur and Navjot Lnu

Ethereum 105
Components of Ethereum 106
Smart Contracts 108
The Ethereum Virtual Machine 109
Characteristics of Smart Contracts 109
Limitations of Smart Contracts 109
Consensus Mechanism 109
Limitations of Ethereum 110
Ethereum Use Cases 110
Hyperledger 111
Hyperledger Frameworks 111
Benefits of Using Hyperledger 112

The Design Philosophy of Hyperledger 113
Tools of Hyperledger 113
Hyperledger Smart Contracts 114
The Architecture of Hyperledger 114
Consensus in Hyperledger 115
Limitations of Hyperledger 115
Hyperledger Use Cases 115
Corda 116
Components of the Corda Platform 117
Corda Smart Contracts 119
Corda's Consensus Mechanism 120
Corda Use Cases 120
User Privacy in Corda 121
Limitations of Corda 121
A Side-by-Side Comparison of Ethereum, Hyperledger and Corda 121
Decision Tree to Choose the Right Framework for a Business Use Case 122
Summary and Conclusions 123
Core Concepts 124
Activity for Better Understanding 124
Reflective Questions (Supplementary) 125
References 125

Chapter 8 Designing a Blockchain Application 127

Mihir Mashilkar, Kashish Patel, Divy Patel and Het Raval

Blockchain Design Guiding Principles and Considerations 127
Do We Need a Blockchain Solution in the First Place? 127
Does the Application Need to Be Feature-Heavy or Feature-Light? 128
What's More Important – Collaboration or Security? 129
Design for Security and Privacy 129
Will the App Have Consistency or Specialization? 129
Will Support Be Centralized or Decentralized? 130
The Mechanics of Decentralization in DApp 130
Monolithic or Modular? 131
Personas 131
Persona Example 1 131
Persona Example 2 132
User Stories 132
Prof. Shaun's User Story 132
Dr. Anika's User Story 132
Functional Requirements 132
Technical Requirements and Tasks 132
Popular Application Design Approaches 133
Ancile 133
Ontology-Driven Approach 134
Software Engineering Strategies for DApps 134
The Model-Driven Approach 134
The User Design Approach 134
The Design Process 134
Blockchain Application Design: Good Practices and Considerations 136

Summary and Conclusions 137
Core Concepts 137
Activity for Better Understanding 137
References 138

Chapter 9 Blockchain Application Development 141

Gagandeep Singh, Manpreet Kaur, Shival Kashyab and Sunil Kajla

Fundamental Differences Between Software and Firmware 141
Popular Application Development Frameworks 142
Ethereum 143
Hyperledger 143
Corda 144
Ripple 144
Quorum 144
Layers of Blockchain Application 144
Tools for Blockchain Applications 146
Integrated Development Environment 149
Good Practices for Blockchain Development 149
Summary and Conclusions 151
Core Concepts 151
Activity for Better Understanding 152
References 153

Chapter 10 Testing and Auditing Blockchain Applications 155

Karen Akshatha Franklin, Philip Samuel Panneer Selvam and Samhitha Keshireddy

Why It Is Critical to Test a Blockchain Application 155
Popular Testing Tools for Blockchain 156
Challenges in Testing Blockchain Implementation 156
Phases of Blockchain Testing 157
Testing Models: Functional Testing 157
Blockchain Bug Management Considerations 158
A Four-Step Bug Management Strategy 159
Blockchain Bug Categories 159
Tools for Testing 161
Test Plan Strategy for Blockchain Applications 161
Opportunities to Enhance Testing Strategies 163
Regression Testing 163
Automation Testing 163
Test Management Considerations 164
User Acceptance Testing 164
Blockchain Application Auditing Considerations 165
A Framework for Auditing Blockchain Solutions 165
A Proposed Audit Checklist for Blockchain Audits 165
Summary and Conclusions 165
Core Concepts 167
Activity for Better Understanding 168
Reflective Questions 168
References 168

Chapter 11 Blockchain System Implementation 171

Bharghava Sai Nakkina, Deepthi Gudapati, Naga Venkat Palaparthy and Sai Sreenath Sadupally

Introduction 171
Disaster Recovery 171
Elements of an Effective Disaster Recovery Plan 172
The Use of Blockchain Technology in Disaster Recovery 173
Verification of Data Integrity 173
Availability 174
Memory Correction 174
Cloud Storage 174
Contract Management 174
Fraud in Contract Management 174
Blockchain as a Tool in Contract Management 175
Product Distribution/Monetization 178
Supply Chain Management 178
Blockchain's Value in Today's Supply Chains 179
Asset Management Using Blockchain 179
Blockchain as a Solution in Asset Management 179
Use of Blockchain for Data Control, Security, Legal Compliance and Assurance 180
Data Control 180
Security 180
Legal Compliance 181
Assurance 181
Blockchain Implementation Challenges 181
Using a COBIT 2019 Approach for Blockchain Implementation 182
A Three-Layer Implementation Approach 183
The COBIT 2019 Implementation Approach 183
Phase 1: What Are the Drivers? 183
Phase 2: Where Are We Now? 184
Phase 3: Where Do We Want to Be? 184
Phase 4: What Needs to Get Done? 185
Phase 5: How Do We Get There? 193
Phase 6: Did We Get There? 193
Phase 7: How Do We Keep the Momentum Going? 193
Summary and Conclusions 193
Core Concepts 201
Activity for Better Understanding 202
References 202

Chapter 12 Blockchain Risk, Governance Compliance, Assessment and Mitigation 205

Bikramjit Pandher, Manoj Kumar Nagavamshi, Poojaben Prajapati and Vijay Kundru

Blockchain Technology Risks 205
Ledger Transparency Risks 206
Blockchain Security Risks 206
Operational Risks 207

Blockchain Application Development Risks 208
Cryptocurrency and Payment Considerations 209
Blockchain Regulatory Compliance Risks 209
Regulatory Compliance Considerations 209
Payment Card Industry Data Security Standards 209
Blockchain Considerations Related to Payment Card Processing 210
Health Insurance Portability and Accountability Act 210
Blockchain Considerations Related to HIPAA 211
General Data Protection Regulation 211
The Personal Information Protection and Electronic Documents Act 213
The COSO Framework 213
A Proposed List of Blockchain Good Practices Based on COSO 214
Conclusions and Recommendations 219
Exercise for Better Understanding 220
References 220

Chapter 13 Blockchain User, Network and System-Level Attacks and Mitigation 223

Nishtha Baria, Dharmil Parmar and Vidhi Panchal

Introduction 223
Blockchain User, Network and System-Level Attacks and Mitigation 224
User-Level Attacks 224
Stolen Private Key 224
Mitigation 224
Malware 224
ElectroRAT 225
PCASTLE 225
Lemon Duck 226
Ransomware 226
Mitigation 226
Implementing System Updates 227
Node-Level Attacks 227
Mitigation 228
Blockchain Network Attacks 229
Flawed Network Design 229
Poor Overall Network Security 229
51% and Double Spending Attacks 229
DoS Attacks 231
Eclipse Attacks 231
Replay Attacks 232
Routing Attacks 232
Sybil Attack 233
Blockchain System-Level Attacks 233
Integer/Buffer Overflow Attacks 233
Mitigation 234
Time Stamp Attacks 234
Mitigation 235
Buffer out of Bounds Attacks 235
Mitigation 235

Race Condition Attacks 235
Mitigation 236
Blockchain Security Best Practices 236
Inherent Security Measures 237
Ethereum 238
Ethereum Smart Contracts 238
Ethereum Security Measures 238
Hyperledger Fabric 238
Hyperledger Smart Contracts 239
Hyperledger Security Measures 239
Corda 239
Corda Smart Contracts 240
Corda Security Measures 240
Summary and Conclusions 240
Core Concepts 241
Activity for Better Understanding 241
References 242

Chapter 14 Smart Contract Vulnerabilities, Attacks and Auditing Considerations 245

Maheswar Sharma, Keerthana Kasthuri, Parvinder Singh and Nynisha Akula

Smart Contracts Security Considerations 245
Reentrancy Attack 246
Real-Life Case Scenario 246
Access Control 246
Real-Life Case Scenario 247
Arithmetic Over/Underflows 247
Real-Life Case Scenario 248
Unchecked Return Values 248
Real-Life Case Scenario 248
DoS Attacks 249
Real-Life Case Scenario 250
Bad Randomness 250
Real-Life Case Scenario 250
Race Condition 251
Real-Life Case Scenario 252
Short-Address Attack 252
Timestamp Dependency 253
Real-Life Case Scenario 253
Smart Contract Auditing Considerations 254
Smart Contract Auditing 254
Control Flow Analysis Tools 254
McCabe IQ 254
Ethereum Virtual Machine 254
Taint Analysis Tools 254
TAJ 254
DYTAN 255
Dynamic Code Analysis Tools 256
MAIAN 256
ContractLarva 256

Vulnerability-Based Scanning Tools 256
Mythril 256
Securify 256
SmartCheck 257
Symbolic Execution Tools 257
DART 257
Manticore 257
Oyente 257
Summary and Conclusions 258
Core Concepts 258
Activity for Better Understanding 258
References 259

Chapter 15 Blockchain-as-a-Service 263

Ramya Bomidi, Srija Guntupalli, Sanober Mohammed and Bhargav Putturu Theja

Introduction 263
A History of Cloud Computing 264
Cloud Computing at a Glance 264
Cloud Computing Attributes 264
Cloud Deployment Types 264
Cloud Computing Service Models 265
Cloud Computing Roles and Responsibilities 266
Benefits of Cloud Computing 267
What Is BaaS? 267
How Does the BaaS Model Work? 268
Advantages of BaaS 268
BaaS Challenges 268
Current and Anticipated Future Interest in BaaS 269
How Is BaaS Different from Serverless Computing? 270
BaaS Server-Side Capabilities 270
How Is BaaS Different from PaaS? 270
How Do BaaS Applications Run? 270
Consensus Mechanism 272
Criteria for Selecting a Blockchain as a Service Partner 273
BaaS Platforms 274
IBM BaaS 274
Oracle 274
Microsoft Azure BaaS 274
Amazon AWS BaaS 275
Alibaba 276
Accenture 276
Baidu 276
Huawei 276
SAP BaaS 277
BaaS Business Use Case 277
Food Traceability with Amazon Managed Blockchain: Nestlé 277
Royalties Information for Publishers with Azure Blockchain Service 279
Global Shipping Business Network Oracle Blockchain-as-a-Service: CargoSmart 279

Proposed BaaS Platforms 280
Functional Blockchain-as-a-Service 280
NutBaaS 280
Full-Spectrum Blockchain-as-a-Service 281
Novel Blockchain-as-a-Service 281
Unified Blockchain-as-a-Service 281
Public Blockchain-as-a-Service 281
BaaS Governance 282
Permissioned Chains 282
Off-Chain Control 282
Summary and Conclusions 282
Core Concepts 283
Activity for Better Understanding 284
References 284

Index 287

Foreword

In a world where businesses rely heavily on information technology (IT) and data, there is no secret that the role of information security and assurance professionals increases every day. Concordia University of Edmonton's (CUE) Master in Information Systems Assurance Management (MISAM) program focuses on the business aspects of IT, namely financial and technology-related audits. The program follows the Information Systems Audit and Control Association model curriculum, and consequently CUE MISAM graduates have excellent skills in both technology and accounting/finance. It is also common for MISAM students to pursue Certified Information Systems Auditor, Certified Information Security Manager or Certified Information Systems Security Professional certifications while completing their studies at CUE. In the MISAM program, graduate students learn while doing, and this book is a product of that educational environment.

The book is result of CUE professor, Dr. Shaun Aghili's, innovative teaching at CUE in the MISAM program. Shaun provides his personal motivation for this compilation in 'The Story Behind this Work.' Having 48 graduate students work on projects that not only provide background to the topic of blockchain, but in more applied settings, is noteworthy. The breadth of topics presented in the book allow readers to learn about various areas, which will appeal to both novices and experts in the field. Shaun relied on students to work on their research topics as they completed their studies. He also received copy-editing support from his graduate assistants, Mr. Aman Dev Singh and Ms. Yetunde Akinsanmi. The book is a team effort and I commend Shaun and all those who contributed in one way or another to the book. Knowledge is meant to be discussed and shared; this book on blockchain is notable for sharing classroom discussions to a wider audience beyond our university campus in Edmonton, Alberta, Canada. By reading this book, may your understanding of blockchain develop and strengthen.

Alison Yacyshyn, PhD
Dean, Mihalcheon School of Management
Associate Professor, Faculty of Management

Preface

THE STORY BEHIND THIS WORK

Circa 2018, I studied a few articles about the distributed, decentralized, peer-to-peer network facilitating transactions for Bitcoin and other cryptocurrencies. While at that point, I was somewhat skeptical about the significant increase in value and future sustainability of Bitcoin (and the flurry of other cryptocurrencies making their debuts at that time), the network infrastructure supporting Bitcoin quickly became an area of interest for me. This network structure, called blockchain, was being examined by a number of researchers and organizations as a potential gamechanger due to its immutable, cryptographically secure and decentralized nature. As a matter of fact, a few commentators were predicting that within the next decade or so, the incorporation of blockchain technology as a consumer-friendly, transparent and decentralized tool was likely to affect the field of information technology in the same way as the popularization of the Internet in the 1990s.

The casual study of blockchain technology led me to the decision to further deepen my understanding of it. I spent the next year taking online blockchain-related courses and studying for several brand new blockchain certification exams dealing with blockchain fundamentals, use cases, architecture and security considerations. This undertaking enabled me to gain a much better understanding of the current state of this technology, its limitations and issues and its potential use cases in supply chain management, decentralized finance, accounting and auditing, as well as its use in government operations, just to name a few.

In January 2021, I put together a research team comprised of second-year graduate students from our master programs in information system security and assurance who were at the time enrolled in their research methodology courses. For our final group research project, I came up with the initial draft of a table of contents for a book on auditing blockchain technologies; a topic that had not received much attention at all until that time. Next, I divided my research students (53 of them!) into small teams of two to four and assigned a chapter topic to each group. For the next two academic semesters (a total of 26 weeks), each team researched and wrote their assigned chapter under my close coaching and supervision. At about the same time, while my students were busy reading, understanding and analyzing reference material pertinent to their assigned chapters, I started to create a comprehensive book proposal around our new research project. During my book proposal research, I quickly found that while there was a large amount of blockchain-related material on the Internet, there were relatively fewer research papers and published books focusing specifically on blockchain technology, despite having found a multitude of books that focused on various cryptocurrencies, especially Bitcoin. My proposal for a book on blockchain auditing was subsequently submitted to CRC Press for review, and much to my delight, I received the green light from CRC Press a few months later!

The book that you are currently reading is the result of this effort. As you study the various chapters of this book, you will come across the names of the authors of each chapter; a total of 53 bright and highly motivated graduate students from our information systems security and assurance programs at Concordia University of Edmonton in Alberta, Canada.

WHO SHOULD READ THIS BOOK?

The idea behind this work is to create a comprehensive study and reference guide for IT assurance specialists who have little knowledge of the emerging technology. As such, this book may be of great interest to a number of readers as follows:

- IT and business readers interested in gaining a solid understanding of blockchains' types, components, challenges and potential use cases;
- Undergraduate and graduate students interested in understanding blockchain technology for research purposes;
- Various business leaders interested in gaining an in-depth understanding of the various aspects of this emerging technology for evaluation and/or implementation purposes;
- Undergraduate and/or graduate-level IT faculty looking to adopt a comprehensive textbook title for courses focusing on blockchain technology;
- Accountants and external auditors looking to deepen their understanding of the technology and its potential use cases in business, accounting and auditing;
- Internal and information systems auditors looking to gain competency in blockchain;
- Individuals looking for a concise, yet comprehensive, reference book on blockchain;
- Individuals looking to efficiently learn all the essentials of blockchain technology as a first step to entering the lucrative blockchain-related job market.

HOW DOES THIS BOOK BENEFIT AUDITING PROFESSIONALS?

As the title clearly suggests, this work was primarily designed for assurance professionals. It not only lays out a solid foundation around auditors' understanding of the mechanics and major components of blockchain technology, but also discusses the current state challenges of this emerging and potentially disruptive technology, such as its risks, vulnerabilities and implementation barriers. Furthermore, the book also discusses a number of potential use cases in sectors such as supply chain management, financial services and accounting/auditing. There are also many other potential use cases, especially in government, that are being investigated currently. However, as the research project advisor and supervisor, I decided to limit our use case studies to the fields and sectors that I have had prior non-academic work experience. As such, it is my sincere belief that both external and internal auditors can greatly expand their blockchain knowledge base and incorporate greater effectiveness into their audit plan that involves this rapidly growing technology.

HOW TO READ THIS BOOK

This work is designed to be both a comprehensive read on blockchain technology, as well as a valuable reference tool for IT and assurance professionals. While this book can be read or studied with a great deal of flexibility, it is important for all readers to start by studying the first three chapters of this work in the order they appear. IT professionals with a great deal of knowledge and experience in networking and cybersecurity will perceive at least some of the concepts discussed in these first three chapters as review. However, as I always remind my information systems security and assurance students, *the first condition to become an expert in a field is to solidly master its fundamentals. Once you develop a solid foundation, adding to it will not be an overly difficult task!*

Once the first three chapters are studied, readers have the option to pick and choose which remaining chapters are of the most interest and use to them. A quick look at this work's table of content should signal to readers that in addition to covering the fundamentals of blockchain, this work also discusses the architecture, security aspects, several potential use cases and implementations considerations related to the technology, in addition to several blockchain-specific auditing tools and checklists.

HOW IS THIS BOOK ORGANIZED?

Each and every chapter in this work follows a similar, overall organizational structure as follows:

- The names of the authors/researchers for each chapter are disclosed at the beginning of their assigned chapter. These individuals were in charge of searching, compiling and analyzing dozens of references during the first part of the research project for each chapter. In the latter part of the research project, the team was responsible for writing the content of the chapter with editing help from me and copy-editing support from my two trusted graduate assistants, Mr. Aman Dev Singh and Ms. Yetunde Akinsanmi. Once the Spring 2021 semester was completed at Concordia University of Edmonton, I spent several months doing a final deep edit of each chapter, before the manuscript was finally ready to go to the CRC editors with some help this time from my wife, Susan.
- Every chapter begins with a brief *Abstract*. Similar to their use in research papers, the abstract enables readers to quickly get an idea in terms of the content and scope of every chapter before delving into it. The rationale behind this is to allow readers the option of skipping some chapters that may not be of interest to them. For example, it is possible that some readers may wish to skip over the more technical chapters, while more technically oriented readers may want to focus primarily on such chapters and skip over chapters focusing on some use cases. As mentioned before, while Chapters 1–3 should be read by all readers, the remaining chapters can be studied in any order, and some chapters may even be skipped by some readers.
- While all chapters always begin with an abstract section, every chapter also concludes with two sections entitled *Summary and Conclusions*, followed by a list of *Core Concepts*. The rationale for including these two sections is two-fold: (a) it enables readers with adequate knowledge, vis-à-vis the chapter content, to forgo reading the entire chapter and instead quickly review some of its core concepts; and (b) these sections can help readers to quickly review and remember some of the more important chapter concepts, as needed.
- Finally, the last section in every chapter is a section entitled *Activity for Better Understanding*. This final section includes several reflective questions aimed at enabling IT and business professionals to start thinking about potential blockchain implementations for their own organization. These reflective questions may also be used by IT and business faculty, as seed for developing test questions or class discussion points. This section also includes one or more learning activities for readers interested in digging a bit deeper into the concepts addressed in the chapter. These learning activities can also be used by IT and business faculty for the creation of short labs aimed at deepening students' understanding of the chapter content.

In conclusion, blockchain technology is a rapidly emerging technology with financially lucrative potential for those willing to take the time to gain expertise in various aspects of it, such as blockchain governance, use-case implementation, security, assurance or application development. Let this book be the first step toward the achievement of your blockchain-related objectives and ambitions.

I wish you all the very best in your blockchain technology careers…

Shaun Aghili
Winter 2022

1 Blockchain Technology
Creating Trust in a Trustless Environment

Harriet Tenge and Maureen Okello

ABSTRACT

This chapter is the first of three aimed at familiarizing readers with blockchain technology's fundamental concepts. Chapter 1 introduces the reader to the foundational concepts of blockchain, such as the Byzantine Generals Dilemma, its history, its basic components, as well as its basic types and attributes. Furthermore, this chapter also discusses how blockchain technology differs from database technology and how blockchain may be a suitable complement to the current and increasingly popular zero-trust architecture.

Topics discussed in this chapter include:

- Blockchain history;
- Blockchain database vs traditional database;
- CRUD vs read and write functions in blockchain;
- The Byzantine Generals Dilemma;
- Blockchain types (public, private, hybrid, consortium);
- Blockchain components (network, blocks, shared ledger and chaining blocks);
- Blockchain attributes and benefits;
- Blockchain as a ZTA tool;
- ZTA and blockchain best practices.

BLOCKCHAIN DEFINED

Blockchain may be defined as a tamper-evident and tamper-resistant, peer-to-peer distributed digital network and decentralized shared ledger. Blockchains are tamper-evident because transactions recorded in its ledgers cannot be changed without easy detection. Tamper-resistant, on the other hand, means that the technology is extremely robust as it is built around cryptographic keys, thereby making it impossible for the blocks to be modified. Users are at liberty to record transactions and share digital files with each other, thereby eliminating the need for an intermediary, such as a bank, or an escrow or a regulatory entity. Blockchain networks are heavily dependent on cryptographic algorithms using public-key and private-key encryptions [42]. A blockchain's decentralized, peer-to-peer network structure helps prevent any single point(s) of failure, as the network can continue to operate efficiently even if a vast number of its nodes fail to work due to some type of network failure or attack [35]. Chapter 2 will cover the various components of the above-mentioned attributes in more detail.

DOI: 10.1201/9781003211723-1

Despite its infancy, blockchain technology has already gone through a number of improvements and transformations since its debut in 2009 (the start of the Bitcoin network). The genesis of the technology is discussed below [28].

BLOCKCHAIN HISTORY

Before blockchain technology gained prominence in 2009, an individual (or perhaps a group of individuals) using the pseudonym *Satoshi Nakamoto* published a whitepaper entitled '*Bitcoin: A peer-to-peer electronic cash system*'. The paper proposed the use of electronic cash as a means of payment in a peer-to-peer network addressing the threat of double spending, and further explained how blockchain technology employs the functionalities of timestamping, hashing and group consensus to accommodate an electronic cash system without the need for a central intermediary [28].

According to Rose (2015; entitled *The Evolution of Digital Currencies: Bitcoin, A Cryptocurrency Causing A Monetary Revolution*) the Bitcoin network was launched by Satoshi Nakamoto in 2009 who later disappeared from the public forum. The first Bitcoin commercial transaction took place on May 22, 2010, at a time when Bitcoin was about a year old. *Laszol Hanyecz,* a programmer living in Florida bought two pepperoni pizzas for the sum of 10,000 BTC, the equivalent of USD $40 at that time [32].

In 2013, after the Bitcoin genesis block was mined, Ethereum founder, *Vitalik Buterin*, a Russian-Canadian programmer presented a paper headlined '*A Next-Generation Smart Contract and Decentralized Application Platform*' introducing the *Ethereum* blockchain, a platform with smart contract capabilities [8].

In 2014, the Republic of Estonia, was the first country to launch e-residency services for its citizenry using Estonian *eID*; the first platform to enable individuals from any nation in the worlds to become 'digital residents' of another nation [21].

The next year (2015), the *Linux Foundation Project* unveiled its *Hyperledger Fabric*, a private, permissioned blockchain framework [26]. In same year, the Ethereum Blockchain came into existence and the first version was dubbed '*Frontier*'. The technology has since been upgraded with new versions released over time such as *Homestead, Metropolis Byzantium* and *Ether 2.0* [5].

Blockchain popularity was further enhanced in 2016 when *Walmart* in collaboration with *IBM* applied the technology to help better monitor its food supply chain. The Walmart Food Safety Collaboration Center in China had also reported at that time an intention to intensify research into food security globally with a commitment project fee of about $25 million spanning five years. This system will track food products such as pork from China, grain exports from Brazil and grapes shipped from Africa, in addition to mango and egg sales and supplies in Walmart US stores using a system of RFID tags and LoT sensors [17].

In addition to Estonia, a number of governments are currently striving to also incorporate the benefits of blockchain into their administrative processes. Tamil Nadu, a state in India, was the first to declare its plan to roll out to its citizens e-governance services using blockchain technology. The project is dubbed *Tamil Nadu Blockchain Policy 2020*, and the plan will utilize the state's family database for its development [9]. The United Arab Emirates (UAE) government planned to go live by the year 2021, leveraging blockchain technology to transition about half of the government services. The UAE government is aiming to save 11 billion United Arab Emirates Dirham (AED) in transactions and document costs, as well as 398 billion AED in print costs and over 77 million labor hours annually through a number of initiatives as follows [38]:

Smart Dubai – Transforming Dubai city digitally and enhancing the living experiences of citizens and visitors;

Dubai Health – Improving healthcare and providing a safe environment for the people;

TABLE 1.1
Blockchain versus Traditional Databases

Domain	Blockchain Technology	Traditional Database	Informative References
CRUD functions	Allows only two function: reads and writes of transactions.	Allows users to perform the four CRUD function; create, read, update and delete.	[31, 37]
Architecture	Peer-to-peer, decentralized architecture. Also may rely on Interplanetary File System (IPFS): a peer-to-peer distributed ledger that contains a hash table and the BitTorrent protocol.	Uses client–server-based architecture.	[4, 37]
Data Integrity	Offers great data integrity due to its tamper-resistant cryptographic attributes	Data may be erased and/or changed due to error or as part of a malicious act.	[37]
Authority	Decentralized system lacking central authority.	Centralized system managed by a selected administrator.	[37]
Speed and Performance	Slower as compared to database technology since it requires signature verification, consensus mechanism and is redundant in nature.	Faster and can process many transactions at a given time. Also offers great scalability.	[37]
Cost	Generally more expensive to implement and harder to maintain.	An established technology that is less challenging to implement and maintain.	[37]

Ajman – Building a cohesive society possessing a positive spirit and attitude through the expansion of a green economy designed to achieve prosperity for its citizens;

Sharjah – Promoting Sharjah (a city in UAE), as a tourism hub using innovative tourism strategies.

BLOCKCHAIN TECHNOLOGY VERSUS TRADITIONAL DATABASE SOLUTIONS

Traditional databases use a centralized server that allows users to change the stored data. The management of the database is under the control of an administrator with the power to grant users the authority to access data. A weakness in the database system is that data may be manipulated in case the credentials of the network administrator are compromised [14]. On the other hand, as mentioned before, blockchains offer a decentralized network structure that may be set up without the need for a central authority or administrator, as is the case with most public blockchains. Table 1.1 summarizes some key features and differences between blockchains versus traditional database systems.

In addition to the various considerations outlined in Table 1.1, the following decision tree reproduced from the National Institute of Standards and Technology (NIST) document NISTIR 8202 is also a useful tool for enterprises to help determine whether a blockchain implementation may be a good use case for various entities [42] (Figure 1.1).

THE BYZANTINE GENERALS DILEMMA

The Byzantine Generals Dilemma (BZD) is a concept that was first discussed by Lamport, Shostak and Pease in a 1982 research paper. BZD denotes a logistical dilemma that revolves around overcoming communications challenges in a trustless environment in an attempt to achieve a common objective. Solving the Byzantine Generals Dilemma revolves around the very same principles upon which blockchain technology was based on; namely, striving to maintain secure communication via

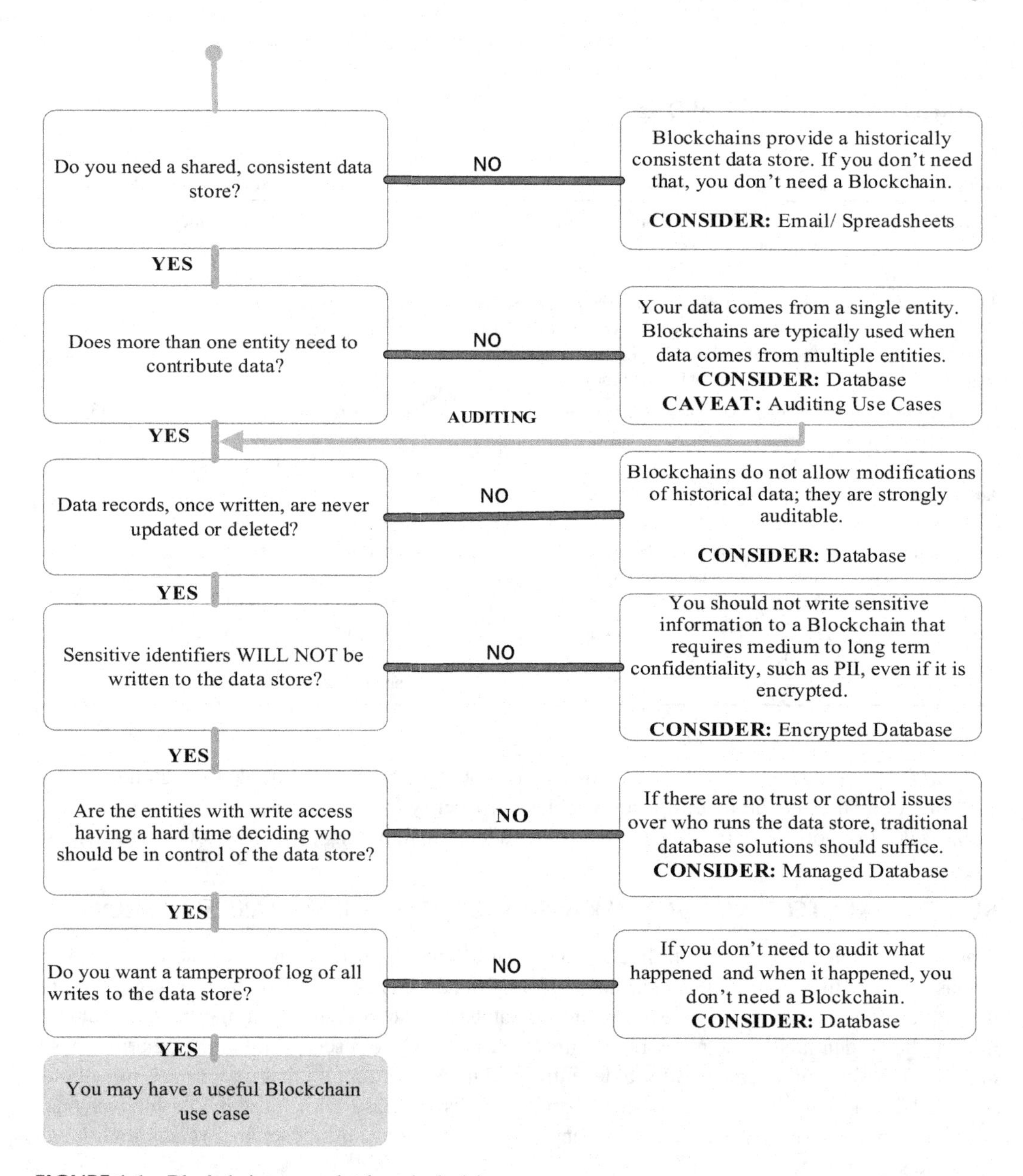

FIGURE 1.1 Blockchain versus database is decision tree.

Source: [42].

the use of cryptography, and achieving consensus among various involved groups in order to achieve a common goal.

To elaborate further, the BZD refers to a hypothetical situation whereby several divisions of the Byzantine army are camped outside a besieged city ready to attack. Before attacking, the generals must all have agreed on a strategy through mutual consent. For this consensus to happen, the generals must communicate with one another. Each general must send plans and instructions to the other generals via messengers in a secure manner; a challenging task in a trustless environment, as messages carried by these messengers may be intercepted and captured by spies loyal to the defending army in the besieged city. An additional concern in this trustless environment is that it is

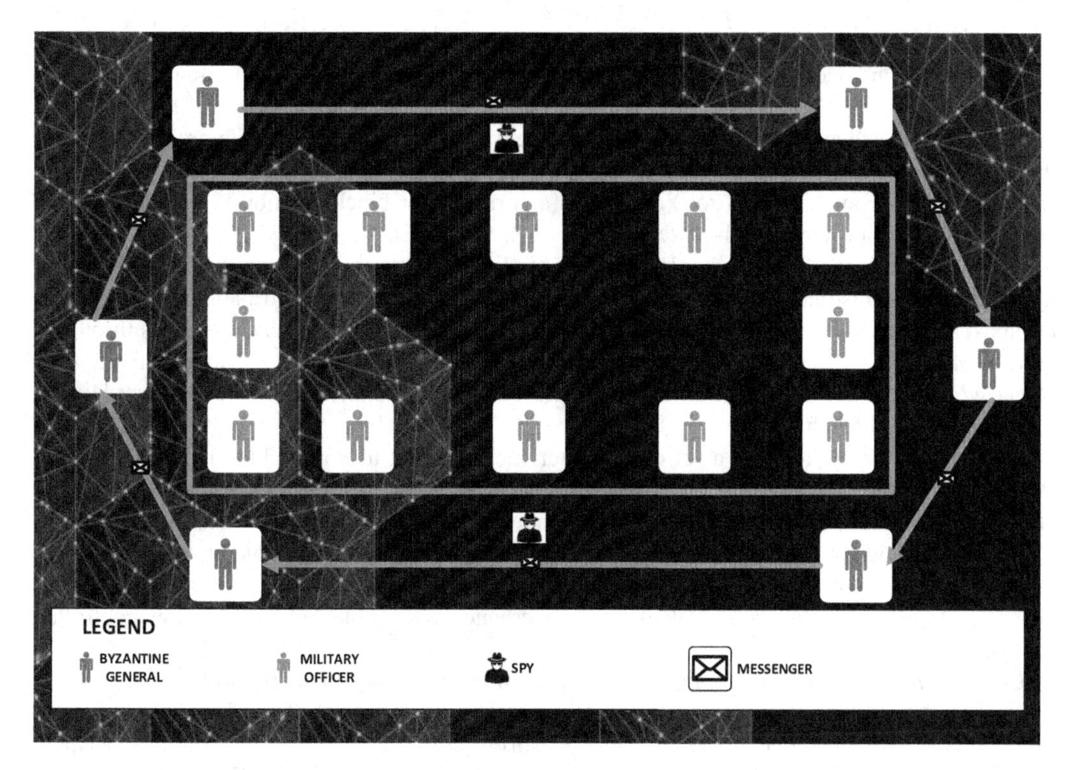

FIGURE 1.2 Byzantine generals dilemma illustrated.

suspected that at least one of the Byzantine generals leading the attack may be a traitor and thus may not execute the agreed-upon attacks in the agreed-upon manner [3].

In order to carry out a successful attack against the besieged city it is imperative that every Byzantine general consents to a common attack strategy through the implementation of a secure communication protocol among the Byzantine generals. Additionally, given the 'traitor general' problem, the common attack strategy must be carried out by a minimum number of the attacking generals (for example: 67%, or four out of six generals in Figure 1.2 for a successful execution).

Despite its fictional and hypothetical nature, the Byzantine Generals Dilemma is a very relevant concept in trustless computer network environments. In other words, just as is the case with the attacking Byzantine generals, a peer-to-peer network represents a trustless environment where various network nodes must also have secure communication protocols for reaching the needed consensus required to carry out the network's objectives. A related concept born out of the Byzantine Generals Dilemma is a concept known as Byzantine Fault Tolerance (BFT). BFT refers to a condition in which some of the nodes in the network can continue to operate even if other network nodes fail to perform. A BFT system can continue to function even if some of the network nodes fail or act fraudulently [3]. The next section will briefly discuss the different kinds of blockchain.

BLOCKCHAIN TYPES

Blockchain architecture may be broken into four main categories; namely, public, private, hybrid and consortium [2]. The following sections provide the reader with an introduction to these various blockchain types. Public and private blockchains and their various sub-categories (i.e.: open or closed) will be discussed in more detail in Chapter 2.

Public blockchains are permissionless, shared and synchronized ledgers that any user can access, join or use with no access control restrictions. The Bitcoin exchange is a well-known instance of a public blockchain where anyone may view the blockchain and conduct buy or sell transactions on it. Public blockchains typically require the consensus of a vast number of the processing nodes during the course of its block creation [24]. Currently, most public blockchains rely on a proof of work (PoW) or proof of stake (PoS) consensus algorithms for block creation [2]. Block creation consensus algorithms are also discussed in Chapter 2.

Private blockchains, on the other hand, are centralized, permissioned networks controlled often by a central administrator (the sponsoring organization). The permissioned nature of these types of blockchain means that read and write restrictions are placed on the blockchain users through a process of identity management and access and control accommodated by a central authority [2]. *Hyperledger Fabric*, *Hyperledger Sawtooth* and *R3 Corda* are examples of private networks [16], as discussed in detail in Chapter 7.

Hybrid blockchains refer to a set of characteristics inherent to both public and private types. While these hybrid types are typically not open to everyone, they still offer the desirable blockchain attributes of integrity, transparency and security. In a hybrid blockchain application, members are in charge of deciding who can participate, and what transactions may be viewable by the public. This combination of controlled access and freedom provides a hybrid blockchain application with a great deal of customizability. In terms of blockchain governance, hybrid blockchains are often controlled by a single entity [11].

Consortium-type blockchains are a type of semi-private blockchain (hybrid) that are often governed by not a single entity, but a group of companies. These types of blockchain applications often work across a number of like-minded sponsoring organizations pursuing common business objectives [41]. Consortium blockchains represents a compromise between fully open, completely decentralized and centrally controlled blockchain applications.

Key characteristics of public and private blockchains are summarized in Table 1.2.

BLOCKCHAIN COMPONENTS: A PRIMER

A blockchain platform consists of a string of blocks created through a consensus agreement by the mining or validating nodes and securely linked together in a chain-like configuration.

Computer networks may be categorized into the following three configurations. Each network type has its own advantages and applicability [25]. Generally speaking, in most use cases, blockchains are decentralized, distributed, peer-to-peer networks.

TABLE 1.2
Public vs Private Blockchain

Domain	Public Blockchain	Private Blockchain	Informative References
Membership	Open to everyone.	Controlled by an organization. Authorized users need permission to participate.	[24]
Permission to read	Participants can read.	Users can usually read, but sometimes an authorization to read records is required.	[41][24]
Permission to write	Everyone can write on a block.	Administrator has the exclusive right to write on the block.	[40]
User Identity	User identity is anonymous.	Participant identity is known.	[24]
Centralized authority	Decentralized network.	Centralized.	[24, 40]
Consensus Protocol	Permissionless.	Permissioned.	[24, 40]

Centralized Networks

A centralized network is controlled by an authorized central administrator who makes decisions about the entire network. A single server in the network performs the processing functions of the network. The server stores all records in a single location from which the server manages the network and allows other computers to connect and transmit requests. The central server is mainly used for large-scale services such as social media, instantaneous messaging and search engines [25] (Figure 1.3).

Generally, this type of network is of great value if the network needs to function on a central servers' updates and enhancements, without having to update each workstation individually. Central servers are fast, easy to develop and easy to patch in case of a found performance issue or security flaw.

Although a centralized network offers many advantages, this type of network also faces certain constraints. The main server controls the entire network and can sometimes serve as a single point of failure when a component of the server malfunctions. Similarly, centralized networks are also exposed to malicious activity, making the server susceptible to hackers who can access transactions and the credential details of peers. In circumstances where the bandwidth of the server is unable to sustain events due to simultaneous requests from members, the system's performance may also be affected [25].

Decentralized Networks

The decentralized network concept was introduced by Paul Baran in 1964 to help improve the telecommunication infrastructure and to overcome the failures of the central servers prone to cyber threats [23]. The network interconnects individual nodes with different servers, through what is known as a peer-to-peer structured system. Moreover, the structure promotes fault tolerance by allowing an undamaged section of the system to continue operating even if a component within

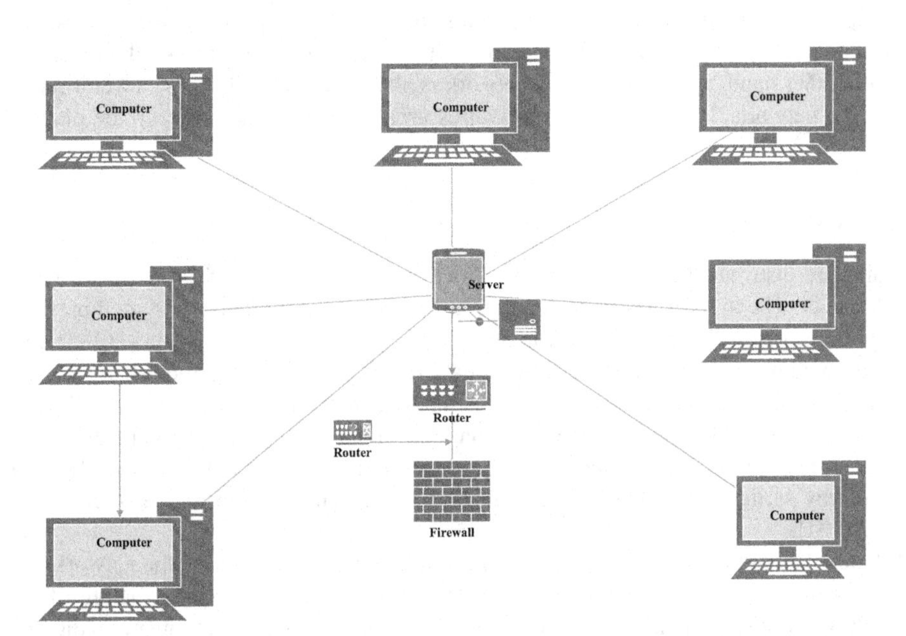

FIGURE 1.3 Centralized networks topology.

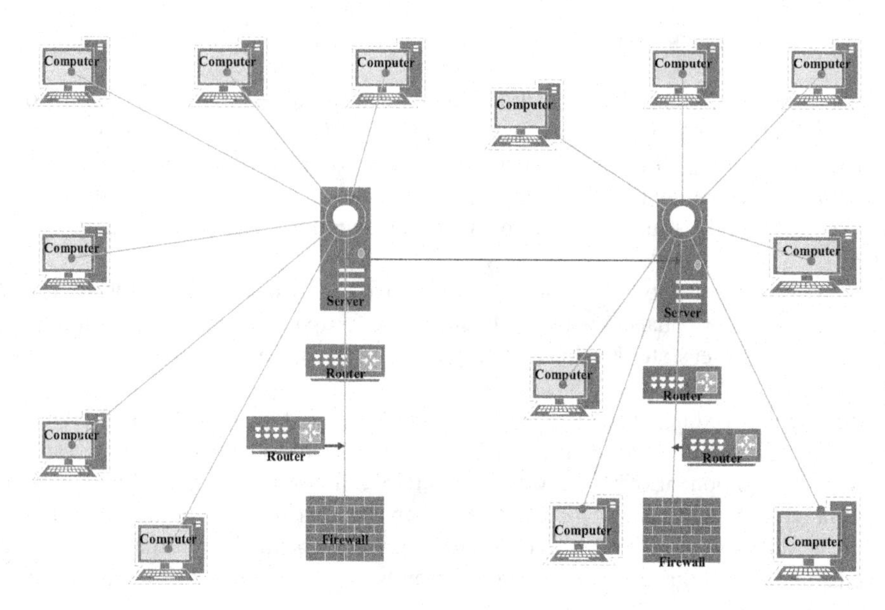

FIGURE 1.4 Decentralized networks topology.

the network malfunctions. Members operating a business in the network are involved in decision-making. This gives members more power and autonomy as compared to the centralized networks (Figure 1.4) [25].

Decentralized networks are considered more reliable and secure because the system is more diversified; thereby allowing information to be stored on independent computers instead of storing data in a central location (higher redundancy). The resilience and secured nature of these democratized platforms makes it difficult for hackers to gain a stronghold in the network. Since network power and control are in the hands of network participants, the system may be considered more trustworthy [25]. On the other hand, the decentralized network is also more costly because several servers participate in the network. Thus, the network requires greater maintenance to keep the platform efficient [1].

Distributed Networks

Blockchains are distributed peer-to-peer networks that have the highest fault tolerance (Byzantine Fault Tolerance). The system has nodes operating within; hence, no node is treated preferentially, as every node can participate in the transaction process and keep track of data [26]. The network is structured with many computers that prevent data from being altered without harmony in nodes. Distributed systems maintain a database that is shared, replicated and synchronized among the peer-to-peer members in a decentralized manner. Additionally, network participants manage and agree by consensus on any updates on transactions stored in the immutable ledgers. The network is administered by participants; therefore, intermediaries or a central authority are often not required (Figure 1.5) [24].

Distributed networks have higher latency in responding to actions because the networks are often spread geographically, thereby creating less feedback time. The tamper-proof nature of the technology makes data stored on the ledger immutable, and data can be shared among peers digitally [10].

One of the challenges of the distributed ledger structure is the difficulty in achieving consensus, as the majority of the network participants must come into an agreement before any decision is made. The use of absolute time to log events depends on the occurrence of those events, but that

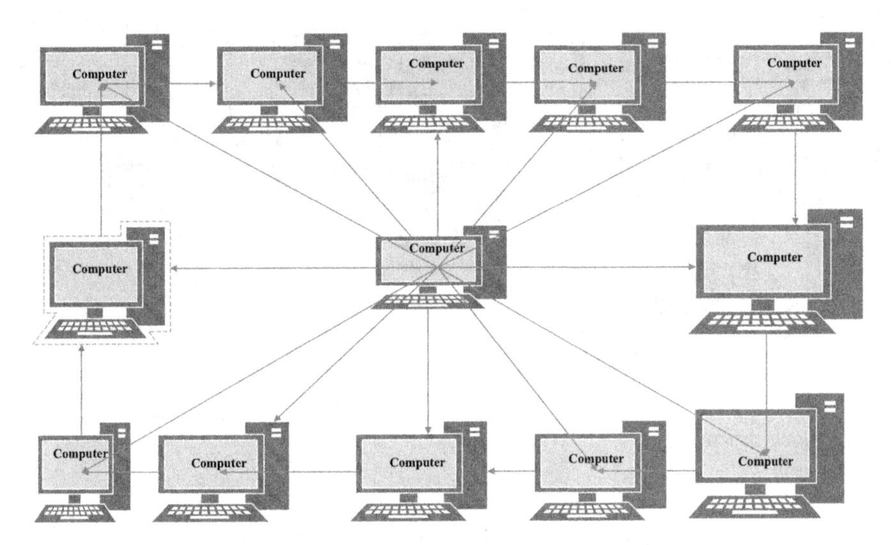

FIGURE 1.5 Distributed networks topology.

seems impossible on a distributed structure. Furthermore, the design and fixing of algorithms on a distributed platform is hard to implement because the nodes lack a common clock; thus, no temporal ordering of commands can take place. The nodes have different latencies which must be kept in mind when designing such a set of rules. There is some difficulty for a node to obtain a global view of the system and hence the system depends on other nodes to make an informed decision. Networked computers which are widely distributed have numerous participants that share resources on the web. The process of computing is also known as grid computing and turns the system into a powerful supercomputer [10].

The characteristics of blockchain are the next topic to be considered.

The Block

A blockchain platform is comprised of a series of transactions securely bundled together in form of blocks. Usually, each block consists of a cryptographic hash of records that serve as a link to the next block except for the genesis block – the root of the blockchain, that contains the information known by all the nodes. Each block structure comprises a block header and a block body [35]. Generally, the header is comprised of six elements as depicted in Figure 1.6.

As can be seen in Figure 1.6, the block header refers to the block metadata which incorporates the timestamp, a hash representation of the block data, the hash of the previous block's header and a nonce. The hash of a previous header is a cryptographic method used in calculating a unique output known as the hash digest (a message digest). A time stamp ensures timeliness and passes through some tests before the blocks can be considered valid. An example of the timestamp is denoted as 2018-02-24 21:27:29. The term nonce is short name for: Number Used Once. Its purpose is to provide a unique random block identifier to help avoid duplicate blocks. The hash of a block is a hash representation of the block data. Finally, the term block data refers to a list of transactions, ledger events and additional data included within the block [39, 42].

Shared Ledger

The blockchain ledger is a collection of transactions that enable a community of peers to share transaction records without the need for a central authority. As mentioned before, data on the blockchain

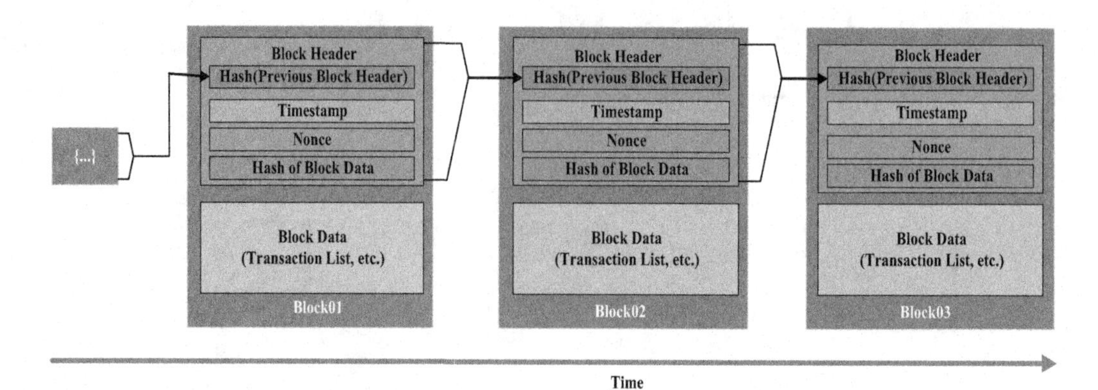

FIGURE 1.6 Block structure.

Source: [42].

ledger are immutable and do not allow transaction records to be changed once published. In the case of a public blockchain use case, the transactions stored in the ledger are publicly available.

Chaining Blocks

A chaining block is linked together with a hash digest of previous block headers to form the blockchain. In a situation where any information related to a previous block is changed by a malicious agent, the hash of all subsequent blocks will also be changed, resulting in a state that will be easily detectable on the blockchain. Chaining blocks are made of transaction lists, a timestamp, nonce, hash of the previous block and the block header that helps to identify and deny any changed block [42].

BLOCKCHAIN ATTRIBUTES AND BENEFITS

Blockchains possess important attributes and benefits as follows: *Consensus* enables blockchains to function in a trustless environment through the consent of each participant to reach an agreement [24]. There are several consensus mechanisms that are being developed, with the two major ones used in a distributed network to reach a consensus agreement being PoW (first use was with Bitcoin), and PoS – PoW's most common replacement requiring much lower energy consumption. Proof of stake (PoS) is a type of consensus model where the nodes validate the blocks to earn transaction fees (also see Chapter 2) [29]. The existence of multi-signature in the consensus process confirms most peers' agreement to a transaction as valid. The Practical Byzantine Fault Tolerance (PBFT) is an algorithm that is proven to help settle disputes among computing nodes when some of the nodes fail to produce the same outcome as the rest of the nodes in the network. *Provenance* allows participants to know the origins and ownership changes that have occurred to the data every single time [15]. The idea originated from the art world and provided written evidence to prove that the artwork had not been altered, forged, copied or stolen [6]. Provenance enables users to track the data source of any transaction conducted within the digital ledger [24]. *Immutability* or *Finality* refers to the concept that data stored in blockchains cannot be modified or expunged unless many of the nodes agree to the changes [6]. The immutability nature of the blockchains secures and protects recorded transactions from being tampered with by participants. In situations where a transaction error occurs, a new transaction must be used to reverse that error. As such, both transactions should be visible [15].

In addition to the attributes discussed, blockchains also offer the following benefits: Blockchain platforms enable users to stay *anonymous* when transacting on the network. Maintaining anonymity means the user's personal information is disguised, or undisclosed [35]. Participants do not know the actual names and addresses of members on the network. Each blockchain network user (operator) is identifiable to a system-generated address [24]. In the fiat currency world, the banking system with the support of the central banks' approved agencies grants permission before payments are made. Contrary to this conventional model, intermediaries are not required on public blockchains due to their decentralized nature [43]. Trust is distributed among users by sharing and storing data on the nodes based on the consensus set rules for ensured security [20]. Blockchains also offer great *transparency* as blockchain applications employ a timestamp indicating when the time blocks were created, registered and verified [13]. Transparency occurs when each node keeps a copy of the transaction record [36]. Anyone on the network can view the stored transaction at any point in time publicly [39]. The assistance of a centralized authority is not needed to improve the transparency of the system [36].

As an effective means to function in a trustless environment, blockchain network participants do not need to know or trust each other when conducting business [42]. Nodes added to the blockchain platform can harmonize and authenticate the entire content of the blockchain in a distributed fashion without a central authority [35]. Trustless can further be explained as a system that is not dependent on an intermediary or agent to function [20]. As discussed in future chapters, blockchain supply chain platforms can provide highly effective quality control through the tracing and tracking of food products in the supply chain. For instance, Walmart in collaboration with IBM piloted food tracking services using an e-certificate connected to product packages through a QR code to manage the food storage and supply chain network. The software supports the procurement supervisors in tracking expired food and storage temperature by viewing detailed information based on the origin of the farm, shipping information, store temperature, batch codes, quality of soil, fertilizer type and a host of others remotely. The system can easily detect and eliminate any food defect or contaminated products while saving cost on product-calls [17]. Blockchain-based supply chain tracking systems also reduce transaction costs related to cross-border transactions use cases [37]. Chapter 6 discusses the use of blockchain for supply chain management in detail. Blockchain systems can accommodate faster funds clearing and settlement. The digitalized network also tends to be more timely in settling transactional disputes and reconciliations. To further explain, transactions processed on the blockchain are verified swiftly; from within a few seconds to some minutes [35]. Due to its swiftness, blockchain is used in digital assets, remittance and online payments [43]. Chapter 5 discusses the blockchain-based transaction settlements in more detail.

Blockchain applications can continue to operate efficiently even if any component of the system fails to work due to its decentralized, peer-to-peer network configuration. As such, single points of failure are avoided [42].

The introduction of blockchains has also caused a paradigm shift from the traditional, fiat currency banking system to the world of cryptocurrencies. *Tokenization* represents an encapsulation of value in tradeable units of accounts known as coins or tokens in a blockchain. In other words, tokenization is a method of changing both physical and non-physical assets into digital tokens. The use of tokens may further enhance the circulation and allocation of rewards and benefits to its participants within the blockchain platform. Examples include voting rights, copyrights and utility [9]. For example, companies can tokenize assets such as real estate or artwork to help solve some inherent liquidity problems affecting such asset types [19]. Finally, blockchain can provide *enhanced security*. Blockchain's consensus method ensures transactions that are recorded on the block are agreed upon by members, thereby enhancing data integrity. Security is improved using hashing methods, where each transaction is encrypted using a hash algorithm and is properly linked to the previous transaction. Each node holds a copy of the transactions performed on the network, meaning data once written cannot be converted on these immutable blockchain networks [36].

BLOCKCHAIN AS A ZTA TOOL

Blockchains share certain noteworthy similarities with a model known as *Zero-Trust Architecture (ZTA)* [34]. ZTA is a security concept centered on the belief that an organization should not automatically trust anything within or outside its perimeters [18]. No implied trust exists in ZTA, which ensures that the trust level is explicitly and dynamically calculated based on context. The ZTA concept draws on technologies such as multifactor authentication, encryption, file system permission, information about users' locations and applications users seek to access for calculating trust. Trust is required to extend or access capabilities that otherwise would not be possible in a trustless environment. Thus, trust is an indication of the relative strength of the assurance of the belief [27].

Table 1.3 compares some key characteristics of ZTA against some blockchain best practices.

CONCLUSIONS AND RECOMMENDATIONS

This introductory chapter examined the fundamentals of blockchain, the types and its components. The goal of this chapter is to provide readers with basic knowledge of blockchain. Topics covered included blockchain origins, the problem of Byzantine Generals and the differences between blockchain and conventional databases. Chapter 1 further highlights the basic attributes and benefits of blockchain. Chapter 1 also discussed the place of blockchain technology in a ZTA environment. Blockchain technology is here to stay, and the technology has numerous potential use cases as explored in future chapters.

CORE CONCEPTS

The top core concepts from this chapter are highlighted below:

- Satoshi Nakamoto is an individual (or perhaps a group of individuals) who became a household name after a publication on Bitcoin's peer-to-peer automated cash system in 2008. The paper proposed the use of electronic cash payment as a better alternative to the traditional banking system. Nakamoto further launched the premier Bitcoin in 2009 and since then, the technology has evolved over the past decade and a half.
- Blockchain is a decentralized, distributed, peer-to-peer network with immutable records that are not alterable. The tamper-proof and tamper-resistant architecture makes it an especially robust technology. Each transaction in the network is protected with a cryptographic hash algorithm that prevents data from being deleted or altered. Hence, members are required to validate block transactions.
- The attributes of blockchain include consensus, provenance and immutability.
- Blockchains have various components, such as blocks, chaining blocks and shared ledgers.
- BZD discusses the theory of Byzantine generals in a fortified city operating in a trustless environment. The Byzantine Fault Tolerance is relevant in blockchain technology: it is a related concept born out of the BZD and the system can withstand the class of failures in a network.
- Blockchain allows users to read and write transactions only. A traditional database, on the other hand, supports users to utilize all the four functions: create, read, update and delete (CRUD).
- Public blockchains are permission-less, shared and synchronized ledgers that any user can access, join or use with no access control restrictions.
- Private blockchains are permissioned networks controlled often by the sponsoring organization. Hybrid blockchains combine the features of both public and private types of blockchain. Consortium blockchains are normally hybrid blockchains where various organizations participate in managing the network.

TABLE 1.3
ZTA and Blockchain Best Practices

Domain	ZTA Best Practice	Informative References	How Blockchain Accommodates Zero-Trust Best Practices	Informative References
Identity and Access Management (IAM)	Encourages the principle of identity and access management of cyber security with network security.	[34]	Provides a secure access and identity management in private blockchains.	[42]
Privilege Identity Management	Focuses on the principles of least privilege to grant access for the completion of tasks.	[33]	Applies the use of hash functions and digital signatures as a zero-proof knowledge principle to verify and restrict users and resource privileges to the ones that require it.	[42]
	Calls for governance strategies, such as restricting resource access based on the principle of least privilege.	[30]	Permissioned blockchains can be designed to only allow certain users to record transactions on a blockchain which any member can read.	[42]
Strong authentication	Recommends encryption of all communications on the network to maintain confidentiality, integrity and provide source authentication in a zero-trust environment Recommends the use of strong authentication to mitigate lateral movement and risks associated with Wide Area Network communications, instead of simply checking the source address of the request.	[41] [12]	Adopts cryptography concepts such as asymmetric key cryptography and symmetric key encryption for encrypting processes.	[42]
System Monitoring Security	Recommends monitoring and measuring the integrity & security posture of all owned and associated assets to detect any consistent activity and any changes in the process.	[33]	The blockchain mining/ validator nodes in blockchain helps publish new blocks in the chain. The mining or validating nodes check the validity of all the transactions in the published block and rejects a block that contains an invalid transaction.	[42]
Security controls on data	Endorses building of security controls around data, or resource and that host-based firewalls do not conform to the zero-trust model, but leveraging on automation helps in understanding the requirements of each application, and thereby codifying them in the local firewall.	[12]	Each block created is hashed to create a digest that represents the block. The hash digest helps to protect the block from any changes by protecting transaction copied in the nodes, each node within the chain will have a copy of the block's hash which detects any changes made.	[42]

- Zero Trust Architecture (ZTA) is based on the concept that organizations should not automatically trust anything in the external and/or internal environment. ZTA is a defense model that considers the most critical aspects of the network's peripherals, the data, applications, assets and services. ZTA eliminates trust in all the perimeters of the network. Its objective is to minimize the risk exposures to an organization in a trustless environment. Thus, ensuring the principles of confidentiality, integrity and availability is attained. The use of blockchain technology is compatible with ZTA principles.

ACTIVITY FOR BETTER UNDERSTANDING

The exercise is designed to equip readers with a better appreciation of Satoshi Nakamoto's white paper on Bitcoin as a peer-to-peer, online payment system. Pay particular attention to the questions below as you review the white paper which can be retrieved at: https://bitcoin.org/bitcoin.pdf

- What is the role of timestamping in blockchain?
- What is the proof-of-work system and how does it work?
- What is the consensus agreement about and what purposes does it serve?
- How does blockchain earn cryptocurrency participants on the platform Bitcoin?
- How does blockchain technology address the double-spending threat with electronic payment?

REFERENCES

[1] Accountlearning. (2020). *Advantages and Disadvantages of Decentralization.* Accountlearning.Com. Retrieved October 27, 2020, from https://accountlearning.com/advantages-and-disadvantages-of-decentralization/

[2] Anwar, H. (2020, April 26). *4 Different Types of Blockchain Technology and Networks.* Retrieved October 12, 2020, from https://101blockchains.com/types-of-blockchain/

[3] Binance Academy. (2020, January 19). *Byzantine Fault Tolerance Explained.* Retrieved October 12, 2020, from https://academy.binance.com/en/articles/byzantine-fault-tolerance-explained

[4] Chainerz, B. (2019, May 7). *Everything You Need to Know about Blockchain and Database.* Retrieved November 11, 2020, from https://yourstory.com/mystory/everything-you-need-to-know-about-blockchain-and-d

[5] Das, V. (2020, September 17). *Ethereum: New Updates, Changes, and Delays.* Retrieved October 9, 2020, from http://www.blockchain-council.org/ethereum/ethereum-new-updates-changes-and-delays/

[6] Devan, G. (2018, November 8). *How Blockchain Technology Is Revolutionizing Data Provenance.* Retrieved October 12, 2020, from https://medium.com/blockpool/how-blockchain-technology-is-revolutionizing-data-provenance-e47610019390

[7] ETGovernment. (2020, September 29). Tamil Nadu Sets 'High Goals' for Strengthening e-Governance and IT Industry Using Blockchain Technology – ET Government. Retrieved October 13, 2020, from https://government.economictimes.indiatimes.com/news/technology/tamil-nadu-sets-high-goals-for-strengthening-e-governance-and-it-industry-using-blockchain-technology/78386382

[8] Ethereum. (no d, p.). *Ethereum Whitepaper.* Retrieved October 11, 2020, from https://ethereum.org/en/whitepaper/

[9] Freni, P., Ferro, E., & Moncada, R. (2020, July 1). Tokenization and Blockchain Tokens Classification: A Morphological Framework – IEEE Conference Publication. *IEEEXplore.* Retrieved October 24, 2020, from https://ieeexplore.ieee.org/document/9219709/authors#authors

[10] GeeksforGeeks. (2019, April 30). *Comparison – Centralized, Decentralized and Distributed Systems.* Retrieved October 29, 2020, from https://www.geeksforgeeks.org/comparison-centralized-decentralized-and-distributed-systems/

[11] Geronion, D. (2018, October 6). *Hybrid Blockchain: The Best of Both Worlds.* Retrieved February 1, 2021, from https://101blockchains.com/hybrid-blockchain/

[12] Gilman, E. (2017, June 28). *4 Easy Ways to Work toward a Zero Trust Security Model.* Retrieved October 13, 2020, from https://opensource.com/article/17/6/4-easy-ways-work-toward-zero-trust-security-model

[13] Golosova, J., & Romanovs, A. (2018, November). Overview of the Blockchain Technology Cases. *ResearchGate*. Retrieved October 10, 2020, from https://www.researchgate.net/publication/329396515_Overview_of_the_Blockchain_Technology_Cases.

[14] Grimshaw, J. (2020, May 21). *IBM Blockchain: What Is Blockchain Technology? Technology: Supply Chain Digital*. Retrieved October 12, 2020, from https://www.supplychaindigital.com/technology/ibm-blockchain-what-blockchain-technology

[15] Gupta, M. (2020). *Compliments of IBM Blockchain for Dummies, 3rd IBM limited edition* (3rd Edition). John Wiley & Sons.

[16] Jessel, B. (2020, January 23). *Hyperledger Fabric, Quorum, Sawtooth, Besu, Corda, and Multichain Saw a 12-Fold Increase in Engineers in the Last Three Years*. Retrieved February 1, 2021, from https://www.forbes.com/sites/benjessel/2020/01/23/hyperledger-fabric-quorum-sawtooth-besu-corda-and-multichain-saw-a-12-fold-increase-in-engineers-in-the-last-three-years/?sh=1823b7b27216

[17] Kamath, R. (2018, July). Food Traceability on Blockchain: Walmart's Pork and Mango Pilots with IBM. *ResearchGate*. Retrieved October 11, 2020, from https://www.researchgate.net/publication/326188675_Food_Traceability_on_Blockchain_Walmart%27s_Pork_and_Mango_Pilots_with_IBM

[18] Keary, E. (2020, February 4). *Security Think Tank: Facing the Challenge of Zero Trust*. Retrieved October 26, 2020, from www.computerweekly.com/opinion/Security-Think-Tank-Facing-the-challenge-of-zero-trust.

[19] Kelley, J. (2020, February 25). *How Tokenization and Digitized Assets Can Help Investors Unlock Trillions from the Economy*. Blockchain Pulse: IBM Blockchain. Retrieved October 24, 2020, from https://www.ibm.com/blogs/blockchain/2020/02/how-tokenization-and-digitized-assets-can-help-investors-unlock-trillions-from-the-economy/

[20] Knirsch, F., Unterweger, A., & Engel, D. (2019, March 11). Implementing a Blockchain from Scratch: Why, How, and What We Learned. *Springer*. Retrieved October 11, 2020, from https://link.springer.com/article/10.1186/s13635-019-0085-3#citeas

[21] Kotka, T., Castillo, C. & Korjus K. (2015, September). *Estonian e-Residency: Redefining the Nation-State in the Digital Era*. Retrieved October 11, 2020, from https://www.raulwalter.com/prod/wp-content/uploads/2015/10/Working_Paper_No.3_Kotka_Vargas_Korjus.pdf

[22] Kramer, M. (2019, May 29). An Overview of Blockchain Technology Based on a Study of Public Awareness. *SSRN*. Retrieved October 11, 2020, from https://papers.ssrn.com/sol3/papers.cfm?abstract_id=3381119

[23] Kumar, R. (2020, June 22). *Your IT Organizational Structure: Should You Centralize or Decentralize?* Software Advice. Retrieved October 27, 2020, from https://www.softwareadvice.com/resources/it-org-structure-centralize-vs-decentralize/

[24] Lastovetska, A. (2020, May 8). *Blockchain Architecture Explained: How It Works and How to Build*. Retrieved October 10, 2020, from https://mlsdev.com/blog/156-how-to-build-your-own-blockchain-architecture

[25] Loki. (2020, January 6). *Centralized vs Decentralized Networks*. Loki. Retrieved October 27, 2020, from https://loki.network/2019/12/05/centralized-vs-decentralized-networks/

[26] Massessi D. (2019, January 6). *Blockchain Consensus and Fault Tolerance in a Nutshell*. Retrieved from https://medium.com/coinmonks/blockchain-consensus-and-fault-tolerance-in-a-nutshell-765de83b8d03

[27] Mital, R. (2020, June 10). *Improving Trust in a Zero Trust Architecture (ZTA)*. Aerospace. Retrieved October 28, 2020, from https://aerospace.org/story/improving-trust-zero-trust-architecture-zta

[28] Nakamoto, S. (2008). *Bitcoin: A Peer-to-Peer Electronic Cash System*. Retrieved October 11, 2020, from https://bitcoin.org/bitcoin.pdf

[29] Naqvi, S. J. (2017, November 9). *Dangers of a Pure Proof of Stake Blockchain Bitcoin*. Retrieved October 24, 2020, from https://medium.com/karachain/dangers-of-a-pure-proof-of-stake-blockchain-439c5dace336bitcoin.org/bitcoin.pdf

[30] Pratt, M. (2018, January 16). *What Is Zero Trust? A Model for More Effective Security*. Retrieved September 22, 2020, from https://www.csoonline.com/article/3247848/what-is-zero-trust-a-model-for-more-effective-security.html

[31] Ray, S. (2018, February 10). *Blockchains versus Traditional Databases – Towards Data Science*. Retrieved November 10, 2020, from https://towardsdatascience.com/blockchains-versus-traditional-databases-e496d8584dc

[32] Rose, C. (2015, July). The Evolution of Digital Currencies: Bitcoin, a Cryptocurrency Causing a Monetary Revolution. *ResearchGate*. Retrieved October 2, 2020, from https://www.researchgate.net/publication/297750676_The_Evolution_Of_Digital_Currencies_Bitcoin_A_Cryptocurrency_Causing_A_Monetary_Revolution

[33] Rose, S., Borchert, O., Mitchell, S., & Connelly, S. (2020, August 11). *Zero Trust Architecture*. Retrieved September 21, 2020, from https://csrc.nist.gov/publications/detail/sp/800-207/final

[34] Shala, B., Trick, U., Lehmann, A., Ghita, B., & Shiaeles, S. (2019, May 11). Novel Trust Consensus Protocol and Blockchain-Based Trust Evaluation System for M2M Application Services. *ScienceDirect*. Retrieved September 23, 2020, from https://www.sciencedirect.com/science/article/pii/S2542660519301234

[35] Shrestha, R. et al. (2020). A New Type of Blockchain for Secure Message Exchange in VANET. Retrieved October 24, 2020, from *Digital Communications and Networks*, vol. 6, no. 2, May, pp. 177–186, https://www.sciencedirect.com/search?qs=10.1016%2Fj.dcan.2019.04.003

[36] Singh, N. (2019, November 04). *Benefits of Blockchain Technology*. Retrieved October 10, 2020, from https://101blockchains.com/benefits-of-blockchain-technology/

[37] Singh, N. (2020, September 15). *Blockchain vs Database: Understanding the difference Between the Two*. Retrieved November 11, 2020, from https://101blockchains.com/blockchain-vs-database-the-difference/#prettyPhoto

[38] UAE Government. (2019, September 17). *UAE Future: 2021*. Retrieved October 11, 2020, from https://u.ae/en/more/uae-future/2021

[39] Vidrih, M. (2019, February 22). *What Is a Block in the Blockchain?* Retrieved October 10, 2020, from https://medium.com/datadriveninvestor/what-is-a-block-in-the-blockchain-c7a420270373

[40] Wang, H., Wang, Y., Cao, Z., Li, Z., & Xiong, G. (2019). An Overview of Blockchain Security Analysis. Retrieved October 10, 2020, from *Communications in Computer and Information Science Cyber Security*, 970, 55–72.

[41] Yafimava, D. (2019, January 15). *What Are Consortium Blockchains, and What Purpose Do They Serve?* Retrieved February 1, 2021, from https://openledger.info/insights/consortium-blockchains/

[42] Yaga, D., Mell, P., Roby, N., & Scarfone, K. (2018, October 3). *Blockchain Technology Overview*. Retrieved October 9, 2020, from https://csrc.nist.gov/publications/detail/nistir/8202/final

[43] Zheng, Z., Xie, S., Dai, H.-N., & Chen, X. (2017, June). (PDF) An Overview of Blockchain Technology: Architecture, Consensus, and Future Trends. Retrieved October 24, 2020, from https://www.researchgate.net/publication/318131748_An_Overview_of_Blockchain_Technology_Architecture_Consensus_and_Future_Trends

2 Blockchain Architecture, Components and Considerations

Aafreen Fathima Altaf Hussain, Temitope Ipentan, Mahakpreet Singh and Grace Moyo Adeyemi

ABSTRACT

This chapter is the second of three aimed at familiarizing readers with blockchain-related fundamentals and components. While Chapter 1 introduced readers to the foundational concepts of blockchain, its history, as well as some of its basic attributes, this chapter aims at providing readers with more detailed information about blockchain components and foundational concepts; especially as it pertains to its cryptographically secure attributes and mechanisms.

Topics discussed in this chapter include:

- Blockchain participants and roles
- Expanded discussion of the various types of blockchain (public/private vs open/closed)
- Consensus algorithms
- Cryptographic and hashing attributes and considerations
- Merkle trees
- Digital signatures
- Access control/identity management/membership service provider considerations
- Crypto wallets
- Interplanetary File System (IPFS)
- Smart contract basics

BLOCKCHAIN PARTICIPANTS AND ROLES

Participants can be considered the next most important component of a blockchain network after the blockchain itself. Blockchain participants interact with each other in various capacities to help generate various blockchain transactions [49].

Participants may be divided into three major categories – individuals, organizations and systems. Individuals are the identifiable group of participants having platform access and using the blockchain network. Examples include employees, teachers, students or an organization's customers. Organizations determine the business strategy, product or services offered as well as a blockchain use case. Individuals act on behalf of the organization based on the organization's authority and business mandates. Systems or devices are the technological elements that make up the business network and can be addressed as a kind of individual participants on their own. These systems are agents of the business networks, performing on behalf of the business and in charge of most of the work that needs to be done. System participants deserve to be regarded separately because more and more systems are autonomous, relying increasingly on artificial intelligence capacities [9].

DOI: 10.1201/9781003211723-2

A more granular categorization of blockchain participants and roles is discussed below [16]:

Blockchain users are participants that have the right to join a blockchain network and perform transactions with other blockchain participants.

Blockchain processing nodes are also users of a blockchain whose main function is to verify transactions and create blockchain blocks, based on an agreed-upon consensus mechanism. It is important to note that while all blockchain processing nodes are users, not all users are processing nodes.

Regulators are authoritative blockchain network participants that have the power to oversee transactions within the network for regulatory compliance purposes, but may not be able to perform a transaction.

Blockchain developers are programmers that create applications enabling users to perform transactions on the blockchain network. Such applications serve as an interface through which users and blockchain communicate.

Blockchain network administrator: Every business running on a private blockchain network must have a network administrator. In its oversight role, the blockchain administrator has the right to create, monitor and manage transactions on the blockchain network.

A certificate authority is the entity that manages the required certificates needed to access a 'permissioned' blockchain or verify various transactions on the blockchain platform.

Blockchain Types

Public Blockchains

A blockchain is referred to as 'public' if it is accessible to any participant wishing to use the blockchain platform. Since a public blockchain provides open access, there is no need for a trusted third-party central registry to provide access control services [10]. Public blockchains allow participants to remain anonymous and have pseudonymity, which is an important privacy attribute to ensure both the confidentiality and security of various transactions. However, the information associated with the transactions carried out and recorded in the public blockchain can be seen by all blockchain users.

Whereas unrestricted access and participant anonymity may be desirable attributes for certain blockchain use cases, public blockchains tend to experience issues related to scalability and privacy.

Scalability and performance problems: The characteristics of a public blockchain where anyone can join the blockchain platform as a user may cause scaling issues and poor performance when compared with the traditional systems for transactions.

Privacy challenges in terms of the visibility and sharing of information: In a public blockchain, there is a lack of an access control mechanism. Therefore, such public blockchains are subject to privacy issues [55].

Private Blockchains

A private blockchain is a platform only open to an affinity group, a consortium, or organizations with a need to share a blockchain's ledger among themselves [10]. To be a part of a private blockchain, authorization needs to be given by the current participants or the administrator of the network. Since identities need to be verified before granting access, the participants' real identities are known. These attributes facilitate better security and control features over the information being shared. The performance and scalability are also much more manageable, as the size of the blockchain can be controlled by the sponsoring organization [55]. Other attributes of a private blockchain include the following:

Limited participation: Only a specific number of participants can take part in the consensus process.

User privileges: Write access is provided by the organization, while the read access is either restricted or made publicly available [35].

Need for access management: Access control is administered by a trusted central authority.

Different levels of access may be required for different roles in usage. For example, activities such as participating in block creation consensus, reading or creating a transaction and executing a smart contract may require the authorization of participants based on different sets of permissions. Therefore, a central authority or group may be needed to perform the review and approval process [24].

Open versus Closed Blockchains

Besides the private and public distinction, blockchains may also be further subcategorized as being 'open' or 'closed'. In an open blockchain, any user can read the stored data, while in a closed blockchain only permissioned individuals can have read privileges.

Public and open platforms: Public and open platforms grant anyone access and read rights to written data as there are no user access restrictions. An example of a public/open blockchain is Bitcoin.

Public and closed platforms: Public and closed platforms allow anyone to write data, but only permitted participants can read the blockchain's data. For example, in a poll or election campaign, all voters may cast their vote (write) to the blockchain but only authorized participants – such as ballot box creators – can have access to (read) the voting or polling results.

Private and open platforms: Private and open platforms allow everyone to read the data, but only permitted individuals can write on the private and open blockchain. It is typically used in supply chains where only vendors can write the supply status to the blockchain, and any private blockchain user can monitor the status and data stored on the blockchain.

Private and closed platforms: Private and closed platforms grant only accredited members read and write privileges of blockchain data, as it may contain private or sensitive information, such as health-related, financial or sensitive government records.

Table 2.1 summarizes the read and write privileges of each type of blockchain.

TABLE 2.1
Read and Write Privileges among Blockchain Types

Public and Open	**Public and Closed**
Everyone can read the data. Everyone can write data. Example: Cryptocurrency platforms, gambling, or video games.	Only allowed participants can read the data. Everyone can write data. Example: Voting platforms/records or whistleblower reporting platforms
Private and Open	**Private and Closed**
Everyone can read the data. Only the administrators can write data. Example: Supply chain monitoring, such as Walmart's DL Asset Track.	Only allowed participants can read the data. Only allowed participants can write data. Example: Military or law enforcement use cases and financial records.

Source: [70].

CONSENSUS

'Consensus' in simple terms means 'a general agreement'. Similar to its general meaning, the consensus mechanism of a blockchain is the heart of the entire system. Consensus drives the blockchain application infrastructure and manages the trust in transactions among processing nodes and the overall blockchain technology [10, 79]. It helps processing nodes to maintain consistency in blockchain transaction data and ensures that blockchain copies contain identical information [38].

Consensus Algorithms

Blockchain networks employ various algorithms. These algorithms are what determines the speed, reliability and scalability of the blockchain. These algorithms include, but are not limited to, the following:

Proof of work (PoW) is the first and also the most secure and trusted of all consensus models. It is stable and has a great ability to ensure unbroken transaction chains. The proof of work concept simply implies that before a processing node can add a transaction into a blockchain network, the processing node must show proof that it has done some work which mostly includes solving a mathematical puzzle using computational power. Solving these puzzles is always difficult, but validating the work done by other processing nodes is relatively easy. PoW relies on proof of resources spent before a value for acceptance is proposed [15]. Proof of work is a hard problem to solve because, as stated earlier, it requires solving a mathematical puzzle, which is further explained in Figure 2.1. In a nutshell, PoW is akin to trying to open a locked door by trying keys from a box full of various keys; yet, once the key that opens the door is identified through a trial-and-error process, it is easy for others to confirm the key required to unlock the door.

The derivation of the nonce value based on the hash requirements is the basis of the proof of work consensus algorithm. To further explain, a nonce should produce a hash output with a mandated

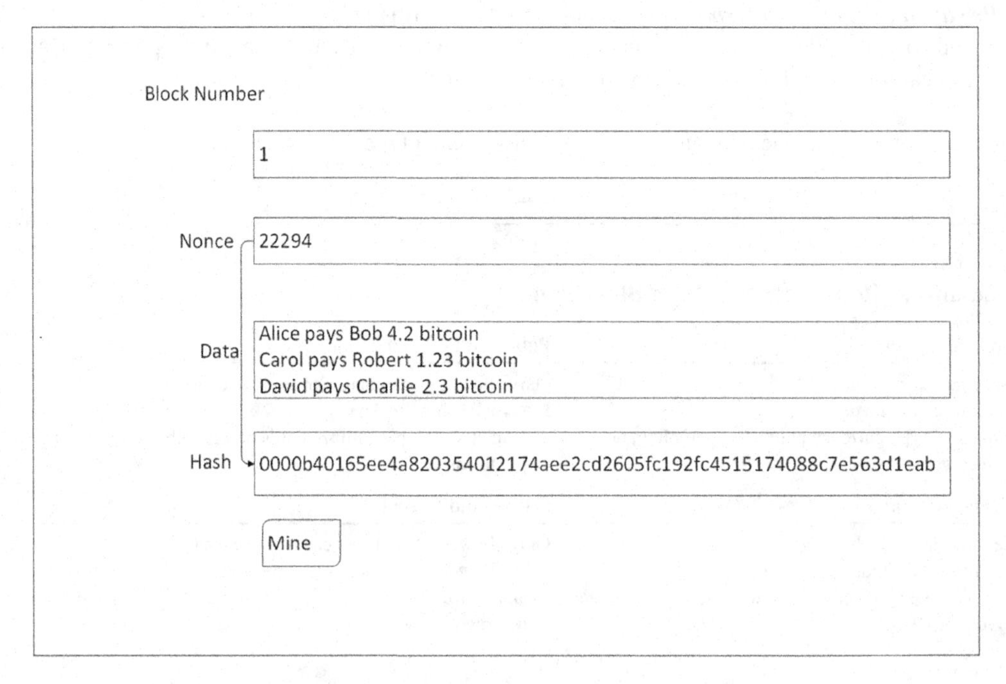

FIGURE 2.1 Illustration explaining a nonce derivation for PoW.

number of zeros at its beginning (for example: four zeros). The work is in how difficult it is to derive the nonce value, compared to the ease of verification because the result is transparent and easily validated by the other nodes. However, this process does not make a block more secure once it gets added to the blockchain. A major disadvantage of the algorithm is the increased energy consumption rate, adversely affecting the environment and validation time for transactions. This validation process in the blockchain network requires several servers verifying the transactions on the blockchain; for example, users making payments and those receiving payments. The high energy consumption in the process of verifying transactions has been said to be as high as the total energy consumption of many smaller nations [83]. The algorithm is also highly vulnerable to attacks in small public blockchains [50].

Figure 2.1 is an illustration of the direct relationship between a nonce and the resulting hash. As can be seen, although it is quite difficult to arrive at the right nonce (the only strategy to solve the puzzle is a process of trial and error), it is easy to confirm and verify the hash simply by inputting derived nonce to confirm the required hash result (requiring four zeros for its first four digits).

Proof of stake (PoS) refers to a mechanism responsible for proving that a node has adequate stake invested in the system. The proof of stake algorithm was created to address some of the perceived inadequacies of the proof of work algorithm. The proof of stake algorithm functions by making validators (miners) stake cryptocurrency as a participation requirement, before permitting validators to add blocks to the blockchain network for which a transaction fee is earned. If a dishonest validator adds an invalid transaction into the blockchain network, the stake would be lost as a penalty. The proof of stake functions as an alternative to the proof of work algorithm [79]. Unlike the proof of work algorithm, PoS is significantly more energy efficient. However, people with few coins may not be able to participate in the block creation process too often. In other words, the higher the stake held in the network, the higher the probability of being selected to validate blocks in the network [15].

The *delegated proof of stake (DPoS)* algorithm is a higher degree of the proof of stake algorithm. To further elaborate, participants with tokens can vote for 'producers', and those with the highest number of votes are the producers that will be permitted to produce blocks. The delegated proof of stake is more advantageous than the proof of stake, as it ensures separation of concerns between the participants and the producers, as well as ensuring proper governance within the network [50]. The delegated proof of stake is heavily reliant on active users in the voting community of the blockchain [79].

Proof of burn (PoB) is often used when starting a new project to maintain fairness in an initial coin distribution. The proof of burn is an alternative to the proof of work algorithm but without energy waste. The native currency or the currency of another blockchain is sent to a wallet address. The blockchain miner can then get an equivalent amount of the currency of the blockchain they are in [27].

The *proof of elapsed time (PoET)* algorithm uses a trusted execution environment in providing randomness and safety through a guaranteed wait time. Nodes are allowed to validate transactions depending on how long they have been waiting since their last opportunity to create a block [79]. Proof of elapsed time relies on a specialized piece of hardware known as the Software Guard Extension (SGX), which allows applications to run trusted codes in a protected runtime environment. It enables the allocation of enclaves (private regions of memory) by user-level codes and protects these codes from processes with higher privileges.

Proof of importance (PoI): Unlike the proof of stake algorithm, PoI does not only rely on the volume of stakes a user has in a blockchain network. Instead, it establishes trust and importance through close monitoring of the movement of the tokens and their usage. Because it determines which nodes within the network are eligible to raise new blocks to the blockchain, it is quite similar to the proof of stake. However, the proof of importance also considers how important the nodes are by using a graph theory to determine each node's rating [79]

The *proof of storage (PoS)* mechanism has resulted in several other variations, such as proof of space, proof of space-time, proof of replication and the like. It allows storage capacity to be outsourced as a condition to participate in the network [15].

Proof of authority (PoA) is designed to be used mainly in a permissioned network by selecting transaction validators based on their reputation and identity in the blockchain. Interested validators must provide the necessary documentation to reveal their identities and reputation [50].

Proof of activity (PoACT) selects stakeholders in a uniform but pseudo-random manner through a combination of the proof of work and proof of stake algorithms for maximum security. This hybrid combination also ensures energy efficiency.

Proof of deposit (PoD) simply refers to the evidence of payment made by other nodes that want to participate in a blockchain [8]. The proof of deposit payment is referred to as a security deposit before any nodes can create new blocks [3].

Ordering-based consensus (OBC) is a similar concept to proof of authority since the ordering-based consensus mechanism is also mainly used in a permissioned network. However, it selects nodes to participate in the network based on their identity service [21]. Identity service refers to a system whereby a processing node is recognized and considered reliable as a result of the node being a member of a verified body.

Proof of capacity (PoC) is also referred to as hard drive mining because hard drive space is utilized for mining. PoC is a consensus mechanism that allows processing nodes in the network to validate transactions and to determine mining rights based on available hard drive space. Nodes that have more unoccupied space on their hard drives can mine available crypto-coins [13].

The Practical Byzantine Fault (PBFT) was first introduced by Miguel Castro and Barbara Liskov to achieve consensus in distributed systems with a voting mechanism if the state changes by providing both safety and liveness [63]. However, it is not scalable due to the number of messages that need to be sent which will increase exponentially as the network grows. This algorithm works in asynchronous systems – like the Internet – and it optimizes the system such that it is more efficient and effective [21].

The maximum number of messages is calculated as follows:

$$M = 1 + 3F + 3F + (2F)(3F + 1) + 3F + 1$$

$$F = (n-1)/3$$

where M = Maximum number of messages
F= Maximum number of allowable faulty nodes and
N = Number of active consensus nodes

PBFT should be used with a consortium of enterprise organizations, where each organization would represent a node on the network. In PBFT, the block creator may not be any special miner, but the proposed block which is committed to the chain would be the most agreed upon block [47].

Simplified Byzantine Fault Tolerance (SBFT) is an improvement on the Practical Byzantine Fault Tolerance (PBFT). It requires about a third of the messages used in PBFT and achieves consensus much faster – most times within 0.2 seconds. Each node in an SBFT consensus can act autonomously without the need to communicate with other nodes [62].

The message count for SBFT is calculated by:

$$M = (3f + 1)(2f + 1) + 1$$

where M= maximum number of messages
F = number of allowable faulty nodes

Delegated Byzantine Fault Tolerance (DBFT) was made popular by a cryptocurrency called NEO. It uses 'bookkeepers' – a group of nodes – to vote for a consensus through a proxy to generate more blocks in the network. Consensus through proxy refers to a mechanism whereby one of the bookkeepers chosen as the speaker proposes his block for verification. The speaker then sends the block to the other delegates in the network to confirm that the block is valid. When two-third of the delegates have confirmed its validity, the block is then added to the network. Once the consensus is confirmed, it is rarely revoked. The speed at which transactions are generated is also very high – about 1,000 transactions per second [63].

Its fault tolerance is calculated as follows:

$$F = (n - 1)/3$$

where F = number of allowable faulty nodes

Federated Byzantine Fault Tolerance is used in critical financial transactions among countries. Nodes in this protocol retain a group of publicly trusted peers and propagate only those transactions that have been validated by the most trusted nodes [13].

A CRYPTOGRAPHY PRIMER

Cryptography is the science of securing information in the presence of third parties – known as adversaries. The main objective of cryptography is to ensure data security, by providing confidentiality, integrity, authentication and non-repudiation [11]. If two parties (Alice and Bob) want to communicate, Alice – who is referred to here as the sender – uses an insecure communication channel to send a message to the receiver – Bob. The channel could be a computer network or telephone line. If the message contains confidential information, it could be intercepted and read by an eavesdropper, or worse the message could be modified by an adversary, and Bob (the receiver) might not be able to detect the manipulation. One of the objectives of cryptography is to provide methods for preventing such attacks from occurring.

Readers should be familiar with the following cryptography-related terms and concepts:

Cryptosystem refers to the full specifications of the keys and the techniques of the keys used in the encryption and decryption process [49].

Cryptoanalysis is the technique used by attackers when trying to crack cryptosystems. The successful attacks may retrieve the plaintext (or a portion of the plaintext) from the ciphertext, then replace it with parts of the original message, or forge a digital signature. Cryptology is often used as a general term to refer to cryptoanalysis [49].

Confidentiality: Only people with the secret key should be able to access the message. If the message gets intercepted during transmission, it should be protected and not accessible to unauthorized parties. Confidentiality can be achieved using algorithms such as Advanced Encryption Standard (AES) and RC4 [37].

Data integrity: The receiver of the message should be able to confirm that the message has not been corrupted, changed or truncated during transmission, either deliberately or accidentally. No one should be able to change the original message or a portion of it with a false message.

Data authentication: The origin of the message should be verifiable by the receiver of the message. There should be no one sending a message to the receiver (Bob) by pretending to be the sender (Alice). The sender and receiver should be able to identify each other when initiating a communication (entity authentication). The Message Authentication Code (MAC) algorithm can be used to achieve data integrity and authentication [37].

Non-repudiation: There should be no denial from the sender later that they did not send the message. There should be the assurance that the origin of the message cannot be denied.

Block and stream ciphers are parts of a cryptosystem. In a block cipher, the plaintext is split into a fixed chunk called a 'block'. A block is a bit of string, which has a fixed length. Furthermore, encryption and decryption are done one block at a time. On the contrary, stream cipher cryptosystems have their keys built using a keystream, the encryption process constructs ciphertext from plaintext and the keystream, and the decryption is done using the ciphertext and the keystream [73].

The two major categories of modern cryptography are symmetric and asymmetric cryptography.

Symmetric cryptography provides a secret communication channel between trusted parties. Assume Alice and Bob have agreed on a secret key. Later when Alice and Bob decide to send a secret message to each other, the secret key is used to convert the message into a scrambled form that cannot be processed by anyone without the key. The original message is usually known as 'plaintext' while the scrambled message is known as 'ciphertext'. This process is called 'encryption'. The process where the ciphertext is received by Bob and he can use the secret key to convert it to plaintext is known as 'decryption' [68]. In symmetric encryption, a single secret key is used by both communication parties to encrypt the plaintext and decrypt the ciphertext, with the secret key being known to both parties. The encryption and decryption algorithms are known publicly. Hence, the secret key must be kept safe. The main problem in symmetric cryptography is the key distribution problem. The key distribution has to be done before the transmission of information. As such, there is a requirement for a secret channel as well as authentication between the communicating parties (Alice and Bob) for the distribution of key material.

Symmetric Cryptography Algorithms

Advanced Encryption Standard (AES) was recommended as a replacement for the vulnerable Data Encryption Standard (DES) by the US National Institute of Standards and Technology (NIST) in 1998. AES is a symmetric block cipher that uses a variable key length of 128, 192 and 256 bits. AES will perform nine processing rounds if the key length and the block length are both 126 bits; and 11 processing rounds if the key and block length are 192 bits.

If both the key and block length are 256 bits, AES will perform 13 processing rounds [2].

The processing rounds involve four steps:

1. *Substitute bytes* – In this step, a byte-by-byte substitution is performed using an S-box.
2. *Shift rows* – Simple permutation is done.
3. *Mix column* – A substitution method, where the data in each column from the shift row step is multiplied by the matrix algorithm.
4. *Add round* – The processing round key is XORed with the data.

Blowfish is an algorithm designed by Bruce Schneider in 1993. It is a 64-bit block cipher with a variable key length ranging from 32 bits (4 bytes) to 448 bits (56 bytes). It is the most efficient algorithm among all other existing encryption algorithms [22]. The blowfish algorithm has two steps, namely [2]:

1. *Key expansion* – The key expansion step converts 448 bits into 4168 bytes. The P-array contains 18 sub-keys of 32-bit each. There are four 32-bit S-boxes which all consist of 256 entries each.
2. *Data encryption* – The data encryption is done using an XOR operation. A total of 16 rounds of data encryption is performed.

RC4 is a symmetric algorithm developed by Ronald Rivest. RC4 has a variable key length ranging from 1 to 256 bytes. There are two phases involved in the RC4 algorithm, namely key generation and encryption. The variable encryption generated by the encryption key uses two arrays, state and keys and the merging operation results. Pseudo-random bytes are generated to produce a stream, which is XORed for the conversion of ciphertext to plaintext. XOR is a logical operation used in comparing two binary bits. If there is a significant difference, a value of 1 will be produced. If both bits are equal, then the result will be 0 [22].

Various advantages and disadvantages of symmetric key algorithms, as well as their applications, are summarized in Table 2.2.

Asymmetric Cryptography Algorithms

Asymmetric cryptography is also known as public-key cryptography. The revolutionary idea of cryptography was introduced by Whitfield Diffie and Martin Hellman in the 1970s. Diffie and Hellman maintained that there might be possibilities to develop a cryptosystem in which there are two different keys. A public key is used for the encryption of the plaintext, while a private key is used for the decryption of the ciphertext to plain text [73]. In this technique, the sender (Alice) encrypts the secret message with the receiver's (Bob's) public key and the receiver decrypts the message with his private key. The public key can be known to everyone, but the private key is only known to the receiver and needs to be protected. The keys are certified using digital signatures. Asymmetric cryptography is convenient and offers greater security, as long as the private key remains intact. The Diffie-Hellman exchange key is one of the most popular methods of key distribution [58].

The *Rivest Adi Adleman (RSA)* algorithm was invented by Ron Rivest, Adi Shamir and Leonard Adleman in 1978. RSA was introduced as a cryptographic algorithm that replaced the less secure National Bureau of Standards (NBS) algorithm. The RSA algorithm is based on the factoring problem, which has a practical difficulty of factoring the product of two large prime numbers[58]. Different encryption keys are also used; namely, a public key (asymmetric key) and a private key (symmetric key). The RSA algorithm involves three steps, namely key generation, encryption and decryption.

Elgamal encryption was developed by Taher Elgamal in 1984. The asymmetric key encryption was established based on the Diffie-Hellman key exchange principle. Distinct from the Diffie-Hellman algorithm, the Elgamal cryptosystem is a complete encryption and decryption system that depends on the discrete logarithm problem. The Elgamal algorithm consists of three components which are the key generator, the encryption and the decryption [58].

Knapsack algorithm was one of the earliest public key cryptosystems invented by Ralph Merkle and Martin Hellman in 1978. In the knapsack cryptosystem, there are also three performed processes: key generation, encryption and decryption. Some of the public-key cryptosystems have been proposed based on the knapsack (or the subset-sum) problem [58]. Knapsack cryptosystems had been considered as one of the promising cryptosystems due to their NP (nondeterministic polynomial time) completeness nature and the high speed of the encryption and decryption process. However, most knapsack-based cryptosystems are considered insecure, because the encryption function in these cryptosystems is essentially linear.

Table 2.3 highlights the advantages and disadvantages of various asymmetric algorithms, with their applications.

Hybrid Cryptography

Asymmetric cryptography has one main drawback; it is much slower than symmetric cryptosystems. As such, asymmetric cryptosystems are mostly used for encrypting small amounts of data, such as credit card numbers. Fortunately, there is a better way to combine symmetric and asymmetric cryptography in order to get the advantages of both. The technique used is called hybrid cryptography.

TABLE 2.2
Comparison of Symmetric Algorithms

Algorithm	Key Size	Advantages	Disadvantages	Applications
AES	• 128, 192 and 256	• Replacement for DES (Data Encryption Standard), excellent security. • Fast and flexible. • Can be applied on various platforms mostly small devices. • Supports large key sizes [54].	• Difficult to decrypt the data (ciphertext) if the secret key gets lost [54]. • The length of the key size is too long which makes the algorithm complex at times.	• AES is used in Coinbase, an online Bitcoin wallet that provides data security protocol. The data are divided using a redundancy system that is cyphered with AES 256 [57]. • Blockchain.info offers *My wallet service* which provides a secure online account. This application uses AES encryption algorithms to protect wallets from possible theft. The data from the Bitcoins are encrypted with AES-256 before storing it on the server [57]. • Bon Pay: A wallet and platform for the payment of cryptocurrency. Uses the AES-256 encryption standard [57].
Blowfish	• 32 to 448	• Suitable and efficient for implementation on hardware. • Provides excellent security, and highly efficient in software. • Process for implementing blowfish is simple since all the operations performed are XORed, addition and table lookup. • Fast encryption speed of blow as its rate on a 32-bit microprocessor is 26 clock cycles per byte. • Secure and flexible because it has a variable key length of up to a maximum of 448 bits [54].	• May not be appropriate for the encryption of larger file sizes [54]. • If the user decides to use a weak key as the secret key, then the algorithm becomes less secure [54].	Blowfish is used in the following applications [69]: • File and disk encryption • Password management • Backup tools • Database security • Ecommerce software • Email encryption • Secure shell • Steganography.
RC4	• 40 to 2048	• Consumes less memory compared with AES and Blowfish. • Fast cipher in SSL (secure socket layer). • A particular RC4 key can only be used once. • The encryption process is 10 times faster than DES encryption [74].	• RC4 is vulnerable to state table analytic attacks [50]. • RC4 algorithm has weaker keys. Keys may be identified with cryptoanalysis that can find situations where one or more generated bytes are strongly linked with a small subset of the key bytes [74].	• Used in recommended encryption mechanisms in SSL and TLS when encrypting HTTPS connections on the Internet. • Suitable for securing communications between Web browsers and eCommerce sites, as well as for use in wireless web applications [74].

TABLE 2.3
Comparison of Asymmetric Algorithms

Algorithms	Key Size	Advantages	Disadvantages	Application
Diffie Hellman	• 2048	• Algorithm is quite fast because the symmetric key is of very short length (256 bits) [23]. • The shared secret key does not get transmitted over the main communication channel [58].	• Lack of authentication [58]. • Vulnerable to man in the middle attacks [23]. • Vulnerable to quantum computing because the security of the secret key relies on its hardness [80].	• Secure socket layer (SSL): In SSL, the key exchange process uses the D-H algorithm to guarantee the identity of one party to another [3]. • Secure Shell (SSH): Diffie-Hellman is a component of SSH, the D-H algorithm is responsible for the parties agreeing on the keys used by the many primitives later in the SSH protocol [3]. • IP Security (IP Sec): IP Security uses the Diffie-Hellman algorithm for the establishment of identities, the preferred algorithms and a shared secret [3]. • In a blockchain, Stealth Address cryptocurrency is created using the Elliptic-curve Diffie-Hellman (ECDH) protocol. Stealth Address in cryptocurrency is used to hide the recipient's identity on the network [48].
RSA	• 2048 to 4096	• Highly secure algorithm since producing the symmetric key from the asymmetric key and modulus is difficult. Also computing the reverse of the encryption is very difficult for the attackers.	• There are many symmetric key encryptions that are considerably faster than any currently available asymmetric encryption [58]. • Generating the key is complex. The process is quite slow [23]. • RSA is also susceptible to quantum computing because it relies on integer factorization [4].	• Digital signature: Used to prove the authenticity and integrity of a message. • For the secure connection between VPNClient and VPNServer • Traditionally used in Transport Layer Security (TLS) and Pretty Good Privacy (PGP) encryption. • Implemented in TLS and OpenSSL. • Implemented into cryptographic libraries such as WolfCrypt and Cryptlib [52].
ElGamal	• 1024	• The same ciphertext is gotten from the same plaintext (with near certainty) each time it is encrypted [58].	• Need for randomness. • Slower speed and a long ciphertext [58].	• Added to OpenPGP (Pretty Good Privacy) for RFC 2440. • Implemented in GnuPG (GNU Privacy Guard), which allows the encryption and signing of data and communication [18].
Knapsack		• Highly effective protocol for secret key distribution [58].	• Need for randomness. • Slower speed and long ciphertext [58].	Suitable for: • Home energy management • Cognitive radio networks • Resource management in software • 5G mobile edge computing [27].

Assume a sender (Alice) wants to encrypt and send a long message to the receiver (Bob) without having a prior shared secret key. A (fast) symmetric cryptosystem is used by Alice to choose a random secret key and encrypt the plaintext. Alice then encrypts this secret key with Bob's public key. Alice sends the ciphertext and the encrypted key to Bob. Bob uses his private decryption key to decrypt the secret key and then uses the decrypted secret key to decrypt the ciphertext. An asymmetric key is used to encrypt the short secret key while a much faster symmetric key is used to encrypt the longer plaintext. Therefore, hybrid cryptography (almost) accomplishes the efficiency of symmetric-key cryptography. Hybrid cryptography can be used in a situation where there is no predetermined secret key between the sender and recipient [73].

Homomorphic Cryptography

The word homomorphism was first used by Rivest, Adleman and Dertouzous in 1978 as a possible solution to encrypt without the need for decryption. Homomorphic encryption is a cryptosystem where computations can be made by the user on encrypted data. Homomorphic technology provides data privacy because it handles data without data leakage and the data can be accessed only by the person with the private key [1]. Homomorphic encryption is still in its developmental stage.

Elliptic Curve Cryptography

Elliptic curve cryptography is a method in public cryptography that is based on the algebraic formation of an elliptic curve over finite fields [45]. The implementation of the elliptic curve in cryptography was recommended individually by Neal Koblit and Victor S. Miller in 1985 [45].

Elliptic curve keys are calculated using the elliptic curve equation, are smaller in size, occupy less memory space and are faster when compared to RSA cryptography [23, 45]. Elliptic Curve Key Exchange (ECKE) can be used for communication over an insecure channel [16].

The advantage of elliptic curve cryptography is the attainment of security using a 164-bit key, which implies that it is more secure than the RSA and Diffie-Hellman algorithms while less power is consumed, thereby allowing for more efficient utilization of portable device batteries [23]. The drawback of elliptic cryptography is its encrypted message size, the algorithm's complexity and it is more challenging implementation compared to RSA [23].

Zero-Knowledge Proof

Zero-knowledge proof and proof of knowledge are progressive ways to recognize passwords and are powerful tools in cryptography [66]. In the zero-knowledge proof system, a verifier is convinced by a prover that a statement is true without disclosing the actual information, but by simply validating the assertion. To further explain, the word 'proof' or 'proof system' is an interactive protocol within two parties, with one party (the prover) attempting to convince another party (the verifier) that a given statement is true [51]. The prover demonstrates knowledge of a secret value that satisfies a given argument (proof or knowledge property) without revealing the secret to the verifier (zero-knowledge property) [8].

The following criteria must be met to be considered zero-knowledge proof [39]:

Completeness: The verifier will be convinced if the prover's statement is true.
Soundness: The verifier cannot be convinced if the prover's statement is false.
Zero-knowledge: If the information is true, no other information will be learned by the verifier other than that the prover's statement is true.

Table 2.4 highlights the advantages and disadvantages of advanced cryptosystems, along with their applications in blockchain technology.

TABLE 2.4
Comparisons of Advanced Cryptosystems

Cryptosystems	Advantages	Disadvantages	Applications
Homomorphic Cryptography	• Improves privacy in bank transactions, voting computations and cloud computing. • Helps in easily retrieving the private information of users [72].	• Has a complex approach because it uses the formation of different ideal lattices [72].	• Privacy preservation in blockchain technology: Implementing homomorphic cryptography in blockchain will allow the processing of ciphertext, which will eventually allow transaction privacy and confidentiality. • To ensure privacy-proof transactions in Huawei blockchain: Huawei's blockchain solution uses the public keys of the homomorphic encryption to encrypt and protect the user's transaction data [42].
Elliptic Curve Cryptography (ECC)	• Provides the highest strength per bit when compared to other cryptosystems [30]. • In ECC, a 164-bit key provides the same security as RSA with a 1024-bit key. This means that ECC requires lower computation and smaller memory [30]. • The same key size used in RSA provides a more secure cryptosystem in elliptic curve cryptography, implying that ECC needs smaller memory when used [30]. • ECC provides increased speed, although the private key process is complicated, and the public key can be calculated very easily [30].	• The size of the encrypted message is significantly bigger than the RSA cryptosystem [30]. • The complexity of the algorithm and the difficulty involved in implementing the algorithm increase the likelihood of errors while implementing the algorithm [30]. • ECC susceptible to quantum computing because the hardness of the elliptic-curve discrete algorithm problem is linked to the complexity in recovering the symmetric key from the asymmetric key, as ECC is used to generate asymmetric keys and symmetric keys [80].	• Internet of Things (IoT): The main application of ECC is in IoT devices (handheld) because ECC is appropriate for resource-constrained schemes due to the efficiency of its energy consumption and decreased heat production [83]. • Bitcoin: Elliptic Curve Digital Signature Algorithm (ECDA), which is ECC-based, is used in Bitcoin. ECC is used in Bitcoin cryptocurrency to ensure that only the rightful owners can spend their funds. ECC is used by Bitcoin to generate public keys from private keys. SECP256k1 is the particular elliptic curve used, which is the curve over a finite field [26]. • Ethereum: The elliptic curve SECP256k1 used in Bitcoin is also used by Ethereum. As such, the process of getting a public key is identical in Bitcoin and Ethereum. EDSA is applied to the private key to get a 64-byte integer, which is two 32-byte integers that denote the X and Y points on the elliptic curve that are concatenated together [7].
Zero-knowledge proof	• Simple to implement, as there is no complex encryption method involved [68]. • Secure since no participant is required to reveal any information. • Increased throughput and scalability of blockchains [68]. • Inefficient verification and authentication methods are replaced; thus, information security is strengthened [68].	• Significant computer power is required [68].	• Zcash: In blockchain technology, Zcash is a cryptocurrency platform that uses a special iteration of zero-knowledge (known as zk-SNARKS), allowing original transactions to remain fully encrypted while being verified under the network's consensus rules [43]. • SNARKS have been implemented to ensure privacy in a blockchain [43].

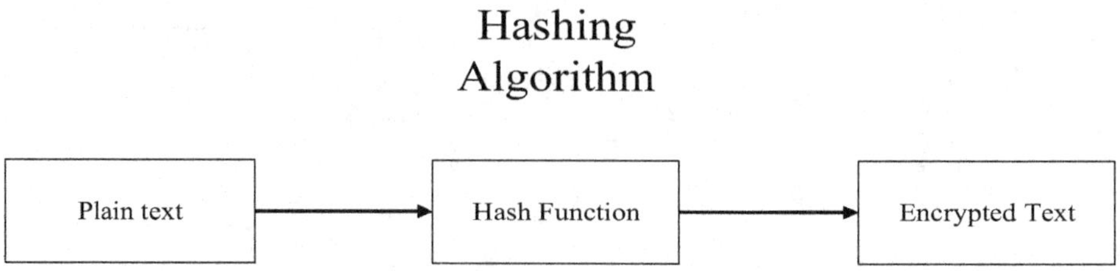

Hashing
Algorithm

Plain text → Hash Function → Encrypted Text

FIGURE 2.2 Working of hash function.

Source: [81].

HASH FUNCTIONS

A hash is the foundation of the blockchain network. A hash function converts an input file, text or image of any size into a fixed-length encrypted output. This technique is called hashing. Hashing lets users randomly select an input data, hash the data and get the same output. Even a minor change to the input would result in a very distinct output [33, 81] (Figure 2.2).

How Does a Hash Work?

A sequence of a fixed length of numbers and letters that do not match the input, resulting in an encrypted value, is referred to as a hash or hash value. The use of fixed-length output increases protection as any person who tries to decode the hash or break the blockchain cannot predict how long or short the input is just by looking at the output length.

Solving a hash begins with the accessible data in the block header, and hence solving a mathematical puzzle. Each block header includes a version number, a timestamp, the previous block hash, the Merkle Root hash, nonce and the target hash. To solve the mathematical puzzle, the block creators may take multiple tries to decide which string of numbers to use as the nonce. The nonce is a series of arbitrary numbers and is applied to the hash of the previous block, which is further hashed. If the hashed output is less than or equal to the target hash, it is approved as the puzzle's solution, and the block will be added to the blockchain. The more complex it is to construct a hash that meets the target hash requirements, the longer it takes to solve the puzzle [33].

Where Are Hash Functions Used?

Cryptographic hash functions are used for many activities on a blockchain. Some of the activities are listed below [81]:

To secure the block header: A publishing node hashes the block header. If the proof of work consensus model is used by the blockchain network, the publishing node hashes the block header with varying nonce values before the puzzle specifications have been met. Then, the hashed output of the current block header is stored within the next block's header, where the current block header's data will be protected.

To secure block data: A publishing node hashes the block data and the hashed output is stored in the block header [81].

For address derivation: An address is a simple, alphanumeric string of characters that serve as the 'to' and 'from' points in a transfer. An address is obtained from the public key of the blockchain network user using a cryptographic hash function with some additional data like the version number or checksums [81] (Figure 2.3).

Public key → cryptographic hash function → address

FIGURE 2.3 Address generation.

Source: [81].

Security Properties of Hash Functions

Pre-image resistance: This implies that hash functions are one-way. For a provided output value, it is computationally very difficult to determine the input value. For example, for a given digest, it is impossible to find the value of x if hash(x) = digest.

Second pre-image resistance: This implies that an input that has a particular output cannot be identified. The cryptographic hash functions are constructed in a way that, for a particular input, it is computationally difficult to determine a second input value that generates the same output. For example, for a given x, it is impossible to find y with any probability of success if *hash(x) = hash(y)*.

Collision resistant: This implies that it is computationally difficult to determine two inputs that generate the same output value. For example, it is impossible to find an *x* and a *y* if *hash(x) = hash(y)* [81].

Non-locality: This implies that similar inputs generate very different outputs in a non-local hash function.

Large state space: A hash collision can be determined only by a brute-force search. As per the Pigeonhole Theorem, brute-force searches involve testing as many inputs as possible. This number should be large enough to make the brute-force search difficult [60].

A successful attack on hash algorithms refers to one that can violate any of the security properties of that hash function. All secure hash functions are prone to a general attack due to the short bit length produced by the hash functions. Attacks on hash functions include brute-force attacks, birthday paradox, differential attacks and length extension attacks [5].

Real-World Hash Applications

Password verification: Most websites save passwords as hashes since saving passwords in plaintext is considered unsafe. Whenever a user enters their password, it is hashed, and the output is compared to the set of hashed values saved on the organization's servers.

Signature generation and authentication: Signature authentication is a cryptographic method used to validate the integrity of digital signatures or messages. A valid digital signature provides the receiver with clear evidence that a known sender generated the message and the message was not modified during transmission.

Data authentication and message integrity: Hashes are also used to ensure that messages and data sent from a source to a receiver are not modified during transmission. To achieve this, the source publishes a hashed version of their data and a key so that the receiver may compare the hash value calculated to the published hash value to ensure that they match [32].

Common Cryptographic Hash Functions

MD4 was designed by Ronald L. Rivest in 1990 [61]. It converts input messages of lengths shorter than 2^{64} bits into a 128-bit output. It was the first cryptographic algorithm to allow efficient use of logic operations and integer calculations on 32-bit processors.

Dobbertin – a cryptographer – attacked MD4 using algebraic and optimization methods that resulted in a 'collision'. The entire message – excluding a few dozen bytes – was under the attacker's control. After the attacks made by Den Boer and Bosselaers, Rivest immediately suggested an improved variant named MD5 [76].

MD5 was developed by Ronald L. Rivest, who also designed MD4 [61]. The MD5 algorithm uses an input message of variable length and produces a 128-bit output [60]. The input message is split into 512-bit blocks. The input message is padded in such a way that the total length is divisible by 512. As many preceding zeros (as needed) are necessary to maximize the message's length to 64 bits less than the 512-multiple. The remaining bits are loaded with 64 bits representing the length of the original message. The MD5 algorithm runs on a 128-bit state, split into four 32-bit words that are mapped to a set of fixed constants. The algorithm then utilizes the 512-bit message blocks to change the state [59].

The MD5 algorithm has fast optimization on 32-bit processors and consists of 64 operations, split into four rounds of 16 operations [61]. These operations are based on a non-linear function, modular addition and left rotation [59]. MD5 is more secure when compared to MD4, but it is slower than MD4 [41]. Boer and Bosselaers proved that MD5 is not collision resistant [76].

Both MD4 and MD5 have wide applications in the industrial and military sectors [41]. Even though MD5 is prone to attacks, it is used as a checksum for data authentication against unintended fraud. MD5 digests are used in blockchain applications to guarantee that a file has been transmitted without any modification during transfer. File servers have an MD5 file checksum so that the user can match the checksum of this file to the received file [59].

SHA-256: SHA-256 was developed by the National Security Agency (NSA). The abbreviation 'SHA' means Secure Hashing Algorithm, and '256' represents a 256-bit output, i.e. 32 bytes [75]. SHA-256 converts input messages of lengths shorter than 2^{64} bits into a 256-bit output [61]. It is a mathematical function performed on digital data. To confirm data integrity, a comparison between the computed hash and the predicted hash is made [75]. This hash function's main characteristic is collision resistance, and to locate a collision, this algorithm would have to be run approximately 2^{128} times [81]. However, neither MD4 nor MD5 are collision-resistant [76].

SHA-256 is used in many blockchain implementations [81]. It is used in various features of the Bitcoin protocol like Bitcoin mining, Merkle trees and Bitcoin address generation [75].

Bitcoin Mining: SHA-256 is used when a block creator has to render the previous block hash [74]. To produce the previous block hash, the miner uses the following formula [75]:

Previous Block Hash = SHA-256 (SHA-256 (Block Header))

Bitcoin address generation: A public key is the foundation of a Bitcoin address. This key is hashed by SHA-256 and RIPEMD160 (Race Integrity Primitives Evaluation Message Digest 160-bit). The formula to create a Bitcoin address is [75]:

Bitcoin Address= RIPEMD160 (SHA256 (Public Key))

SHA-256 is also used to create Secure Sockets Layer (SSL) certificates. These certificates are used to set up and validate secure communications for websites and online services. In SSL certificates, the cryptographic components are created by SHA-256. A secure connection between a server and an Internet Service Provider (ISP) is referred to as an 'SSL handshake'. Credit card numbers, login details and other confidential data are protected with this link using SHA-256. SHA-256 hashes all the user information during the process [75].

MERKLE TREES

Merkle trees were conceived by Ralph Merkle [82]. Robert Merkle patented the Merkle tree in 1979. A Merkle tree records all transactions in a block by creating a digital fingerprint for the whole set of transactions [64]. Merkle trees are like an index portion of a book, as they allow the user to check whether a transaction is included in a block or not. Examples of some hash functions using Merkle trees are SHA-0, SHA-1, SHA-2, RIPEMD and RIPEMD-160 [6].

THE WORKING OF MERKLE TREES

Merkle root or the root hash is the hash left by continuously hashing pair of nodes to generate Merkle trees [64]. Each node in the Merkle tree is a hash value [82]. There are two types of nodes: leaf nodes and non-leaf nodes. The leaf node is a hash of separate transactions, and a non-leaf node is the hash of a previous hash. Merkle trees require an equal number of leaf nodes because they are binary. The last hash will be duplicated if the number of transactions is odd until it produces an even number of leaf nodes [64].

There are four transactions, A, B, C and D, in a block in Figure 2.4. Each transaction is hashed and stored in the leaf nodes as Hash A, Hash B, Hash C and Hash D. Sequential pairs of leaf nodes are summed up by hashing Hash A and Hash B separately and hashing Hash C and Hash D separately, resulting in Hash AB and Hash CD. Hash AB and Hash CD are called parent nodes. The Merkle root is then created by hashing Hash AB and Hash CD called Hash ABCD. The Merkle root is then stored in the block header and detects if any modifications are made in the transaction data, thereby maintaining data integrity [82]. If any modification is made, then the Merkle root also changes [64].

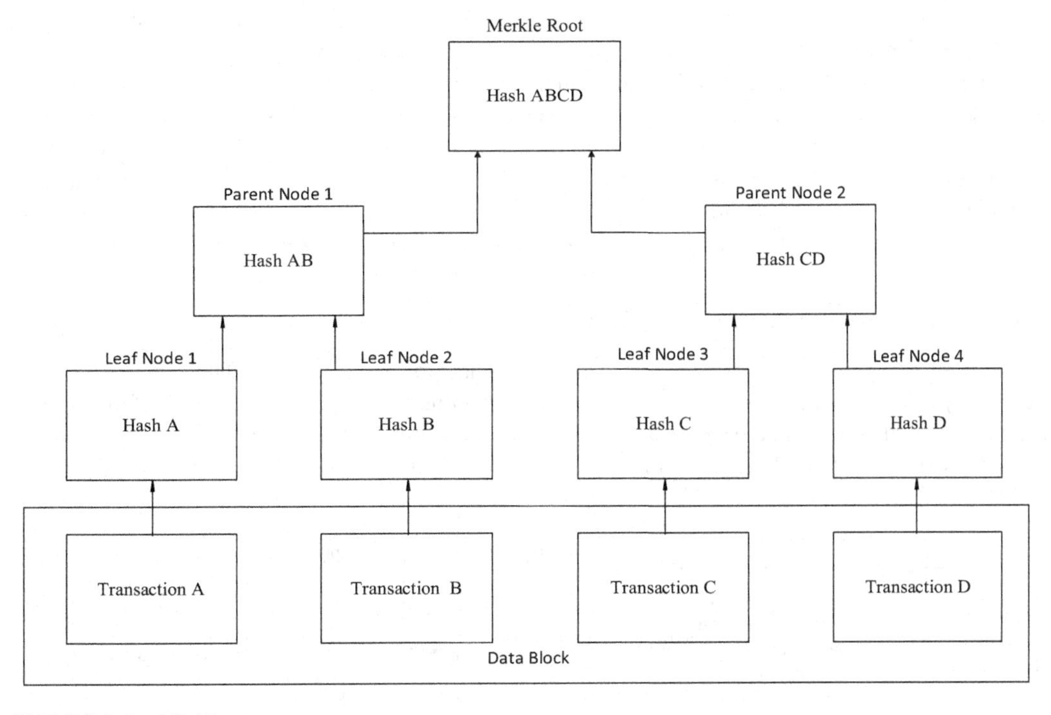

FIGURE 2.4 Merkle tree structure.

Source: [19].

Benefits of Merkle Trees

Merkle trees reduce the need for a large memory space; authenticate data to prove integrity, and entail fast and easy computation. Blockchain miners/validators can compute hashes steadily as they receive transactions from other members. Furthermore, blockchain users can validate transactions in the block. As such, Merkle tree management requires only a small volume of information to distribute across the network [64].

DIGITAL SIGNATURES

A digital signature is an electronic representation of the traditional handwritten signature. A digital signature is used to prove that the data have been signed by the claimed 'signatory' and can be used to determine if any modifications have been made to the data after they were signed. This signature is made up of a series of bits in a computer and can be created on the transmitted data or the data recovered from storage.

Claimed signatory: A signatory is called a claimed signatory before the authentication of the signed information until the signatory's true identity is confirmed.

Key pair owners: Signatories are also called key pair owners because they own a pair of a public and a private key.

Intended signatory: An individual or an organization that intends to generate signatures in the future is called an intended signatory [29].

Digital Signature Generation

The key pair owner uses the private key to generate digital signatures. A message digest is created using a hash function before the digital signature generation process. The message digest is a concise version of the data to be signed and is used as an input to generate digital signatures. After the generation of the message digest, a digital signature algorithm is selected, and other information pertinent to the algorithm is obtained (for example, Elliptic Curve Digital Signature Algorithm). A digital signature is generated by the intended signatory using the private key, message digest, a digital signature algorithm and other details. The signatory may validate the signature using the public key to identify computation errors in the digital signature generation process [29] (Figure 2.5).

Digital Signature Verification and Validation

The verifier performs the digital signature verification and validation process. The verifier uses the public key of the claimed signatory to verify digital signatures. The domain parameters of the digital signature algorithm used in the signature generation process are obtained from the claimed signatory or certificate authority (CA). For a digital signature to be checked and verified, a message digest is

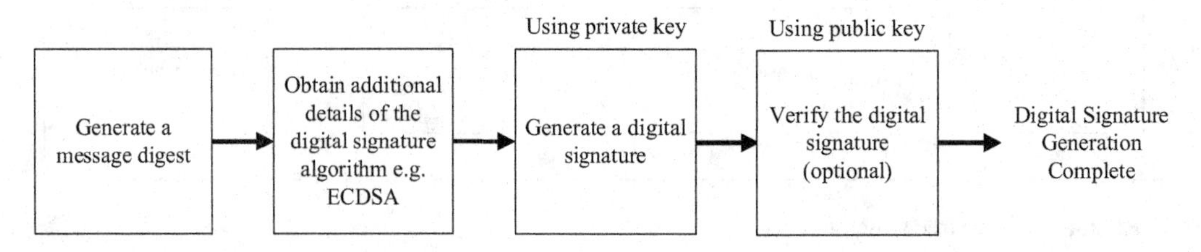

FIGURE 2.5 Digital signature generation.

Source: [29].

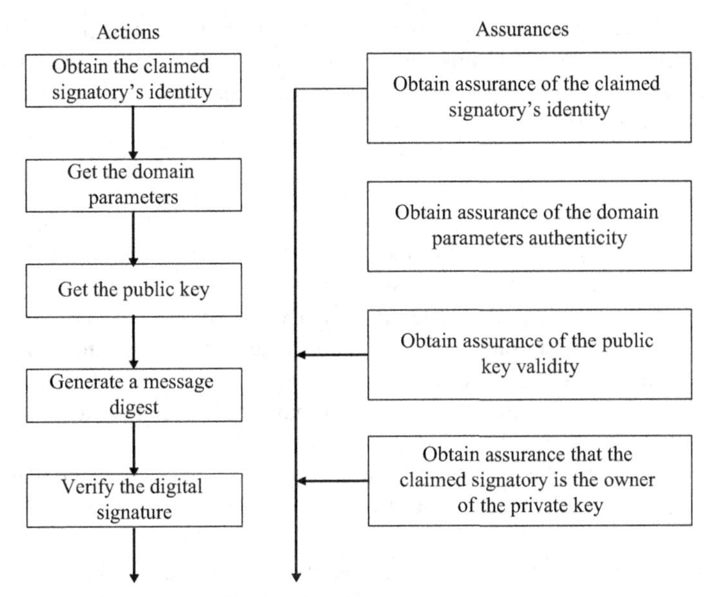

FIGURE 2.6 Digital signature verification and validation.

Source: [29].

generated. This message digest is created using the same hash function as in digital signature generation. The digital signature is verified using the digital signature algorithm, domain parameters, the public key and the new message digest.

Before the verified digital signature is approved, the verifier must have an assurance of the following:

1. Identification claimed by the signatory.
2. The domain parameters authenticity for ECDSA (Elliptic Curve Digital Signature Algorithm).
3. The public key validity.
4. That the private key owner is the claimed signatory.

All the methodologies to procure the above four assurances are given in SP (Special Publications) 800-89.

The digital signature and the signed data will be treated as valid if the verification and assurance processes are successful (Figure 2.6). The digital signature will be treated as invalid if the verification or assurance process fails [29].

ACCESS CONTROL, IDENTITY MANAGEMENT AND MEMBERSHIP SERVICE PROVIDERS

Access control is implemented to prevent attackers from accessing vulnerable devices. The access control model consists of five elements – subject, object, object owner, operation and reference monitor.

The description of each element is as follows:

Subject: A subject is an active entity capable of accessing objects.
Object: An object is an access-controlled element.

Object owner: An object owner owns the object and describes the access policies.
Operation: An operation is how the subject accesses an object. For example, read, write, delete, etc.
Reference monitor/access authority: A reference monitor decides to give or deny the object's access to the subject [65].

Identity Management

Identity management identifies, authenticates and authorizes users to access an organization's services; for example, the level of access, permissions and limitations for each user to validate their identity and access services [77]. A good identity management model helps improve decentralization in the blockchain. Some hierarchical identity management models are the conventional model, centralized model, federated identity model and user-centric identity model. These typical models are vulnerable to attacks because they have a central authority to control identity representing a single point of failure [36].

The self-sovereign identity (SSI) model is a new approach to identity model development in self-sovereign identity. SSI improves security and trust by allowing users to exercise absolute control over their identities [36].

The main characteristics of the self-sovereign identity model are as follows:

Existence: Individuals have a distinct existence and can never exist in digital form alone.
Persistence: Identities last as long as the user wants.
Portability: Identity information must be easily portable.
Control: Users have complete ownership of their identities and should be able to manage those identities.
Access: Users should have access to their identities conveniently without any facilitators and be aware of any changes made to their identity claims.
Transparency: Technologies used to control identities should be transparent about how they are run, handled and reconfigured and should grant open access to all users.
Interoperability: Users can widely utilize their identities as long as they have absolute control over them.
Consent: The identity information can be exchanged or used only with the user's permission.
Minimization: To improve security, minimal information should be used to carry out the necessary work, and data exposure must be limited.
Protection: In the case of a dispute between user rights and network needs, preference should be given to the user, thereby protecting their rights.

A self-sovereign identity model architecture has three roles, as follows:

- *Issuer*: An entity that creates and issues claims is called an issuer. A claim is a statement made by an entity.
- *Holder*: An entity that retains and exchanges the claims is called a holder.
- *Verifier*: A third party that validates the identity of the identity holder is called a verifier [36].

A range of identity management models like *Uport*, *Sovrin* and *ShoCard* has been developed to improve the decentralization functionality based on blockchain and SSI technology as described next.

Uport is an SSI-based open-source platform that allows users to register their own identities on a smart contract. Uport facilitates the one-way exchange (unidirectional) of identities between users. In this model, users need not expose personal details for restricted use. The threats to privacy in a Uport platform are still not clear.

Sovrin is a consortium blockchain. Users can access this blockchain without any prior approval. A Sovrin identity has a group of validator nodes called stewards to attain consensus and uses decentralized identifiers (both unidirectional and omnidirectional). In Sovrin, the users have complete ownership over their identities with the option of selecting the SSI characteristics that they want to share with the parties depending on them. Users communicate with Sovrin via a smartphone application and a software control agent. Despite all the benefits of this approach, the Sovrin platform is not commonly used because the users do not fully understand the functionality of the platform, hence the user interface is still a problem to be tackled.

ShoCard is a platform that facilitates the development of mobile identities to enable a shared user identity across various regions. The infrastructure of ShoCard focuses on the Bitcoin blockchain. This approach integrates mobile applications, blockchain and biometrics in a federated identity model (typical hierarchical model) considering the privacy and security factors. A ShoCard identity uses unidirectional identifiers, and the user has absolute control over their identities. The threats to privacy in ShoCard are still not clear. Users communicate with the ShoCard infrastructure via a smartphone application and use the QR code scanning model for all activities [36].

Membership Service Providers

A membership service provider (MSP) is a decentralized element used in identity management in a blockchain that verifies users who wish to access the blockchain network. Access control functionality is given to users of the blockchain network by the MSP.

The activities of the membership services are as follows:

1. Verification of user identity
2. Registration of users
3. Granting access based on user roles

The users are permitted to access the blockchain network when an identity is issued. Membership services use certificate authority (CA) to assist in identity management and authentication services. The certificate authority identifies all users and applications [12].

CRYPTO WALLETS

A wallet maintains a list of public and private keys that can be easily managed by the wallet user. This can be done in two ways; namely, watching the blockchain local copy or communicating with a copy that belongs to another full node user. In other words, the wallet generates a balance from transactions it can manage. With the use of private keys, transactions can be generated via the wallet when the wallet references a transaction that can be unlocked by a person with private keys, and with the address that the transaction wants to be relocked with. A wallet does not store coins but can reference any transactions on the blockchain. However, a wallet may store a local list of transactions that it has been engaged in, the user preferences and the updated balance that is constantly updated [34].

Types of Wallets

There are three types of software wallets, namely:

Full node wallets, where the entire blockchain is downloaded locally in the full node wallet. Transactions are processed and verified locally and then shared with their peers.

Thin node wallets use a connection to another full node when processing transactions.

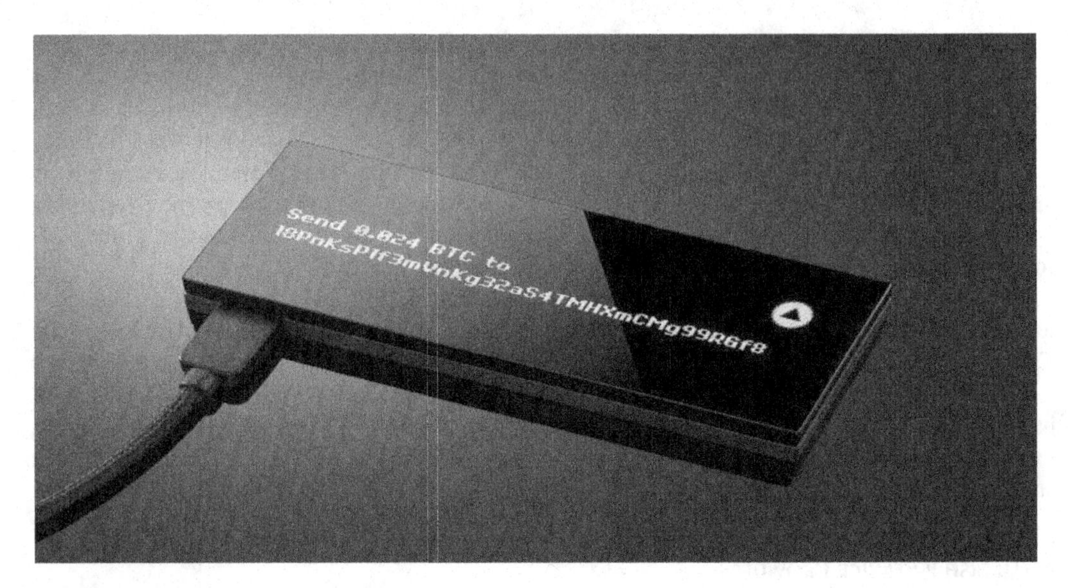

FIGURE 2.7 Example of a hardware wallet.
Source: [46].

Online wallets exist only on an online wallet site. The data transactions do not usually sync with a local client.

Hardware wallets are physical devices for storing private keys and data such as the account balance. A primary example of a hardware wallet is KeepKey, illustrated in Figure 2.7.

Hardware wallets, generally, are the most secure. As such, the owner's cooperation will be required to unlock it, in case of confiscation. However, in the case of losing the PIN or the device, the recovery capability can be used. If the investigator understands the recovery steps, the coins may be accessed without the need to recover the PIN.

Cold Wallet or Cold Storage

Cold storage refers to any key that is kept offline. It could be on a USB key, a paper note or a hardware wallet. Although there may be a key that is stored offline and there would be a need for it to be imported into a wallet to transfer funds out, the offline wallet will still be able to receive coins from the sender. This is because the coin was never received but it only references the address on the blockchain. As such, an address and its private key or a backup stored on a piece of paper in the safe can still get richer while being completely offline.

Setting up a paper wallet is very easy. The private key can be written and stored on a piece of paper and kept somewhere safe like a bank safety deposit box. Public/private key address pairs can also be generated without being online. The paper wallet is very secure because the key pair can be generated, and the private key would never have to appear in a wallet or on a computer until any funds received need to be moved to the public key address.

A useful tool for generating a public/private key pair is a free application called *WalletGenerator*, a universal paper wallet generator for cryptocurrencies [34] (Figures 2.8 and 2.9).

How Are Keys Stored in a Wallet?

Wallets store keys in several ways. These methods describe the ways public keys are created from a single or many private keys. Understanding how keys are created is important because it can help

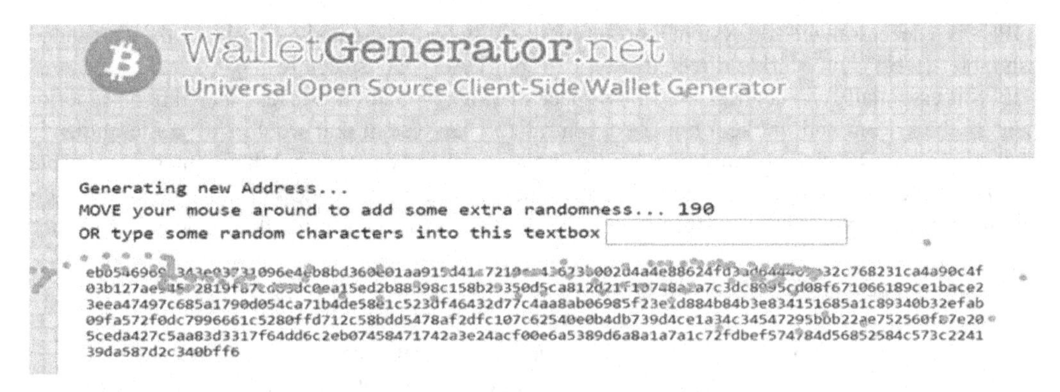

FIGURE 2.8 Example of a wallet generator.

Source: [78].

FIGURE 2.9 A generated public/private key pair ready to print in a paper wallet.

Source: [78].

in tracking wallets used by individuals or organizations with complex transaction needs. These methods are divided into three primary categories:

Nondeterministic: Also known as Type-0 or *just bunch of keys* (JBOK). Nondeterministic keys are stored in a simple list of public/private key pairs. This method means that there are a lot of keys that need to be managed. It also means that a lot of data need to be backed up or kept safe.

Deterministic: Also known as Type-1 or seeded wallets. In deterministic wallets, all the private keys are obtained from a single seed that is based on a random number. This method is considerably better because the seed only needs to be stored and backed up to be able to recover all the private keys generated. This makes managing the wallet much easier.

Hierarchical deterministic: Also known as HD wallets. This is the most up-to-date wallet protocol that is in use and was implemented in the Bitcoin core in 2016. As with the deterministic wallets, all the private keys are obtained from a single seed, but the keys in an HD wallet can generate their own private and public keys in a hierarchical structure. Also, the seed can be backed up, and the backed-up seed can be used to recover the entire structure of the tree. The seed is often represented by a series of 12 to 24 words [34].

INTER-PLANETARY FILE SYSTEMS

An Inter-Planetary File System (IPFS) is a protocol that allows media to be stored on a distributed file system [40]. It generates a unique fingerprint (Hash) when a file is added and removes duplicate files across the network [17].

IPFS is a decentralized file system whose functionality core is created by applying the distributed hash tables (DHT) and Merkle-directed acrylic graph (DAG) data structures [40].

IPFS is essentially a decentralized peer-to-peer sharing application, because there are no superior nodes in the system and unlike a typical decentralized file system that would save a file until it is deleted by a user, IPFS saves only files that are being used and ensures that they remain immutable [62]. Therefore, less used files are automatically deleted from the network. This attribute of the IPFS is a major reason why it is seemingly impossible to have a successful distributed denial of service attack on the network [56].

The sub-protocols of IPFS are as follows:

Identities ensure that processing nodes' identities are generated and verified as genuine.
Networks ensure that there is seamless connectivity to other networks and optionally check the integrity of messages sent in the network.
Routing ensures the exact delivery of information in the network to the intended recipient.
Exchange ensures data distribution is seamless. *BitSwap* is used to serve as a medium for nodes to gain the blocks they need.
Objects are used to represent data structures and communication hierarchy in a network.
Files ensure that there is a clear hierarchical structure in file systems.
Naming ensures the use of the self-certifying mutable name system [17].

IPFS has a name service called *Inter-planetary Name System* (IPNS). IPNS allows IPFS files to be identified by names that are easy to identify and understand by users. This is helpful for users to easily identify a file to be downloaded after an update has occurred.

An advanced feature of IPFS is that it supports file versioning using data structures like GIT, a type of software used for identifying and tracking any changes in files and also to enhance coordinated work during software development among multiple people. IPFS also makes use of directories and files using the mountable file system known as File System in User Space (FUSE) [62].

There are four types of IPFS files as follows:

Blob is a representation of a major part of a stored file in the IPFS.
List is a representation of a complete file of a combination of blobs and other lists designed to compress data in the network.
Tree is a representation of a directory. It holds a combination of blobs, lists, trees and commit.
Commit is a snapshot of the history of any file on the network.

Since every file in the network is identified by its hash, the hash of a category is required to download any file. For example, to download a snapshot of a file, the hash of the IPFS Commit is required [40].

SMART CONTRACTS

Nick Szabo postulated smart contracts in the late 1990s in an article entitled *Formalizing and Protecting Partnerships on Public Networks*. A smart contract is a secure and decentralized program that runs on top of the blockchain. It represents an agreement that is executable and enforceable automatically when certain conditions are met. A smart contract is a decentralized framework. The blockchain mechanism's major privacy benefits made blockchain a standard decentralized model for smart contracts [14].

Smart Contract Lifetimes

Smart contracts are software that executes autonomously on a blockchain once it has been deployed. The storage of smart contracts on the blockchain is done openly, as smart contracts can be read and used by anyone on the blockchain. Smart contracts cannot be altered and are permanent on a blockchain. In the blockchain network, smart contract execution is done by 'workers' (commonly known as miners or validators) who earn cryptocurrency as a payment for smart contract execution [53].

Smart contracts have characteristics that should be considered when designing a distributed system:

Smart contracts are immutable: Smart contracts cannot be modified once they have been deployed. As such, if a contract application logic has a flaw, the flaw will exist forever since there are no ways of making updates.

Smart contracts are open: In blockchain applications, one of the core principles is transparency. In other words, the records stored in the blockchain are accessible to everybody and since the smart contract is a component of the blockchain, the 'source code' of the smart contract can be viewed by anybody. Therefore, methods and algorithms that will be kept secret should not be implemented by smart contracts.

Information that relates to smart contracts is always available: Apart from the code of a smart contract, all blockchain users can view the values that the contract variables hold, the historical data and all the transactions that are related to the contract.

The above-mentioned attributes have implications such as:

1. Smart contracts cannot be used for storing private data such as private keys and protected records.
2. Smart contracts cannot be used in performing operations that require secret information, e.g., in the creation of a digital signature.
3. Smart contracts cannot be used in generating secret information, e.g., in the generation of a secret key [31].

How Developers Ensure Users Cannot Exploit Bugs or Unintended Functionality

Having in mind that the internal code of smart contracts cannot be changed once deployed, developers are well advised to use defensive programming techniques through testing procedures, measuring the test coverage, setting up continuous integration and performing security audits on the contract which will ensure that the users cannot exploit bugs or unintended functionalities [53].

Smart contract developers should consider a 'kill switch', which is a piece of code that will leave the smart contract inoperable permanently. Invoking a 'kill switch' should result in all funds associated with the smart contract being transferred to the contract owner. The kill switch should also result in the prevention of users from interacting with the referred contract in the future. It should be pointed out that even if a smart contract is 'killed', the code and data of the contract will remain on the blockchain [31].

Turing Completeness

Turing completeness is a property that defines programming languages used in simulating a Turing machine. Alan Turing developed the idea of the Turing machine in the 1930s. The Ethereum Virtual Machine is a well-known example of a Turing complete smart contract device [67].

To verify if Turing complete languages are necessary or not, the Ethereum blockchain is evaluated for various control flow mechanisms, such as loops and recursion functions. Only 35.3% of the smart contracts use control flow mechanisms. Therefore, the use of non-Turing complete smart contracts is sensible. Note that the Bitcoin platform uses non-Turing complete languages [44].

Turing completeness introduces an unwanted and burdensome attack surface along with it. A real example is the Decentralized Autonomous Organization (DAO) attack in 2016, which was triggered as a result of Turing completeness. This attack robbed 3.6 M ETH (Ether), equivalent to USD 50M. Thus, Turing complete smart contracts could be high-risk implementations [67].

Untrusted Code

Unexpected risks and failures may occur from calls to smart contracts built using untrusted code. These calls may introduce malicious code into the smart contracts. Interacting with untrusted code is unsafe. It may disrupt the control flow leading to many vulnerabilities.

To minimize the danger, it is recommended to delegate calls to trusted contracts only. When calling an untrusted contract, state changes should be avoided after the call. The use of the 'call()' in the code instead of using the 'transfer()' and 'send()' functions is recommended [25].

Chapter 14 discusses smart contract attacks and best practices in greater detail.

CONCLUSIONS AND RECOMMENDATIONS

This chapter discussed the various blockchain components, cryptographic principles and algorithms, differences between public and private and open and closed blockchains, how Merkle trees function in blockchain, blockchain-related access control concepts, the use of hashing algorithms, smart contracts and cryptocurrency wallets. This information-rich chapter is an attempt to familiarize readers with the architectural components of blockchain platforms.

CORE CONCEPTS

- A blockchain network is created by linking multiple nodes and running software, with transactions being requested through a wallet. This transaction is sent to all processing nodes on the blockchain network with each processing node validating the transaction against predefined rules given by the blockchain network developers. All validated transactions are then stored in a block and secured with a hash. When other processing nodes in the network confirm that the hash is right, the block is added to the blockchain signifying the completion of the transaction and establishing immutability [20] (Figure 2.10).
- There are a number of important blockchain-related roles and responsibilities. These include users, processing nodes, blockchain regulators, blockchain developers, blockchain administrators and a certificate authority. These various participants – although quite different in their network roles – exhibit synergy. These interactions are what enable blockchain transactions to be successfully generated throughout the network.
- Blockchains may be broken down into four main categories: Public, private, open and closed. All users can read the stored data in an open blockchain but only allowed participants can read the data in a closed blockchain.
- Consensus algorithms refer to the various mechanisms that have been developed over time to ensure agreement in blockchain transaction. This is a pivotal aspect in blockchain as it determines the speed, reliability and further development of the blockchain network. Although there are several consensus mechanisms currently available as consensus algorithms, there is room for further development of new mechanisms.

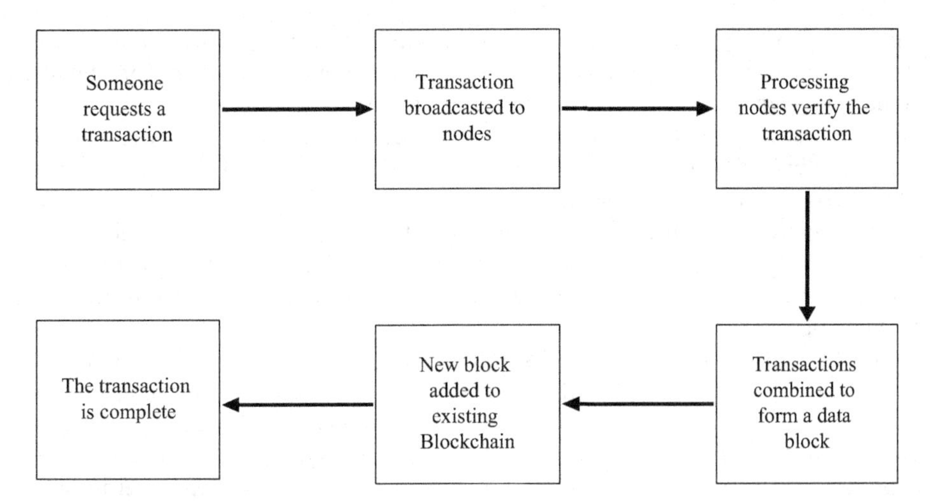

FIGURE 2.10 Blockchain transaction lifecycle.

Source: [20] Created with Visio.

- A nonce is a random numerical value that constitutes the heart of the proof of work consensus mechanism. The processing nodes in a blockchain network aim at producing a nonce that meets the required hash value during a PoW process (for example: a hash that starts out with four zeros). Once the PoW process produces a suitable nonce, it can be easy verified by the other participating nodes in order to reach PoW consensus. Once verified by the other nodes, the node that determined the nonce value first will get to create the corresponding block and earn the block reward; typically a cryptocurrency sum.
- Cryptography is an essential part of blockchain technology. There are two major categories of algorithms in cryptography, namely symmetric algorithms and asymmetric algorithms. In symmetric algorithms, one key is used for encryption and decryption, while in asymmetric algorithms, two keys are used, which are the public key and the private key that are also known as a 'key pair'. The encryption is done using a public key, and decryption is done using the private key. In blockchain technology, cryptography is used to protect the identities of the network users, to ensure the transactions are secure and to protect all the valuable information.
- A hash is an encrypted value of a fixed length of numbers and letters that do not match the input. The security properties of a hash function are collision resistance, one-way functionality, non-locality and large state space. Some examples of hash functions are MD4, MD5 and SHA256. Hash functions are used for password verification, signature generation and authentication, data authentication and message integrity.
- A Merkle tree records all transactions in a block. Merkle trees require less memory space. A Merkle root is the hash left by continuously hashing pair of nodes to generate Merkle trees and is stored in the block header. The Merkle root authenticates transaction data and proves data integrity.
- Digital signatures prove identification. A digital signature generation and verification process is included in the digital signature algorithm. A digital signature is generated using a private key and is verified using the public key. The verifier must have four assurances before the verified digital signature is approved: assurance of the claimed signatory's identity, assurance of the domain parameters authenticity, assurance of the public key validity and assurance that the claimed signatory is the owner of the private key.
- Access control is used to prevent cybercriminals from accessing vulnerable systems. Identity management is used to identify, validate and authorize users to access the resources of an

organization. Compared to the hierarchical identity management models, the self-sovereign identity model improves security and trust by allowing users to have control of their identities. Some other identity models are *ShoCard*, *UPort* and *Sovrin*.

- The inter-planetary file system can accommodate peer-to-peer, decentralized resource sharing. It aims at eliminating the inefficiencies that arise from the absence of files in some units across a network by seamlessly connecting all devices in the network to the required files. It also ensures the security of files shared within the network. The inter-planetary file system uses the name service called Inter-Planetary Name System (IPNS). This is a decentralized form of domain name system (DNS). It helps users to name files in a manner that is easily decoded by humans.
- A Membership Service Provider (MSP) gives users the access control functionality of the blockchain network. An MSP verifies and registers users on the blockchain network using a certificate authority and provides access to services based on user roles.
- There are different types of wallets: software wallets, hardware wallets and cold wallets. It is important to be familiar with the wallet types and their different features in terms of convenience and security.
- Smart contracts have features such as transparency, and protocols for user authentication and token transfer. However, smart contracts cannot be modified once deployed on the blockchain. Smart contracts do not preserve user privacy and should not store or be used to create secret information. Due to the weaknesses and vulnerabilities of smart contracts, developers should use defensive programming techniques and implement a 'kill switch' to render the smart contract inoperable, if warranted.

ACTIVITY FOR BETTER UNDERSTANDING

Using your web browser, go to etherscan.io (The Ethereum Blockchain Explorer). Spend some time exploring this platform to better familiarize yourself with the inner-workings of it; then try to answer the following questions:

1. What is the current price of Ether?
2. How fast are transactions being posted?
3. Click on a transaction link on etherscan.io and determine the ether balance and transactional costs for the transaction.
4. How fast are blocks created?

REFERENCES

[1] Acar, A., Aksu, H., Uluagac, A. S., & Conti, M. (2018). A Survey on Homomorphic Encryption Schemes. *ACM Computing Surveys*, *51*(4), 1–35. https://doi.org/10.1145/3214303

[2] Agrawal, M., & Mishra, P. (2012). A Comparative Survey on Symmetric Key Encryption Techniques. *International Journal on Computer Science and Engineering (IJCSE)*, *4*(5), 877–881.

[3] Ahmed, M., Sanjabi, B., Aldiaz, D., Rezaei, A., & Omotunde, H. (2012). Diffie-Hellman and Its Application in Security Protocols. *International Journal of Engineering Science and Innovative Technology (IJESIT)*, *1*(2), 69–73.

[4] Alderson, E. (2017). *Elliptic Curve Cryptography and Quantum Computing* (thesis).

[5] Al-Odat, Z. A., & Khan, S. U. (2019). Constructions and Attacks on Hash Functions. *2019 International Conference on Computational Science and Computational Intelligence*, 141–142. https://doi.org/10.1109/csci49370.2019.00030

[6] Al-Odat, Z. A., & Khan, S. U. (2019). Constructions and Attacks on Hash Functions. *2019 International Conference on Computational Science and Computational Intelligence*, 139–140. https://doi.org/10.1109/csci49370.2019.00030.

[7] Badretdinov, T. (2018, July 16). How to Create a Bitcoin Wallet Address from a Private Key. Medium. https://medium.com/free-code-camp/how-to-create-a-bitcoin-wallet-address-from-a-private-key-eca3ddd9c05f.

[8] Bangerter, E., Barzan, S., Krenn, S., Sadeghi, A.-R., Schneider, T., & Tsay, J.-K. (2013). International Workshop on Security Protocols 2009: Security Protocols XVII. In *Bringing Zero-Knowledge Proofs of Knowledge to Practice* (Vol. 7028, Ser. Lecture Notes in Computer Science book series, pp. 51–62). Berlin: Springer. https://doi.org/10.1007/978-3-642-36213-2_9

[9] Baset, S. A., Desrosiers, L., Gaur, N., & Novotny, P. (2019). *Blockchain development with Hyperledger: Build Decentralized Applications with Hyperledger Fabric and Composer.* Packt Publishing.

[10] Bashir, I. (2017). Chapter 1 Types of Blockchain. In *Mastering Blockchain: Deeper Insights into Decentralization, Cryptography, Bitcoin, and Popular Blockchain Frameworks* (p. 75). Packt Publishing.

[11] Bashir, I. (2017). Chapter 9 Hyperledger. In *Mastering Blockchain: Deeper Insights into Decentralization, Cryptography, Bitcoin, and Popular Blockchain Frameworks* (pp. 507). Packt Publishing.

[12] Bashir, I. (2018). Hyperledger. In *Mastering Blockchain: Distributed Ledger Technology, Decentralization, and Smart Contracts Explained* (Second, pp. 672, 686). Packt Publishing.

[13] Bashir, I. (2018). *Mastering Blockchain: Distributed Ledger Technology, Decentralization, and Smart Contracts Explained* (Second). Packt Publishing.

[14] Bashir, I. (2018). Smart Contracts. In *Mastering Blockchain: Distributed Ledger Technology, Decentralization, and Smart Contracts Explained* (Second, pp. 95, 402–403). Packt Publishing.

[15] Bashir, I., & Prusty, N. (2019). In *Advanced Blockchain Development: Build Highly Secure, Decentralized Applications and Conduct Secure Transactions* (pp. 34–35). Packt Publishing.

[16] Batten, L. M. (2013). Elliptic Curve Key Exchange (ECKE). In *Public Key Cryptography: Applications and Attacks* (p. 95). John Wiley & Sons.

[17] Benet, J. IPFS Design. In *IPFS- Content Addressed, Versioned, P2P File System.* https://ipfs.io/.

[18] Biglar, A. (2018). Some Applications of Knapsack Problem. *ResearchGate*, 1. https://doi.org/10.13140/RG.2.2.15115.39209.

[19] Bosamia, M., & Patel, D. (2018). Current Trends and Future Implementation Possibilities of the Merkel Tree. *International Journal of Computer Sciences and Engineering*, *6*(8), 294–301. https://doi.org/10.26438/ijcse/v6i8.294301

[20] Botjes, E. (2017, August 11). Pulling the Blockchain Apart: The Transaction Life-Cycle. https://medium.com/ignation/pulling-the-blockchain-apart-the-transaction-life-cycle-7a1465d75fa3.

[21] Castro, M., & Liskov, B. (1999). Practical Byzantine Fault Tolerance. In *Proceedings of the Third Symposium on Operating Systems Design and Implementation.*

[22] Chandra, S., Bhattacharyya, S., Paira, S., & Alam, S. S. (2015). 2014 International Conference on Science Engineering and Management Research (ICSEMR). In *A Study and Analysis on Symmetric Cryptography* (pp. 1–8). Chennai; IEEE.

[23] Chandra, S., Paira, S., Alam, S. S., & Sanyal, G. (2014). A Comparative Survey of Symmetric and Asymmetric Key Cryptography. *2014 International Conference on Electronics, Communication and Computational Engineering*. https://doi.org/10.1109/icecce.2014.7086640

[24] Chuen, D. L. E. E. K., & Deng, R. H. (2018). Blockchain: From Public to Private. In *Handbook of Blockchain, Digital Finance, and Inclusion, Volume 2: ChinaTech, Mobile Security, and Distributed Ledger* (Vol. 2, pp. 145–177). Academic Press.

[25] ConsenSys. (2020, July 10). *Ethereum Smart Contract Security Recommendations*. ConsenSys. https://consensys.net/blog/blockchain-development/ethereum-smart-contract-security-recommendations/.

[26] Cook, J. D. (2020, January 17). Bitcoin Keys and Elliptic Curve secp256k1. Applied Mathematics Consulting. https://www.johndcook.com/blog/2018/08/14/bitcoin-elliptic-curves/.

[27] Daeri, A., Zerek, A. R., & Abuinjam, M. A. (2014). International Conference on Control, Engineering and Information Technology (CEIT'14) Proceedings. In *ElGamal Public-Key Encryption* (pp. 115–117). IPCO.

[28] Dhillon Metcalf, V., Metcalfe, D., & Hooper, M. (2020). *Blockchain Enabled Applications: Understand the Blockchain Ecosystem and How to Make It Work... for You.* APRESS.

[29] Digital Signature Standard (DSS). (2013), 9–14. https://doi.org/10.6028/nist.fips.186-4

[30] Elliptic Curve Cryptography. girlstalkmath.com. (2017, July 3). https://girlstalkmath.com/2017/06/30/elliptic-curve-cryptography/#:~:text=One%20main%20disadvantage%20of%20ECC,used%20form%20of%20cryptography%2C%20RSA.

[31] Fotiou, N., & Polyzos, G. C. (2018). Smart Contracts for the Internet of Things: Opportunities and Challenges. *2018 European Conference on Networks and Communications (EuCNC)*. https://doi.org/10.1109/eucnc.2018.8443212

[32] Frankenfield, J. (2020, February 4). *Cryptographic Hash Functions Definition*. Investopedia. https://www.investopedia.com/news/cryptographic-hash-functions/.

[33] Frankenfield, J. (2020, June 30). *Understanding Hash*. Investopedia. https://www.investopedia.com/terms/h/hash.asp.

[34] Furneaux, N. (2018). Wallets. In *Investigating Cryptocurrencies: Understanding, Extracting, and Analyzing Blockchain Evidence* (pp. 95–107). John Wiley & Sons.

[35] Guegan, D. (2017). *Public Blockchain versus Private Blockchain*, 1–8. https://doi.org/halshs-01524440

[36] Haddouti, S. E., & Ech-Cherif El Kettani, M. D. (2019). Analysis of Identity Management Systems Using Blockchain Technology. *2019 International Conference on Advanced Communication Technologies and Networking*, 2–5. https://doi.org/10.1109/commnet.2019.8742375

[37] Hankerson, D. R., Vanstone, S. A., & Menezes, A. J. (2004). In *Guide to Elliptic Curve Cryptography* (pp. 2–186). Springer.

[38] Hargrave, J., & Karnoupakis, E. (2019). *What Is Blockchain?* O'Reilly Media.

[39] Hellwig, D., Karlic, G., & Huchzermeier, A. (2020). *Build Your Own Blockchain: A Practical Guide to Distributed Ledger Technology*. Springer. https://doi.org/10.1007/978-3-030-40142-9

[40] Hill, B., Chopra, S., Valencourt, P., & Prusty, N. (2018). *Blockchain Developers Guide*. Packt Publishing

[41] Hossain, M. A., Islam, M. K., Das, S. K., & Nashiry, A. (2012). Cryptanalyzing of Message Digest Algorithms md4 and md5. *International Journal on Cryptography and Information Security*, *2*(1), 12. https://doi.org/10.5121/ijcis.2012.2101

[42] Huawei Technologies. (2018). Huawei Blockchain Whitepaper Toward a Trusted Digital World. White paper. https://www.huaweicloud.com/content/dam/cloudbu-site/archive/hk/en-us/about/analyst-reports/images/4-201804-Huawei%20Blockchain%20Whitepaper-en.pdf.

[43] Jagati, S. (2020, July 24). *Zero-Knowledge Proofs, Explained*. Cointelegraph. https://cointelegraph.com/explained/zero-knowledge-proofs-explained.

[44] Jansen, M., Hdhili, F., Gouiaa, R., & Qasem, Z. (2019). Do Smart Contract Languages Need to Be Turing Complete? *Advances in Intelligent Systems and Computing*, 19–26. https://doi.org/10.1007/978-3-030-23813-1_3

[45] Kalra, S., & Sood, S. K. (2011). Elliptic Curve Cryptography. *Proceedings of the International Conference on Advances in Computing and Artificial Intelligence – ACAI '11*. https://doi.org/10.1145/2007052.2007073

[46] KeepKey. (2020). *KeepKey Hardware Wallet*. https://keepkey.myshopify.com/products/keepkey-the-simple-bitcoin-hardware-wallet.

[47] Kent, P., & Bain, T. (2019). *Cryptocurrency Mining for Dummies*. Wiley.

[48] Kenton, W. (2020, November 21). What Is a Stealth Address? Investopedia. https://www.investopedia.com/terms/s/stealth-address-cryptocurrency.asp.

[49] Kilroy, K. (2019). *Blockchain as a Service*. O'Reilly Media.

[50] Kim, S., Deka, G. C., & Zhang, P. (2019). Chapter 7 Consensus Mechanisms and Information Security Technologies. In *Role of Blockchain Technology in IoT Applications* (pp. 181–209). Academic Press.

[51] Kurmi, J., & Sodhi, A. (2015). A Survey of Zero-Knowledge Proof for Authentication. *International Journal of Advance Research in Computer Science and Software Engineering*, *5*(1), 494–500. https://doi.org/https://www.researchgate.net/publication/316492793

[52] Lake, J. (2018, December 10). [web log]. https://www.comparitech.com/blog/information-security/rsa-encryption/.

[53] Margaria, T., Steffen, B., Colombo, C., Ellul, J., & Pace, G. J. (2018). Contracts over Smart Contracts: Recovering from Violations Dynamically. In *Leveraging Applications of Formal Methods, Verification and Validation. Industrial Practice 8th International Symposium, ISoLA 2018*, Limassol, Cyprus, November 5-9, 2018, Proceedings, Part IV (pp. 300–315). Springer.

[54] Moh, H. H. (2014). *Implementation of (Aes) Advanced Encryption Standard Algorithm in Communication Application*. Thesis, UMP, Kuantan, Pahang.

[55] Mohan, C. (2019). State of Public and Private Blockchains. *Proceedings of the 2019 International Conference on Management of Data – SIGMOD '19*. https://doi.org/10.1145/3299869.3314116

[56] Mölken, R. van. (2018). *Blockchain across Oracle: Understand the Details and Implications of the Blockchain for Oracle Developers and Customers*. Packt Publishing.

[57] MTEK Labs. (2019, June 1). AES-256, the Encrypting Algorithm Increasing the Blockchain Security. https://mteklabs.tech/aes-256-the-encrypting-algorithm-increasing-the-blockchain-security/.

[58] Nithya, S., & Raj, E. G. D. P. (2014). Survey on Asymmetric Key Cryptography Algorithms. *Journal of Advanced Computing and Communication Technologies*, *2*(1), 1–4.

[59] Pittalia, P. P. (2019). A Comparative Study of Hash Algorithms in Cryptography. *International Journal of Computer Science and Mobile Computing*, *8*(6), 148–149. https://ijcsmc.com/docs/papers/June2019/V8I6201928.pdf.

[60] Poston, H. (2020, September 29). *Hash Functions in Blockchain*. Security Boulevard. https://securityboulevard.com/2020/09/hash-functions-in-blockchain/.

[61] Rachmawati, D., Tarigan, J. T., & Ginting, A. B. (2018). A Comparative Study of Message Digest 5(MD5) and SHA256 Algorithm. *Journal of Physics: Conference Series*, *978*, 1. https://doi.org/10.1088/1742-6596/978/1/012116

[62] Rahmati, P. (2019, October 19). Simplified Byzantine Fault Tolerance for Blockchain Applications. *Medium*.

[63] Raj, K. (2019). *Foundations of Blockchain: The Pathway to Cryptocurrencies and Decentralized Blockchain Applications*. Packt Publishing.

[64] Ray, S. (2017, December 14). *Merkle Trees*. Hacker Noon. https://hackernoon.com/merkle-trees-181cb4bc30b4.

[65] Riabi, I., Dhif, Y., Ben Ayed, H. K., & Zaatouri, K. (2019). A Blockchain Based Access Control for IoT. *2019 15th International Wireless Communications & Mobile Computing Conference (IWCMC)*, 2086. https://doi.org/10.1109/iwcmc.2019.8766506

[66] Rivest, R. L. (2014). Cryptography. In J. V. Leeuwen (Ed.), *Algorithms and Complexity* (Vol. A, pp. 719–750). Elsevier Science Publisher B.V.

[67] Rolfe, T. (2019, August 11). *Turing Completeness and Smart Contract Security*. Medium. https://medium.com/kadena-io/turing-completeness-and-smart-contract-security-67e4c41704c.

[68] Rousey, M. (2020, February 1). New Technology Explained Zero-Knowledge Proof – Security Enhancing Protocol. https://changelly.com/blog/zero-knowledge-proof-explained/.

[69] Schneier on Security. *Products That Use Blowfish*. https://www.schneier.com/academic/blowfish/products/.

[70] Schurtenberger, M. (2020, March 7). *Public vs Private Blockchains: Why Public Blockchains Are the Future*. Bitcoin Suisse. https://www.bitcoinsuisse.com/outlook/why-public-blockchains-are-the-future.

[71] Security, S. (2020, October 22). *Decoded: Examples of How Hashing Algorithms Work*. Savvy Security. https://cheapsslsecurity.com/blog/decoded-examples-of-how-hashing-algorithms-work/.

[72] Sharma, D., Sharma, P., & Deep, V. (2019). Homomorphic Encryption – Need of the Hour. International Journal of Engineering and Advanced Technology, 8(6S), 748–751. https://doi.org/10.35940/ijeat.f1145.0886s19

[73] Stinson, D. R., & Paterson, M. B. (2019). In *Cryptography: Theory and Practice* (4th ed., pp. 1–186). CRC Press Taylor & Francis Group.

[74] Stošic, L., & Bogdanovic, M. (2012). RC4 Stream Cipher and Possible Attacks on WEP. *International Journal of Advanced Computer Science and Applications*, *3*(3). https://doi.org/10.14569/ijacsa.2012.030319

[75] Sunnatovich, M. I. (2019, May 3). *Use Cases for Hash Functions or What Is SHA-256?* Medium. https://medium.com/@makhmud.islamov/use-cases-for-hash-functions-or-what-is-sha-256-83036de048b4.

[76] Tilborg, H. C. A. van (Ed.). Encyclopedia of Cryptography and Security, 260–261. https://doi.org/10.1007/0-387-23483-7

[77] Tykn. (2010, June 22). *Blockchain Identity Management: The Definitive Guide (2020 Update)*. Tykn. https://tykn.tech/identity-management-blockchain/.

[78] *Walletgenerator.net – Universal Open Source Client-Side Wallet Generator*. https://walletgenerator.net/.

[79] Welfare, A. (2019). Commercializing Blockchain. In *Commercializing blockchain: Strategic applications in the real world* (pp. 205–352). Chichester: Wiley.

[80] Wolchover, N. (2015). *Quantum-Secure Cryptography Crosses Red Line*. Quanta Magazine. https://www.quantamagazine.org/quantum-secure-cryptography-crosses-red-line-20150908/.

[81] Yaga, D., Mell, P., Roby, N., & Scarfone, K. (2018). Blockchain Technology overview, 7–12. https://doi.org/10.6028/nist.ir.8202

[82] Yang, X., Liu, J., & Li, X. (2019). Research and Analysis of Blockchain Data. *Journal of Physics: Conference Series*, *1237*, 3. https://doi.org/10.1088/1742-6596/1237/2/022084

[83] Young, C. (2021, February 11). Why Bitcoin Mining Consumes More Electricity Than Entire Countries. Retrieved March 15, 2021, from https://interestingengineering.com/why-bitcoin-mining-consumes-more-electricity-than-entire-countries

[84] Zahan, A., Hossain, M. S., Rahman, Z., & Shezan, S. K. A. (2020). Smart Home IoT Use Case with Elliptic Curve Based Digital Signature: An Evaluation on Security and Performance Analysis. *International Journal of Advanced Technology and Engineering Exploration*, 7(62), 11–19.

3 Blockchain Tokens and Cryptocurrencies

Lavanya Vaddi, Jaskaran Singh Chana and Gurjot Singh Kocher

ABSTRACT

The first three chapters of this book aim at providing readers with a fundamental understanding of the various attributes and components of blockchain technology. For many public blockchain implementations, the use of tokens or cryptocurrencies constitutes a major benefit. As such, this chapter aims at first familiarizing readers with various blockchain governance types and considerations in order to explain the concept of a 'fork' in a blockchain. It is also important to note that a permanent or hard blockchain fork often results in the creation of a new type of cryptocurrency. The chapter provides several examples of such occurrences.

The various types of blockchain token types such as equity and utility tokens, as well as current token standards, such as ERC 20, 223 and 721 are also discussed in this chapter. The last major section of this chapter provides readers with an overview of some more popular cryptocurrency types such as Bitcoin, Ether, Ripple, Litecoin and Monero, to name a few. Finally, a brief discussion of some major, current crypto exchange platforms, such as Bittrex, Coinbase and Bitfinex, is also provided.

As mentioned in the chapter, as of February 2020, there were close to 5,400 cryptocurrencies created with a total market capitalization of over $200,000,000. While the long-term performance and sustainability of most cryptocurrencies currently on the market remain highly uncertain, it is important for auditors and aspiring blockchain professionals to be familiar with the basic attributes of the more popular cryptocurrencies discussed in this chapter.

BLOCKCHAIN GOVERNANCE

As discussed in previous chapters, blockchain technology offers important advantages and control to its clients with respect to their needs. Organizational governance is defined as a system by which an organization reaches and implements decisions related to the pursuit of its objectives. Governance structure is accepted and followed by all the participants or users. Although governing bodies are typically centralized in nature, blockchain is a peer-to-peer decentralized system that keeps on evolving with time and need.

Blockchain governance is the system via which blockchain could adapt to changing circumstances and requirements. Adaptability and upgradability form the crucial characteristics of blockchain governance. Broadly speaking, blockchain governance generally involves the following four central bodies:

Core developers are responsible for keeping up the fundamental code supporting the blockchain.

Node operators are the set of operators who keep a full copy of the blockchain record. The developers of the code rely upon node operators to accept the features they supply. Node

DOI: 10.1201/9781003211723-3

operators make the decisions related to the implementation of the features offered by the code developers on the nodes.

Token holders include the clients/users of the blockchain token. Investors form a major portion of this group of stakeholders who possess the right to vote with respect to feature implementations, upgrades, etc.

Blockchain team consists of a firm or non-government organization (NGO) primarily responsible for fundraising and project development tasks. These teams consist mostly of investors and supporters who negotiate with the node operators and code developers regarding the functioning of the blockchain [31].

Popular Blockchain Governance Strategies

Off-chain governance maintains a balance between the users, miners and organizations. This governance system has a centralized nature where the main parties who developed a blockchain make various governance-related decisions. The parties who were not involved in the core development of the blockchain lack the knowledge and expertise to make governance-related decisions; therefore, this technique follows a direct governance approach in which everyone votes directly on every decision. Bitcoin and Ethereum are examples of these governance techniques. To make up for the drawbacks of off-chain governance, new techniques were developed in the on-chain governance system discussed later.

Benevolent dictator for life is the simplest strategy of governance in which the monopoly of a single person (original creator/lead developer) like Vitalek Buterin, the creator of Ethereum, prevails in terms of the power over blockchain governance decisions.

Core dev team governance is often employed in situations where the most prominent core developers are responsible for making decisions and the structure map of blockchain. Although any client has the power to request a feature, the final decision stays with the core developers.

Open governance is a technique used by Hyperledger and Corda where a team is elected by the mutual decision of the client community in a democratic manner. This elected team is responsible for making various governance-related blockchain decisions.

On-chain governance may be preferred in blockchain-based systems to help ensure that none of the users or a group of users could enforce their will on a blockchain. Interaction rules between the users are defined in this kind of governance. The interactions are completely based on rules of the blockchain infrastructure, refereed as '*rule of code*'. Rule sets could be layered as one rule set could be the base for other rules. When the rules and the processes of decision-making are encoded into the infrastructure of the blockchain system, the governance model is referred to as on-chain governance. In on-chain governance, consensus is reached via a direct voting scheme. The outcome of the voting is calculated algorithmically and run according to the blockchain protocol [28]. Involvement of the node user is not necessary, because the process follows the on-chain process. This results in more effective decision-making, thereby reducing the possibility of a hard fork. There are also some challenges related to on-chain governance, such as scalability. As a blockchain network continues to grow, it will become increasingly more difficult to enforce governance rules in an effective and efficient manner [31].

BLOCKCHAIN FORKS

A cryptocurrency could go through a substantial impact due to a blockchain fork. As a matter of fact, a permanent fork could be the cause for the creation of a new cryptocurrency. For example, in August 2017, the Bitcoin community had a disagreement which led to a *Bitcoin Cash* fork on the Bitcoin blockchain. In July 2016, a major part of the Ethereum community modified the consensus rules as a response to the DAO attack which resulted in two chains with two cryptocurrencies; one

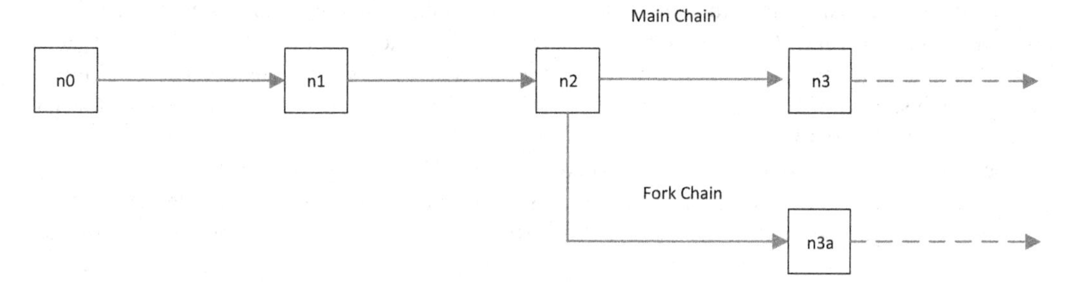

FIGURE 3.1 Blockchain forks illustrated.

is *Ethereum Classic* and the other is *Ether.* A fork could be defined as the modification of consensus rules which might or might not result in split of a blockchain network and a new version of the blockchain emerging. Decentralization is one of the distinct characteristics of blockchains. Blockchain forks are one of the implications of decentralization (Figure 3.1) [33].

Why Forks Happen

Due to disagreement of some users on the present blockchain, the blockchain splits into different instances. Occasionally, where different blocks are generated at the same time, the short-term uncertainty is noticed. At other times, it is seen as a modification in the consensus rules which might cause a new version of the blockchain. Forks might pave the way to the new attacks and menace data protection using cross-chain analysis. Overall, any ensuing fork compromises trust and induces uncertainty [33].

Types of Forks

Blockchain forks are mainly divided into two types; process-based and protocol-based forks. As the fork is the state of a network where there would be two or more active chains, many forks will have an instant effect and dissolve quickly, while some forks split permanently. As the consensus rules change, some followers of the older blockchain will decide to stay with the same version, while some choose to upgrade to the new version; therefore, the blockchain splits and results in two versions of that blockchain. The fork is called process-based, if the blocks have the same acceptance set and the state of network is determined by the allotment of the consensus related resources. An acceptance set is the criteria for a block to be approved, so that it can be appended to the blockchain.

The protocol-based fork emerges where there is a difference in the acceptance of new blocks. The differences are caused due to the changes in the consensus ruleset. The process-based fork is unintentional, if two or more consensus-relevant nodes found a new block at the same time. It is referred to as the probabilistic block race where the condition gets resolved in a short time. The intentionally caused process-based forks are a little complex and difficult to resolve. They emerge if the consensus-relevant nodes successfully create a new block while holding the information of the existence of it. This phenomenon is called *selfish mining*. Process-based forks may lead to the record or completely delete the transactions in the dominant version. The protocol-based fork might also emerge unintentionally, if there are bugs in the system. If the users have bugs in their accounts, the block might get appended to the blockchain even if it is rejected by other users, leading to the creation of a new blockchain. The intentional protocol fork emerges if a part of the users chooses to modify the ruleset and continue with the new version of the protocol. As mentioned before, this results in the generation of a new cryptocurrency in many cases.

Persistence analysis refers to the existence and continuity of the blockchain, despite being dissolved due to a fork. The persistence analysis and the outcomes of the process-based forks depend on the allotment of the consensus-related resources and block propagation speed. The persistence analysis and the outcomes of the protocol-based forks depend on the additional factor that is the change in the protocol.

The protocol-based fork is divided into three types, namely soft, hard and forced forks.

Hard forks are converse to soft forks. The rules to accept a block become more lenient. That is why the blocks generated with the new protocol would normally be rejected by the old protocol implementation.

A soft fork is described as the modification of the consensus protocol where the ruleset becomes stricter. The blocks which are under the new protocol will definitely be approved by the old protocol implementation.

A forced fork will lead the way to a chain split. The new and old rules are completely different. The certain risk in the forced forks is that new blocks will be approved only by a subgroup of the clients [33].

The DAO Hack

Circa 2016, a Decentralized Autonomous Organization, named DAO, was deployed on Ethereum. Initially, DAO received a good margin of crowdfunding; approximately equivalent to 150 million US dollars of ether within 28 days of release. It soon reached a market capitalization of 1 billion USD. Structurally, it possessed a code-based internal capital DAO; an autonomously operated system that depended on the users for some tasks which it could not perform. At the same time, vulnerabilities in the smart contract that governed the DAO were exploited by an attacker and approximately 30% of the funds were stolen. The attacker performed multiple withdrawal transactions and transferred them into another DAO.

The proposals for a response to attack were presented that suggested freezing the ether in the DAO and its child (developed by the attacker). Other proposals included a hard fork and moving the funds to a new DAO. Amidst these discussions, the attacker released an open letter that according to the principal code of law in the DAO smart contract, the ether was acquired legally. Finally, a controversial hard fork of Ethereum was declared as everyone was not in agreement and two chains, Ethereum and Ethereum classic, were born. The fork was successful as major stakeholders followed the change [29].

The child DAO that the attacker developed was known as *dark DAO* and, although most of the funds were recovered, the attacker still netted about $8.5 million USD [8].

TOKENS IN BLOCKCHAINS

Tokens are issued by various cryptocurrencies and can be considered as a unit that has some value to it. Tokens have numerous use cases in terms of incentives for validating a block, usage of blockchain platform, etc. They are also used to represent the ownership of an asset or profit-sharing, etc. [24].

Table 3.1 represents the benefits of a token with respect to its role, purpose and benefits.

Types of Tokens

A token is a programmable resource constrained by a smart contract in a decentralized network on account of Ethereum [10]. These can also be looked at as value compartments used for moving the responsibility of a resource to the other node regulators on the equivalent blockchain [24]. An *initial coin offering* (ICO) is a technique for collecting assets for a task through a very large number of small sum contributions from numerous funders on the web. This is how a new digital currency gets its start. An ICO takes into account a large number of protocols to create the tokens on the

TABLE 3.1
A Quick Guide to Crypto Tokens Usage and Value

Roles that tokens can play	For the purpose of	Benefits it provides
Rights	Bootstrapping engagement	Product usage, voting, governance, ownership, contribution
Value exchange	Economy creation	Work rewards, buying, spending
Tolls	Skin in the game	Running smart contracts, security deposits, usage fees
Function	Enriching user experience	Joining a network, connecting with the user, incentive for usage
Currency	Frictionless transaction	Payment unit, the transaction unit
Earnings	Distributing benefits	Profit sharing, benefits sharing, inflation benefits

Source: [31].

available blockchain [25]. The primary distinction between a crypto token and a crypto coin is that the coins have their own blockchains, whereas the tokens do not. The tokens are set up on an existing blockchain [10]. Broadly speaking, cryptocurrency tokens can be classified into five different kinds, depending upon the sectors that they cover [4]

Security tokens are the components of a resource or an asset with inherent value. For instance, backed by the gold reserves of the UK Royal Mint and in collaboration with the Chicago Mercantile Exchange, a new gold-based cryptocurrency was developed utilizing the blockchain technology from two different startups transacting with tokens called *Royal Mint Gold.* A single Royal Mint gold token depicts 1 gram of real gold digitally. Users could purchase and trade tokens on the institutional platform and all the transactions are recorded on a blockchain [12].

Payment tokens are responsible for providing payment services. These tokens are generally generated on credit card information to prevent third-party attacks. For instance, Ripple and Utility Settlement Coins (UTC) are considered payment tokens. Various algorithms have been designed to generate payment tokens like mathematically reversible algorithms, static tables or one-way non-reversible cryptographic functions.

General platform tokens permit clients to create 'apps' and automate exchanges in the event that they meet certain principles. These are used on platforms used to purchase various items, such as clothes, concert tickets, books etc. [10].

Equity tokens are defined as the digital depiction of the privilege to engage in liquidation and viable rights, like dividends and voting. Equity tokens grant the advantageous interests of the share to the token holder which can be passed to any other user without the approval of the official titleholder on the share [17]. Security tokens have the attributes of both fungible and non-fungible tokens. The token standards involved are ERC-1400, R-token [25].

Utility tokens distribute the computerized control to the service. There is no necessity for any endorsement if the tokens were working when received [10]. The client who has utility tokens would have the option to utilize various capacities stated by the business to exchange the tokens [25]. Utility tokens are responsible for providing access to services or products. Examples include *Filecoin* and *Golem.*

Token Standards

Majorly accepted token standards in Ethereum include ERC-20, ERC-721, ERC-777 and ERC-1155.

ERC-20 is the only widely accepted ERC standard as of the time of this writing. ERC-20 is the conventional standard that accommodates the primary functions for the token transfer. ERC-20 grants tokens authority, which can then be used by third-party users on the chain. ERC-20 is a protocol interface which requires six compulsory functions, three secondary functions, and two events to be implemented with an API [10]. ERC-20 protocol defines a transfer method for transferring the token from one wallet to another [34]. Typical functionalities include trading, transferring, sharing and exchanging the tokens. All such functionalities are obtained through the interpretations of functions, which emulates a digital token using a smart contract [21]. Events are the logs with which the token could be tracked. The six functions are total supply, balance of, transfer, transfer from, approve and allowance; and two events: transfer and approval. Initially decentralized applications (DApps) had their own token standards. Later, with the introduction of the ERC-20, all the DApps started using ERC-20 [25]. The transaction will have a call to the data section of the transfer method which contains the amount being transferred and the recipient wallet. ERC-20 causes a token balance update which is kept in smart contracts storage. Additionally, the transfer method is followed by a transfer event which keeps the log. The logs are parsed in the following sections: contract address, value, sender and receiver address. Overall, ERC-20 depicts a social behavior with respect to tokens [34].

ERC 223 is backward compatible with ERC-20. ERC-223 is an improved version of ERC-20, rectifying its drawbacks. ERC-20 standard's main drawback was that it could not support some receive mechanisms of tokens. Therefore, a new function was introduced called *tokenFallBack* [26].

ERC-721 is a non-fungible standard. Here, every token is specific and facilitates the tracking of different assets. These tokens are controlled by smart contracts with 10 mandatory functions and three events [10]. As mentioned earlier, functions use smart contracts to obtain the functionalities of the digital token. The compulsory functions are named as: *symbol*, *name*, *totalSupply*, *balanceOf*, *ownerOf*, *approve*, *takeOwnership*, *tokenOfOwnerByIndex*, *transfer* and *tokenMetadata*. The events are the logs to track the tokens [21]. There are three events, namely, *Transfer*, *ApprovalForAll* and *Approval*. The term non-fungible means that the token is unique and cannot be replaced with another item. The non-fungible tokens are distinctive in nature. They cannot be divided into smaller parts and are non-interchangeable [25]. ERC-721 standards can be used to detect recently popular *deepfake* videos. Deepfake videos are harmful because they have the power to create a fake reality and thus confuse viewers. As such, it is crucial to identify and defy deepfake videos. To achieve this, the users must be provided with the access to the trusted data origin of the digital content. There should be a possibility of tracing back the authenticity of the content. The proof of authenticity using blockchain technology could provide a solution to this issue. Typically, public blockchains are more suitable for deepfakes [16]. NFTs provide the digital certificate as the notation of ownership which gives the proof of authentication. NFTs do not provide any copyrights or the original copy to buyers which is why NFTs are regarded as having 'bragging rights' [37].

ERC-777 adds up many progressive features and is backward compatible with the ERC-20 standard. There are 13 ERC-777 compulsory functions and five events to be implemented. ERC-1155 is a multi-token standard. Various standards of tokens can be sent at once in a single contract. It needs the implementation of six compulsory functions and four events [10]. ERC-1400 is a confined standard. ERC-777 is partly fungible and cannot be exchanged [24].

ETHEREUM GAS AND EIP 1559

When it comes to Ethereum blockchain platforms, in order to effectively deal with an exchange or execute an agreement, a transaction or processing fee is charged. This transaction fee is referred to as *gas*.

Ethereum has been developing more effective solutions to address the rapidly increasing gas charges on the Ethereum platform. As of 2021, Ethereum has implemented its newest gas-related protocol named Ethereum Improvement Protocol 1559 (EIP 1559) [32]. EIP 1559 is a gas-related Ethereum network upgrade that will require gas fees payable in Ether, the network's original cryptocurrency, to be destroyed every time ether is used for gas [13]. Furthermore, the Ethereum network has set a base gas expense for each transaction completed on the Ethereum network in an attempt to offer expedient service to all users [13]. Clients who wish to go through with their transaction quicker than the standard processing times may add a *tip* amount for the validator to fast-track their transaction. In simple terms, this tip can be considered equivalent to a tip given to waiters in a restaurant for fast and efficient service. The major difference in case of gas (with respect to the restaurant's case) is that the tip is provided before the receipt of service.

BLOCKCHAIN TOKEN USE CASES

The replacement of sensitive information with tokens is referred to as tokenization. Any unique information like a primary account number (PAN), security code, expiration date, etc. of a card can be tokenized. To understand the concept of tokenization, transactions involving credit or debit cards are good examples. Whenever an individual swipes their debit or credit card, the PAN is tokenized by generating random alphanumeric IDs. In this way, there is a record of the transaction but without the real PAN number being stored in the system. This token is sent to the processor which in turn detokenizes the ID prior to authorizing the transaction. With respect to the validity of the token, it is valuable to a particular individual only; thus, it is of no use to any third parties.

Blockchain tokens find their usage in a variety of areas ranging from gambling, gaming and loyalty points to various marketplaces [10]. In addition, tokens also find their application in casino chips, sale and purchase of drinks at cultural events, access approval and laundry coins [24, 27]. Below are a few use case examples:

Real estate assets can be tokenized using blockchain-based tokens. In 2018, an asset management firm based in New York, *Elevated Returns*, was able to do real estate tokenization for the first time. An offering was made on the Ethereum blockchain worth $18 million for a resort (St. Regis) in Aspen, Colorado. The initial plans were to sell 50% of resort as a single asset; but later, it was adjusted to 18.9% of ownership via sale of tokens. The sale was completed through *Indiegogo* with *Templum Markets LLP* as partners.

Artwork tokenization focuses on generating tokens for the artworks of popular artists. Typically, a particular organization is responsible for the ownership of a limited numbers of prints; as such, the limited prints can be tokenized accordingly. In the next step, the company offers the prints to the public in exchange for tokens. Finally, in the case of an outright sale, the print is delivered physically to the mailing address once the required number of tokens is redeemed. Tokenizing art pieces also enable individuals to have a partial ownership of a painting alongside other individuals.

Gold assets tokenization provides for a partial ownership of gold in a digital token format. Let's assume that gold is owned by Company A; and, the vault in which it is kept is owned by company B. A contract between A and B is maintained for the ownership registry of the gold. For every token sold, the amount of gold ownership by A is transferred to the vault company B. A digital signature or certificate can be used by a token holder to provide their ownership for redeeming the gold. JP Morgan announced that they will be tokenizing the gold bars on their blockchain network known as *Quorum.*

Cargo tokenization focuses to eradicate issues related to delays in delivery or goods reclamation caused by bill of lading (B/L) issues. To further explain, a B/L acts as a proof of ownership in case merchandise is lost or misplaced during transport. *CargoX* came up with an Ethereum-based open system known as *Smart B/L.* Initially an application will be used by the carrier to create a Smart B/L, subsequently received by an exporter. When payment is received by the exporter, the importer transfers the Smart B/L token's ownership to the new merchandise owner. The importer can claim the goods ownership by presenting the token to the carrier [36].

CRYPTOCURRENCIES

As of February 2020, a total of 106 million users of cryptocurrency were reported. The cryptocurrency exchange noticed an increase of 15.7% in adoption by January [15]. As of April 2020, there were around 5,392 cryptocurrencies which were traded with a total market capitalization of $201 bn [5]. According to the data incorporated from 24 different exchanges, the surge in the price of Bitcoin is the main reason for the increase in the number of users [15]. There are several characteristics that all crypto coins possess. These characteristics are divided into two categories, namely, transactional and monetary characteristics.

The *transactional characteristics* of the cryptocurrencies include *irreversibility*, meaning that once the transaction is completed, it can never be reverted or altered. *Pseudonymity* means the cryptocurrency accounts are not related to the real-time personal information of the user. Another attribute is *security* because only the user with the private key has control over the funds. The final attribute is *permissionlessness* as the cryptocurrency platforms are indeed permissionless. Anyone can download and use them for free [4].

Cryptocurrencies also possess *monetary characteristics* such as the *capacity to control the currency's supply*. Another important distinction is that in a bank account, fiat money represents a debt to the account owner on behalf of funds custodian (the bank); whereas a cryptocurrency represents itself, as it uses no financial intermediary [4].

Broadly speaking, cryptocurrencies can also be classified into two major types: public and private. This classification is based on traceability. To further explain, *public cryptocurrencies* provides user anonymity, yet the transactions are easily traceable. *Private cryptocurrencies* provide privacy, which makes it hard to trace the transactions. Private cryptocurrencies hide transaction details; meaning that although they have a public open ledger, the information regarding transactions being incurred is a bit obfuscated to protect the privacy of the end user.

As their name suggests, *Altcoins* are used as an alternative to other digital currencies. A huge part of altcoins relies upon the same or similar blockchain innovation as Bitcoin, with a point of improvement of certain particular blockchain attributes [30]. Altcoins evolve because of forks or ICOs [9, 14].

Cryptocurrencies work as the medium for exchange in blockchain decentralized networks. Spillover effects assess how a one unit shock to a certain cryptocurrency will affect the other cryptocurrency's price [20, 22]. Spillover effects in the digital currency market and other monetary markets are major topics to be investigated. Due to the relative recency of cryptocurrencies as an asset class, there has not been a great deal of research done regarding cryptocurrency diversification best practices, hedging strategies and/or the correlation of cryptocurrencies vis-à-vis other traditional asset classes, such as stocks or bonds. Cryptocurrencies are the legal tenders not currently taxed; however, this non-taxation trend is increasingly meeting opposition by a number of major governments [19, 20].

POPULAR CRYPTOCURRENCY TYPES

The next section discusses some more popular cryptocurrencies currently available on various exchanges. Readers should note that the following brief cryptocurrency discussions revolve around a limited number of currently popular cryptocurrencies, and are in no way comprehensive.

Bitcoin was established by Satoshi Nakamoos in 2009. It is regarded as an investment asset by many users. As a permissionless digital cash system, Bitcoin has a transaction fee. For a quick transaction, the transaction fee is high. If a new block is created, the miner earns a reward. Bitcoin is sometimes referred to as a deflator currency. The Bitcoin supply is limited to 21 million coins. As more blocks are being created by miners, the reward keeps decreasing accordingly. Due to a fork in August 2017, Bitcoin led the way to the creation of *Bitcoin cash* [4]. Bitcoin is currently the most popular form of cryptocurrency, due to its platform security's track record and its spectacular rise in value over the past decade [3].

Ether is the digital money of Ethereum. Ethereum was conceived in 2015 by the Russian-born Canadian, Vitalik Buterin. Ethereum is a platform where users can develop and execute applications called *decentralized applications* (DApps) [4]. DApps are hosted on Ethereum which can be used to pay for services like social media, gaming and finance. Every DApp has a different token, which can be purchased with Ether. *Decentralized finance*, also known as *DeFi*, came to birth because of Ethereum. DeFi coins are used in which protocols run on smart contracts that run on Ethereum [40]. At the time of this writing, Ethereum can create a new block in just 14 to 15 seconds, with plans for major performance improvements in the near future. A major difference between Bitcoin and Ethereum is that Bitcoin focuses on being a worldwide decentralized digital currency, while Ethereum is a decentralized computing platform. As mentioned before, due to the occurrence of a fork in 2017, the Ethereum platform was split into two: Ethereum and Ethereum Classic. The Ethereum platform follows a peer-to-peer architecture which results in high security and fault tolerance [4, 38].

IoTA has a capital of over 14 billion USD. IoTA was considered to be the sixth largest digital currency in December 2017. It uses *Tangle*, which is an acyclic graph used to categorize the transactions. IoTA utilizes blocks that allow transactions for certain funds when necessary. It also uses an autonomous machine economy to conduct the transactions without a fee. Scalability, decentralization and modularity are some other features of IoTA [14].

Ripple was established by the Ripple Company in 2012. Unlike Bitcoin, Ripple is able to perform complicated block generation tasks that involve high computations by using a minimum amount of energy. In terms of block creation, Ripple is currently about 1000% faster than Bitcoin and 300% faster than Ethereum. Ripple transactional costs are relatively low. Ripple uses the *RippleNet* to communicate payment information to suppliers and banks [3]. The US Security Exchange Commission (SEC) has filed a lawsuit against Ripple and two of its employees for raising the currency up to $1.3 billion by using non-registered security offerings. At the time of this writing, the status of the SEC lawsuit remains unknown [26].

Litecoin was developed to resolve some of the drawbacks of Bitcoin. In October 2011, Charlie Lee founded Litecoin. Litecoin was used to reduce the computing power required for coin mining. Litecoin uses the *Scrypt* algorithm as its consensus mechanism. Litecoin is faster in the creation of blocks; and its reward is scheduled to be halved once it reaches 840 million Litecoins in approximately 4 years. The transaction fees also are much lower when compared to Bitcoin [4].

Monero's chief objectives are privacy and non-traceability. It was designed by Nicholas van Saberhagen in April 2014. Similar to Bitcoin, the transaction fee varies with the traded currency. Characteristics of Monero are *Ring Confidential transactions* (RingCT), *Stealth Addresses* and *Ring Signatures*. For every transaction, the Stealth Address makes sure that the sender uses a onetime address. The feature of identity hiding of the sender and receiver is called Ring Signatures. RingCT defines the way of hiding the transaction money in Monero. RingCT was used for the first time in January 2017; later, it became mandatory for every Monero transaction. In Monero, transactions cannot be tracked. The main advantage of Monero is dynamic scalability [4, 9].

Dash was released in January 2014. It was originally known as *Xcoin* and later branded to *Darkcoin*. Its current name is a mixture of two words, Digital and Cash, combined to form Dash. Falling under the category of altcoins, it is derived from Litecoin, which in itself is a derivative of Bitcoin. Dash's major advantage over Bitcoin is that it provides refined privacy protection with quicker processing of business transactions. In other words, it provides a less expensive, faster, more private and secure way to process transactions [12, 14].

Stellar was created by the Stellar Development Foundation in 2014. The developers of Stellar were Joyce Kim and Jed McCaleb, who was also one of the developers of Ripple. Ripple uses the Byzantine Fault Tolerance (BFT) algorithm for its transactions. Stellar's currency is called *XLM* or *Lumens*. Stellar does not use miners, and it recycles its transaction fees. It also has a distributed exchange [4].

EOS is an open-source platform launched in June 2018. Its crypto coin EOS (Ethereum Operating System) is very similar to Ethereum. EOS was launched with an ICO of 4 billion USD. With its high capacity, EOS can process up to 4000 transactions per second. The EOS tokens are used to purchase the bandwidth and the storage of the blockchain. Users with tokens get the chance to vote on various governance issues [4].

Tether is a controversial cryptocurrency branded as a *stable coin*, meaning that, originally, its equivalency was designed to be measured with respect to real-world assets. Specifically, one tether was designed to equal 1 US dollar at all times. Thether was established under the name *Realcoin* in 2014 by the startup founders Craig Sellars, Brock Pierce and Reeve Collins. There are no transaction fees in Tether. It is one of the most highly traded cryptocurrencies on the market because there is a possibility of buying other cryptocurrencies with the realcoins which constitute USD [4].

Neo, originally known as *Ant shares*, was developed by *Da hongfi*. Neo works using the mechanism of Byzantine Fault Tolerance, which makes it much faster performing than Ethereum in early 2021. While Neo can handle 10,000 transactions per second, Ethereum can only handle 15 per second. Neo is indivisible, so it exists only in form of whole numbers [18].

Toycoins form a category of coins which were developed for fun. Examples include *Jesuscoin, Dogecoin, Shitcoin, Satancoin*, etc. Investors used to not pay much attention to these types of coins as they lack a serious purpose. At some point, Dogecoin managed to receive investments of 450 million USD. Elon Musk, the CEO of *Tesla* and the website *Reddit* were mainly responsible for the popularity of dogecoin and its ensuing stardom [4, 9].

Metacoins are the protocols constructed on the platforms of different cryptocurrencies. *Counterparty*, for example, is a peer-to-peer financial network that offers its very own currency, called *XCP*, and has the same advantage of not requiring an intermediary. Its purpose is to democratize money in the same way that the Internet democratized the generation and exchange of information [1].

CRYPTOCURRENCY EXCHANGES

The following section briefly discusses a handful of more popular cryptocurrency exchanges.

Bittrex is a trading platform for cryptocurrencies. Headquartered in Zurich, Switzerland, Bittrex aims at providing a secure way to exchange cryptocurrencies.

Poloniex is an exchange platform for cryptocurrencies established by Tristan D'Agosta in 2014. With headquarters in Delaware, Poloniex was hacked in 2014. Bitcoins that were on hold on the platform were stolen by hackers. An estimate says that 12.5% of the Bitcoin supply was stolen. To avoid such scenarios in the future, Poloniex has moved to cold storage for preserving their funds. Also, it requires the use of two factor authentication among its users [6]. Poloneix is currently the largest altcoin exchange in terms of daily trading volume [11].

Kraken has been the oldest and largest platform for the exchange of Bitcoin since 2011. It made a name for itself by aiming to provide financial freedom via increased cryptocurrency adoption. There is the versatility of APIs provided by Kraken. The role of its API is to provide access into the Kraken accounts without the need to share sign-in info. These APIs provide limited and controlled access only, and are typically used by experienced traders for their bots to communicate with Kraken accounts [39].

Coinbase has a goal to develop a transparent financial system, that is efficient, fair and accessible since its discovery in 2012 according to the company. Its distinct features include being arguably the most trusted, as well as being the easiest one in use. Coinbase reaches over 7000 institutions that includes over 43 million verified users with over 115,000 ecosystem partners in over 100 countries. Coinbase went public in 2021 [2].

Bitfinex's unique features include peer-to-peer financing, margin trading and over-the-counter (OTC) options. Bitfinex has its own token, UNUS SED LEO, that provides a variety of benefits including P2P lending fees, as well as discounts on trading fees. It provides state-of-the-art digital asset trading services and global liquidity providers [1].

Binance was established in 2017 in China, but later relocated to the Cayman Islands. In 2021, Binance was deemed to be the largest crypto exchange in the world in terms of its daily trading volume. Binance was founded by Changpeng Zhoa, the creator of high-frequency trading software. At the time of this writing, Binance is under investigation by the US Department of Justice and the Internal Revenue Service [35].

SUMMARY AND CONCLUSIONS

Blockchain technology has the potential to transform entire industries from a centralized culture toward a decentralized structure. To keep a check on this characteristic, different governance systems have been developed and many more are being developed. Different tokens have been developed over the years with different standards governing them, with ERC 20 being the most prominent among those at this time.

Cryptocurrencies were introduced to facilitate decentralized financial transactions on various blockchain platforms. Examples include Ether, Bitcoin, Monero, Dogecoin, etc. The last sections of this chapter provided readers with a quick overview of some popular cryptocurrencies and crypto-exchanges.

CORE CONCEPTS

- Blockchain governance is the system by which the blockchain adapts to changing times and requirements. The main characteristics of blockchain governance are adaptability and upgradeability. There are four key bodies, namely, core developers, node operators, token holders and blockchain teams, which are typically involved in blockchain governance.
- A fork is the reason for the generation of new cryptocurrencies. The blockchain splits into separate database instances due to disagreements among some users on the current blockchain. This leads to the evolution of new cryptocurrency, while some users choose to continue with the old blockchain and some decide to follow the new fork.
- The main difference between a crypto token and a crypto coin is that coins have their own blockchain, whereas tokens do not, as tokens are created on blockchains that already exist. Tokens are classified into numerous types based on their intended purpose, some of which include security, payment, utility and equity tokens.
- Token standards can be seen as the interface to create, transfer and perform token operations. The most accepted token standards are ERC-20, ERC-223 and ERC-721, of which ECR-721 is the only non-fungible standard.
- Gas is the transaction fee charged on the Ethereum exchange structure. In 2021, skyrocketing gas fees were revamped with the introduction of EIP 1559. In order to speed up urgent transactions on Ethereum, a tip may be included by users.
- Ethereum is the most prominent cryptocurrency after Bitcoin. The GHOST protocol forms the basis of the Ethereum chain. The major difference between Ethereum and Bitcoin is that Ethereum is a platform, not a currency. Ether is the currency for Ethereum.
- Based on traceability attributes, cryptocurrencies could be classified into public and private types. While the public cryptocurrencies enable anonymity but are still traceable (for example: Bitcoin), private cryptocurrencies are designed to be untraceable.
- There are a number of exchange platforms, including Bittrex, Coinbase and Poloniex. In 2021, a number of cryptocurrencies and exchanges are under investigation or close scrutiny by

various government entities, such as the US Security Exchange Commission, the US department of Justice and the Internal Revenue Service.

ACTIVITY FOR BETTER UNDERSTANDING

Readers interested in learning a bit more about the ongoing cryptocurrency revolution are encouraged to watch the YouTube video entitled: *Inside The Cryptocurrency Revolution* available at: https://www.youtube.com/watch?v=u-vrdPtZVXc in order to get a better cryptocurrency perspective.

After reading the chapter and watching the video, consider the following questions:

- In your personal opinion, within what time frame and in what manner will cryptocurrency transactions be taxed by various countries?
- In your opinion, what are some likely transactional challenges associated with blockchain-based currencies for organizations?
- In what capacities is Vitalin Buterin (founder of Ethereum) likely to influence the growth and popularity of the cryptocurrency markets in general?
- Of the cryptocurrencies discussed in this chapter, which ones are likely to still be around within the next decade?
- Thinking of your enterprise's objectives and long-term goals, which cryptocurrencies discussed in this chapter are most likely to be adopted in the future by your organization, based on their basic attributes?

REFERENCES

[1] *About Us*. Bitfinex. https://www.bitfinex.com/about

[2] *About*. Coinbase. https://www.coinbase.com/about

[3] Antonopoulos, A. M. (2021). *Mastering Bitcoin: Programming the Open Blockchain*. La Vergne: Stanford.

[4] Auwera, E. V., Schoutens, W., Giudici, M. P., & Alessi, L. (2020). *Financial Risk Management for Cryptocurrencies*. New York: Springer.

[5] Bagshaw, R. (2020, April 22). *Top 10 Cryptocurrencies by Market Capitalisation*. Yahoo! https://ca.movies.yahoo.com/top-10-cryptocurrencies-market-capitalisation-160046487.html

[6] Bittrex vs Poloniex: Comparing Safety and Fees to Find the Better One. (n.d.). Retrieved March 4, 2021, from https://www.bitdegree.org/crypto/tutorials/bittrex-vs-poloniex

[7] Bouraga, Sarah. (2020). An Evaluation of Gas Consumption Prediction on Ethereum Based on Transaction History Summarization. https://www.researchgate.net/publication/344406010_An_Evaluation_of_Gas_Consumption_Prediction_on_Ethereum_based_on_Transaction_History_Summarization

[8] cryptopedia (2021). "The DAO: What Was the DAO and How Was It Hacked?," *Gemini*. Retrieved March 1, 2021, from https://www.gemini.com/cryptopedia/the-dao-hack-makerdao

[9] Distributed, D. (2017). *Know Your Coins: Public vs. Private Cryptocurrencies*. Nasdaq. https://www.nasdaq.com/articles/know-your-coins-public-vs-private-cryptocurrencies-2017-09-22

[10] Di Angelo, M., & Salzer, G. (2020). Tokens, Types, and Standards: Identification and Utilization in Ethereum. *2020 IEEE International Conference on Decentralized Applications and Infrastructures*. doi:10.1109/dapps49028.2020.00001

[11] Elendner, H., Trimborn, S., Ong, B., & Lee, T. M. (2018). The Cross-Section of Crypto-Currencies as Financial Assets. *Handbook of Blockchain, Digital Finance, and Inclusion*, *1*, 145–173.

[12] Faden, M. (1970, January 1). *Britain's Royal Mint Trades Gold in Blockchain Transactions*. American Express. https://www.americanexpress.com/us/foreign-exchange/articles/blockchain-transactions-digital-trade-gold/

[13] Benjamin, G. (n.d.). Ethereum to Become a Deflationary Asset and See Its Supply Reduced with Proposed EIP 1559 Upgrade. Retrieved March 22, 2021, from https://blockchain.news/postamp?id=ethereum-become-deflationary-asset-see-its-supply-reduced-proposed-eip-1559-upgrade

[14] Girasa, R. (2018). Technology Underlying Cryptocurrencies and Types of Cryptocurrencies. *Regulation of Cryptocurrencies and Blockchain Technologies*, 29–56.

[15] Hamacher, A. (2021, February 25). There Are Now 106 Million Cryptocurrency Users Globally: Report. Retrieved April 5, 2021, from https://decrypt.co/59267/there-are-now-106-million-global-crypto-users-report?utm_campaign=daily_bundle_template&utm_content=Crypto+user+numbers+top+106+million+worldwide&utm_medium=email&utm_source=sg_email

[16] Hasan, H. R., & Salah, K. (2019). Combating Deepfake Videos Using Blockchain and Smart Contracts. *IEEE Access*, *7*, 41596–41606.

[17] Henderson, A., Burnie, J., & Thornborough, J. (2019). Eversheds Sutherland Occasional Paper: Issuing Equity Tokens on the Blockchain Legal Reimagining. Retrieved from https://www.eversheds-sutherland.com/documents/services/financial/equity-blockchain-2030.pdf

[18] Laura, M. (2021, January 5). *NEO vs Ethereum: Could NEO Be the Next Big Thing after Ethereum?* BitDegree.org Crypto Exchanges. https://www.bitdegree.org/crypto/tutorials/neo-vs-ethereum#:~:text=NEO%20can%20handle%20about%2010%2C000,NEO%20provides%20a%20great%20alternative

[19] Liang, J., Li, L., Chen, W., & Zeng, D. (2019). Towards an Understanding of Cryptocurrency: A Comparative Analysis of Cryptocurrency, Foreign Exchange, and Stock. *2019 IEEE International Conference on Intelligence and Security Informatics*. doi:10.1109/isi.2019.8823373

[20] Liu, J., & Serletis, A. (2019). Volatility in the Cryptocurrency Market. *Open Economies Review*, *30*(4), 779–811.

[21] McKie, S. (2017, October 2). The Anatomy of ERC20. Retrieved May 23, 2021, from https://medium.com/blockchannel/the-anatomy-of-erc20-c9e5c5ff1d02

[22] Moratis, G. (2021). Quantifying the Spillover Effect in the Cryptocurrency Market. *Finance Research Letters*, *38*, 101534. https://doi.org/10.1016/j.frl.2020.101534

[23] Atzei, N., Bartoletti, M., & Cimoli, T. (2017) *A Survey of Attacks on Ethereum Smart Contracts.* In International Conference on Principles of Security and Trust. https://eprint.iacr.org/2016/1007.pdf

[24] Oliveira, L., Zavolokina, L., Bauer, I., & Schwabe, G. (2018). *To Token or Not to Token: Tools for Understanding Blockchain Tokens.* https://www.researchgate.net/publication/328162731_To_Token_or_not_to_Token_Tools_for_Understanding_Blockchain_Tokens

[25] Patel, D., Nandi, S., Mishra, B. K., Shah, D., Modi, C. N., Shah, K., & Bansode, R. S. (2020). IC-BCT 2019: Proceedings of the International Conference on Blockchain Technology (Blockchain Technologies). In M. Shirole, M. Darisi, & S. Bhirud (Eds.), *Cryptocurrency Token: An Overview* (pp. 133–140). Springer.

[26] Press release. (2020, December 22). Retrieved May 15, 2021, from https://www.sec.gov/news/press-release/2020-338

[27] Ranade, A., & Shaikh, Z. (2020). A Survey on Blockchain Technology with Use-Cases in Governance. *SSRN Electronic Journal.* doi:10.2139/ssrn.3568629

[28] Ryan. (2019). *About Us.* Bittrex Global. https://bittrexglobal.zendesk.com/hc/en-us/articles/360009624200

[29] Reijers, W., Wuisman, I., Mannan, M., & De Filippi, P. (2018). Now the Code Runs Itself: On-Chain and Off-Chain Governance of Blockchain Technologies. *SSRN Electronic Journal.* doi:10.2139/ssrn.3340056

[30] Reynolds, B. (2002). Virtual Relationships: Short- and Long-Run Evidence from Bitcoin and Altcoin Markets. *Journal of International Financial Markets*, *52*: 173–95.

[31] Mitra, R. (2020). What Is Blockchain Governance: Ultimate Beginner's Guide, *Blockgeeks*, 24 April. Retrieved March 1, 2021 from https://blockgeeks.com/guides/what-is-blockchain-governance-ultimate-beginners-guide/

[32] S., & Li, S. (n.d.). Ethereum's Berlin Hard Fork Anticipated for April. Retrieved March 22, 2021, from https://blockchain.news/postamp?id=ethereum-berlin-hard-fork-anticipated-april

[33] Schär, F. (2020). Blockchain Forks: A Formal Classification Framework and Persistency Analysis. *Singapore Economic Review*, 1–11. doi:10.1142/s0217590820470025

[34] Somin, S., Gordon, G., & Altshuler, Y. (2018). Network Analysis of ERC20 Tokens Trading on Ethereum Blockchain. In A. J. Morales, C. Gershenson, D. Braha, A. A. Minai, & Y. Bar-Yam (Eds.), *Unifying Themes in Complex Systems IX* (pp. 439–450). Springer International Publishing.

[35] Studio, O. (2021, February 05). The 4 Best Bitcoin Exchanges Reviewed (2021). Retrieved March 4, 2021, from https://observer.com/2021/02/best-bitcoin-exchange/

[36] Taylor, K. (2020, August 6). *Top 4 Use Cases of Tokenization.* HitechNectar. https://www.hitechnectar.com/blogs/use-cases-tokenization/

[37] Vennavally-Rao, J., & Somos, C. (2021, March 06). What Are NFTs? Cryptocurrency Technology Is Driving New Digital Art Craze. Retrieved March 20, 2021, from https://www.ctvnews.ca/sci-tech/what-are-nfts-cryptocurrency-technology-is-driving-new-digital-art-craze-1.5336423?cache=yes%2F7.305626%3Fot%3DAjaxLayout

[38] Vujicic, D., Jagodic, D., & Randic, S. (2018). Blockchain Technology, Bitcoin, and Ethereum: A Brief Overview. *2018 17th International Symposium*. https://doi.org/10.1109/infoteh.2018.8345547

[39] *Why Kraken?* Bitcoin and Cryptocurrency Exchange. https://www.kraken.com/en-us/whykraken

[40] Wise, H. (2021, January 14). Beyond Bitcoin: Number Two Crypto Ethereum Is Climbing Faster. *This Is Money*. https://www.thisismoney.co.uk/money/markets/article-9138547/Beyond-Bitcoin-Number-two-crypto-Ethereum-climbing-faster-given-rise-DeFi.html?ito=email_share_article-top

4 The Advent of the Triple Entry Accounting

Implications for Accountants and Auditors

Ibukunoluwa Mabo, Oluwatobiloba Iroko and Oyegbenga Oyenekan

ABSTRACT

This chapter looks at the implications of blockchain technology in modern accounting and its possible implications for accounting and audit professionals. After a brief historical introduction about the evolution of financial accounting (the journey from single-entry to double-entry and now, triple-entry accounting), the chapter provides suggestions in terms of using blockchain as an anti-fraud control to help mitigate financial statement, cash larceny, inventory and payroll fraud schemes based on the Association of Fraud Examiners (ACFE) fraud tree model. This chapter also includes a blockchain-focused purchase cycle audit template for internal and external auditors.

THE HISTORY OF ACCOUNTING ENTRIES

Single Entry

The first records related to the use of single-entry accounting may be traced back to India. Chanakya wrote a book about finance and management during the period of the Mauryan Empire (321 BC to 185 AD) in order to preserve the ways and methods books of account were handled [4]. Single-entry bookkeeping remained widely used until the middle of the 19th century, even after a superior approach had been developed [2].

Single-entry bookkeeping relied on one side entry of accounting concerning recording of financial events into a book called a cashbook. Small businesses mostly made use of single-entry transactions as they record small transactions; especially as it pertained to the recording of revenue and expenses on a cash basis [26].

An effective accounting system is judged by how well records are stored and how well they can meet the information needs of internal and external decision-makers [28]. As such, single-entry bookkeeping is likely to work for smaller operations with a limited volume of activity. Its advantages included the fact that it did not require complicated computations. The downside of the single-entry system of accounting is that it cannot solve a lot of questions raised in the modern world of accounting where financial fraud needs to be detected through extensive forensic analysis [14].

DOI: 10.1201/9781003211723-4

Double Entry

The double-entry system had its start in the 15th century by an Italian Economist named Luca Pacioli. Double entry means recording transactions on two sides of the accounting ledger by using two types of accounts; a debit and a credit. The practice of Islamic accounting, which the Venetians practiced and applied, may have also influenced double entry [9, 10].

The principle of double entry revolves around the requirement that for every debit entry there must be a corresponding credit entry for the same amount of money or transactions, and for every credit entry there must be a corresponding debit entry for the same amount of money or transactions [35].

In 1494, Luca Pacioli published the first bookkeeping work, which was titled '*Summa de Arithmetica, Geometria, Proportioni et Proportinalita*' in Venice; a book on arithmetic and geometry. The work of Luca Pacioli marks the beginning of the double-entry system of accounting [31]. Factors that led to the emergence of systematic bookkeeping, as identified by Littleton (1927), included increases in literacy, money and trade, as well as, the introduction of Arabic numerals, and the decimal system, and finally a significant increase in private property ownership [27].

Triple-Entry Accounting

The term 'triple-entry' was initially conceived by a specialist in accounting named Yuji Ijiri. Added to the existing debit and credit entries, Yuji recommended the incorporation of a third layer of entries using a collection of accounts to further illustrate variations in income. Another theory of triple entry was coined by a financial cryptographer named Ian Grigg. Grigg wrote an article entitled *Triple-entry Accounting*, which was published on his website in 2005 [16]. He developed a new theory, '*the receipt is the transaction*', where to prevent transaction fraud and minimize unintended errors in internal documentation, a third entry can include a digitally authorized receipt supported by a financial cypher among two parties [5]. The triple entry appeared in Satoshi Nakamoto's 2008 paper '*Bitcoin: A Peer-to-Peer Electronic Cash System*' [6].

Triple-entry accounting is a modern and effective approach to solving the core issues of trust and accountability that impact existing accounting systems [16]. In triple-entry accounting, business organizations need to properly execute transactions internally on a blockchain structure and the equivalent entry will be registered in a publicly shared ledger [5].

Jason Tyra published a brief paragraph in *Bitcoin Magazine* in 2014 endorsing the triple-entry principle suggested by Grigg. Ever since Tyra's article was published, triple-entry accounting has become an increasingly accepted concept [6]. Bitcoin is often used as a model to describe a triple-entry bookkeeping system [17]. A brief article posted by Deloitte in 2016 suggests that the introduction of blockchain triple-entry accounting would be a turning point in accounting [5, 24]. As such, the implementation of blockchain has become necessary with more modern financial events requiring a decentralized and distributed management of trust [9].

Double-Entry versus Triple-Entry Accounting

The implementation of a blockchain-based accounting system where double entry and distributed records are transformed into decentralized public records could offer high levels of certainty as to the integrity of the audited entity records [6]. Dai and Vasarhelyi define a possible future blockchain-based accounting environment in which a third, cryptographically protected yet accessible entry is incorporated into the blockchain, over the standard two entries used to record a transaction in a double-entry ledger. The blockchain-based accounting environment defined by Dai and Vasarhelyi would potentially prevent the potential for data manipulation and other types of financial fraud, encourage a transition to population versus sample-based assessment and facilitate the production of accounting reports effectively in real time with the implementation of other innovations, including smart contracts [8].

There is a possibility that a greater number of transaction recordings and related data collection will be done on an automated basis due to the ability to incorporate smart-contract and machine-learning technologies into blockchain-based data collection systems. For example, it has been projected that applying common credit scoring to financial information would likely lead to the development of different decisions if they were based on real-time financial information from a blockchain-based accounting system rather than on periodic reports presently available [6].

Triple-entry accounting can be achieved using blockchain technology, as multiple accounting transactions like payments, management of information related to (smart) contracts, transactions regarding the exchange of goods and services, and many others can be carried out [9]. In addition, blockchain technology can meet all accounting legal requirements and accounting standards [11]. Some of the components of blockchain technology that make it a suitable architecture for the realization of triple-entry accounting include the immutability of the register, the transparency of transactions and the consensus mechanism [10].

In triple-entry accounting, the shared entry is the fully reliable source of the transaction record. However, a local copy of the shared transaction repository may be incorporated as a sub-ledger to the general ledger of each transactional party [17]. The public register can be accessed by a large community, and transactions can be traced and known by public users and stakeholders who have an interest in the performance of the company. It is important to note that, once added, transactions cannot be eliminated from the blockchain. If an incorrect block needs to be modified, the node must insert a new block with a correcting entry to correct the previous one [10].

Some steps need to be taken to update the shared record between parties on the blockchain system with a new transaction record such as having two parties involved. One party initiates a transaction entry called 'request', 'offer' or 'transaction draft' and the other party accepts it. There is a signature-gathering process, in which one party adds their signature to the transaction entry draft and the counterparty accepts by countersigning before the entry is processed by the shared transaction repository. The shared transaction repository checks to determine if the signatures are valid and signs the entry after confirmation of validity. This generates a hashed triple-signed receipt, whereby all the parties hold the same data that cannot be misplaced or changed [17].

THE EFFECT OF TRIPLE-ENTRY ACCOUNTING FOR ACCOUNTANTS

The long-run application of triple-entry accounting is likely to have various effects on accountants and the accounting profession [36]. Triple-entry accounting recognizes the automation of recognition, measurement, presentation and disclosure, which replaces some of the traditional accounting jobs such as recording and preparing financial statements, thereby increasing the job opportunities for accountants to create and enforce rules and guidelines to guide financial accounting using blockchain technology [37]. Triple-entry accounting is also effective in assuring the authenticity of source documents and the genuineness of smart contracts [36]. In addition, blockchain use facilitates the use of smart contracts which allow for efficient and effective contract management, reduced tax violations and support for accrual basis accounting [21]. Finally, the cost of financial audits could potentially be reduced under the triple-entry accounting system as this is a new and more efficient approach to tackle the issue of underlying trust and openness that affects the existing accounting systems [36, 37].

THE IMPLICATIONS OF BLOCKCHAIN FOR AUDITORS

Blockchain technology offers a wide range of opportunities and challenges to the audit and assurance profession. While the significance of traditional audit and assurance remains the same, it is important to note that the auditor's approach may have a significant influence at deriving more effective and efficient audit outcomes. The advent of triple-entry accounting through blockchain technology raises some essential questions in the auditing occupation such as: What are the possible blockchain-related skillsets auditing professionals need to learn and adopt in order to remain relevant? [3].

One of the advantages of the use of triple-entry accounting in the audit and assurance profession is its further development of control test levels, thereby increasing the validity of audits, and improving the internal structures and the execution details of the assessment. Having the right set of controls reduces the level of error and fraud induced in the financial statements and records, thereby helping auditors produce better audit reports. As time passes, the audit profession is very likely to move from individual assessments to developing testing algorithms – like smart contracts – which are automated and operational through the decentralized ledger [8].

Also, near real-time auditing and financial reporting on a blockchain network could allow auditors the flexibility of focusing on complex and risky transactions. Accessing information for an audit becomes easier because of the transparency of the blockchain technology and audit evidence, which makes it simple to extract. The ability of each piece of data to be stored in blocks emphasizes transparency and security which remains free from manipulation from outsiders. It increases trust in audits because the data are more secure [37].

Blockchain technology is also likely to reduce audit preparation time. Due to the decentralization of the network, transactions on a blockchain network could reduce the gap between transactions occurrence and verification dates. However, a change to auditing on blockchain networks means that the control risk will decline while the inherent risk may increase. The characteristics of blockchain technology make audit evidence immutable, thereby lessening and improving the work of auditors, which eventually leads to quicker auditing and financial reporting [3].

It is important to also note that since audited financial statements are a required tool for large entities, blockchain technology will not replace financial reporting and financial statement auditing. Financial reporting and auditing play a key role in debt and equity funding, involvement in capital markets, mergers, and acquisitions, and more [8]. Financial statements reflect management assertions, including estimates, many of which may not be easily summarized on blockchains [37].

FINANCIAL FRAUD MITIGATION USING BLOCKCHAIN TECHNOLOGY

Fraud prevention is an essential requirement for the success of organizations all over the world. According to the 2020 *Association of Certified Fraud Examiners'* (ACFE) *Report to the Nations* (RTTN), the total loss caused by fraud in 2,504 cases from 125 countries in the world was more than $3.6 billion; translating to a loss of 5% of revenue to fraud each year [31]. Carrying out transactions on blockchain technology could greatly reduce occupational fraud. A description of ACFE fraud schemes is included below.

Financial Statement Fraud

Financial statement fraud is a scheme conducted mainly by management that entails a deliberate misstatement or exclusion of material data in the organization's financial statements in order to deceive financial statement users [13].

Financial statement fraud could be significantly mitigated when a distributed digital ledger is used by an organization to perform transactions and store data. The digital records provide increased visibility and transparency of the transactions made between members of a business network in a supply chain. As mentioned before, blockchain records cannot be altered or deleted and, therefore, hiding illegitimate financial activities becomes practically impossible [37]. As such, financial statement fraud could be significantly mitigated when a distributed digital ledger is used by an organization to perform transactions and store data.

Some items on the balance sheet cannot be fully quantified, such as intangible assets. For instance, goodwill, intellectual property and a company's reputation may be hard to explicitly quantify on the blockchain using the triple entry accounting for balance sheet [36].

Skimming

Skimming is a scheme that occurs when payments are stolen before they are recorded. In a skimming scheme, employees try to hide theft by deleting invoices or forging a credit memo [13]. According to the RTTN, skimming fraud amounted to an average of $47,000; and accounted for about 11% out of all occupational frauds. In 2020, the skimming fraud scheme was rated as a medium risk [31].

If transactions are carried out on blockchain technologies, it would be increasingly difficult to steal payments, as most employees will not have the encryption keys to access payments and, if they do, they will have to provide proof of identity and their access to the system would be visible. Also, since transactions are recorded in real time on blockchain systems, it is almost impossible to steal payments before they are recorded [37].

Cash Larceny

Cash larceny occurs where payment is taken from an organization once it has been entered into the organization's accounts and records during a specific period. Cash larceny can be done by reversing transactions, altering cash counts and forging cheques [13].

In 2020, cash larceny fraud amounted to 8% of all reported cases of occupational fraud. It caused a high median loss of $83,000. In 2020, the cash larceny fraud scheme was rated as a high risk [31].

Due to the immutable characteristic of blockchain technology, records stored in the ledger cannot be manipulated or deleted. Every transaction is timestamped and embedded into a block after verification and cryptographical hashing. The hashing process includes the hash from the previous block, which links the block to an unbreakable chain in the network. If an employee alters the recorded transaction, the chain of blocks will break, and the cause will be easily noticed [37].

Register Disbursement Schemes

Register disbursement schemes occur when employees create bogus entries on a cash register to fraudulently remove cash. An example is a fraudulent refund [13]. In 2020, the register disbursement fraud scheme amounted to 8% of all reported cases of occupational fraud. It caused a median loss of $20,000. In 2020, the register disbursement fraud scheme was rated as a low risk [31].

Register disbursement schemes can be easily mitigated in blockchain technology because blockchain's decentralized network provides a transparent platform in which fraudulent data and transactions can be easily identified and flagged. Transactions recorded on the chain can only be accessed by authorized users who have a user-specific encryption key [37].

Check Tampering

This is a scheme in which the individual steals his or her employer's funds by diverting, forging or modifying a check or electronic payment meant for a third party for his or her benefit [13]. In 2020, the check tampering fraud scheme amounted to 10% of all reported cases of occupational fraud. It produced a median loss of $110,000. In 2020, the check tampering fraud scheme was rated as a high risk [37]. Blockchain-enabled smart contracts could go a long way in eliminating cheques as a payment tool.

Billing Schemes

A billing scheme occurs when an individual causes his or her employer to make a payment by tendering invoices for fabricated goods or services, inflated invoices or invoices for personal purchases [13]. In 2020, the billing fraud scheme amounted to 20% of all reported cases of occupational fraud. It produced a median loss of $100,000. In 2020, the billing fraud scheme was rated as a high risk [31].

A blockchain-based expense and billing system can be used by organizations to validate expenses and issue payments. In this system, expenses are created and recorded using blockchain-validated records created by the vendor [37]. Due to the immutable nature of blockchain technology, the transactions in this system cannot be altered by employees.

Payroll Schemes

Payroll schemes occur when an individual causes his or her employer to release payment by creating false compensation claims [13]. In 2020, the payroll fraud scheme amounted to 9% of all reported cases of occupational fraud. It produced a high median loss of $62,000. In 2020, the payroll fraud scheme was a high risk [31].

Blockchain's finality and immutability allow for a permanent, time-stamped audit trail for every stage of a business transaction and process. This could help management to eliminate the risk of issuing payments for false claims [37].

Expense Reimbursement Schemes

In expense reimbursement schemes, an employee claims reimbursement of fictitious or inflated business-related expenses [13]. In 2020, the expense reimbursement fraud scheme amounted to 9% of all reported cases of occupational fraud. It produced a high median loss of $100,000. In 2020, the expense reimbursement fraud scheme was rated as a high risk [31].

Blockchain technology allows an independent verification of third parties. It creates a verifiable record of every transaction that has been made; and those transactions can only be created and/or altered with consent. A blockchain-based expense system can be used by organizations to validate expenses and make payments. In this system, expenses are charged directly to the organization by the vendor and are validated on the blockchain network. The entire transaction is authenticated and carried out on the network. Due to the immutable nature of blockchain technology, the transactions in this system cannot be altered by employees [37].

Misuse of Inventory

Misuse of inventory occurs when an employee abuses the non-cash assets of the target organization. Blockchain's decentralization and data authentication can prevent participants from misusing inventory assets. Similarly, theft of inventory occurs when an employee steals or misappropriates the non-cash assets of the victim organization [13]. If an organization's inventory is digitized on a blockchain network, there will be better traceability and transparency of the movement and use of inventory, since transactions are immutable on a blockchain network, and are authenticated and updated by consensus between participants in the network.

OTHER NON-ACFE-RELATED BLOCKCHAIN THREATS

Double-Spending Attacks

Double spending is a problem that arises when transacting digital currency that involves the same tender being spent multiple times, resulting in an interruption of the blockchain network and stolen cryptocurrency [20]. A double-spending attack occurs when a block is duplicated to steal credible information or currency with the ability to make it look legitimate [19]. This is one of the most difficult attacks that can occur in blockchain due to the amount of time and the rigorous process required for manipulating complex algorithms in order to make a fraudulent transaction appear legitimate [7].

A double-spending attack involves spending a cryptocurrency for which delivery of goods or services has already been accomplished in a previous transaction. In a double-spending attack, an attacker uses a fraudulent chain to alter the target transaction in a way that deceives the victim and benefits the attacker [18]. The records of payment are composed of the transaction and shared in a network through the status-quo chain. Double-spend attackers change the chain status within the network with fake blocks after the receipt of a service or product(s). Therefore, a double-spend attack – if executed successfully – compromises the immutability and trustworthiness of a blockchain [15].

The double-spending attack can be resolved using a P2P distributed timestamp server to generate computational proof to utilize consensus mechanisms that verify transactions with certainty [29]. The consensus mechanisms are alternatively known as proof-to-work [2]. Practically, the mechanism ensures that each participant node verifies the transaction [29]. The higher the computing power utilized by an attacker, the higher the likelihood of the success of a double-spending attack [32].

Exchange Hacks

Due to their large crypto holdings and lack of full-proof security, cryptocurrency exchange hacks have been popular among hackers over the past decade. To avoid exchange wallet attacks, it is safer to keep cryptocurrencies in a hardware or paper wallet, which will use fewer Internet touchpoints to keep money secure from malicious online hackers [22]. Chapters 2 and 3 provide more discussion regarding crypto wallet types and safety features.

Social Engineering

Blockchain-focused social engineering takes several different forms, but the goal is always the same; to get a user's private key or, more specifically, the user's cryptocurrency. One of the most popular types of social engineering is phishing, used with emails. To avoid being a victim of a social engineering attack, it is important to make sure to refrain from publicly disclosing login credentials or private keys to anyone [22].

Malware

Due to the unauthorized and unnoticed takeover of a device, crypto-jacking creates performance problems, increases energy consumption, and opens the door for other hostile codes. To prevent malware attacks, it is important to run comprehensive malware and other routine security checks to help ensure that user computers have not been infected with malware [33].

A PROPOSED PURCHASE CYCLE BLOCKCHAIN AUDIT CHECKLIST

The implementation of the proposed blockchain audit checklist below (Table 4.1) allows for reduced issues relating to fraud schemes in the completion of a purchase cycle.

FRAUD SCHEMES ASSOCIATED WITH PURCHASE CYCLE

The purchase cycle may be the target of many with fraud schemes which, if not checked using good practices and effective controls, may affect the purchasing decision. The immutable, decentralized, distributed, tamper-resistant and consensus-mandated nature of the blockchain technology backed

TABLE 4.1
Purchase Cycle Blockchain Implementation Audit Checklist Template

S/N	Blockchain-Purchasing Cycle Risks to Consider	Yes	No	N/A	Comments / Action Required
1	Does the creation of blocks have documented approvals? If yes; What level of authorization exists for the creation of block or initiation of the transaction?				
2	Is there a blockchain control mechanism in place between initiation and approval?				
3	Is there segregation between internal and external cross-checking of blocks?				
4	Is a blockchain consensus protocol used in choosing a block creator?				
5	Are procedures adequate for recording, evaluating, resolving, or escalating problems?				
6	Were validators evaluated for competency and trustworthiness?				
7	Is there a validation check between the purchase order input and output between parties on the block? How often is the validation done?				
8	Are end-to-end procedures executed properly? If yes, what control exists?				
9	Are controls in place to ensure that there is no conflict of interest and not collusion among validators?				
10	Are blocks created internally reviewed/audited? If yes, How often?				
11	Is a mechanism in place to prevent the creation of blocks for duplicate transactions; or, processing the same transaction twice?				
12	Are there aspects of the purchase cycle that cannot be modeled into the creation of blocks? If yes, how are they treated?				
13	Did an expert in smart contracts conduct a block review? If so, what frequency?				
14	Do the digital signatures guarantee transaction authenticity and integrity? If yes, how do you protect your private keys?				
15	Are there private keys recovery steps and procedures? Is it an escrow account? How can it be accessed? Who can access it?				
16	Are the blockchain access rights issued on a need-to-know basis?				
17	Is the access control list efficient and effective in avoiding conflict of interest and mitigating fraud?				
18	Is the storing of confidential data on the blockchain adequately controlled and structured for business continuity and disaster recovery?				

Source: [34].

with best practices and controls around this technology makes it a great tool for mitigating fraud in the purchase cycle. Obviously, the effect of the purchasing decision not being finalized in an expedient, accurate and fraud-proof manner will most likely result in an organization's ability to attain its set goals and objectives.

Additionally, it is evident from Table 4.2 that from the very first stage in the purchase cycle, different fraud schemes may occur, resulting in sometimes significant monetary and reputational losses.

By providing appropriate and convenient payment alternatives, and improving the overall purchase cycle by utilizing accurate and real-time data, the threat associated with billing schemes and placing superfluous orders should be greatly mitigated.

TABLE 4.2
Mapping of Fraud Schemes Associated with Purchase Cycle with Good Practices and Controls

S/N	Stages in the Purchase Cycle	Fraud Scheme	Good Practices and/or Controls
1	Order requisition	Billing scheme Unnecessary order	Offering multiple payment options. Upgrading and embedding the billing software into the blockchain network. Only use accurate and real-time data.
2	Selecting supplier	Price fixing Conflict of interest	Verify the identity of registered vendors on the block. Ensure all parties relating to the transaction are aware of all policies and standards embedded in smart contracts. Create suitable policies for adding and removing suppliers from the blockchain network. Report any form of suspicion or concerns to the company or available hotlines. Smart contracts should be designed with access controls that permit only approved suppliers and users to create transactions.
3	Quotation and tendering	Bid rigging Price fixing	Avoid carrying out transactions outside the blockchain network. Further scrutinize bid prices or proposals that look suspiciously low or high. Do not split the contract between two entities with identical bids. Provide staff training on how to follow policies and controls guiding the purchase cycle on the blockchain network.
4	Issuing a purchasing order	Kickbacks and bribery	Frequent update of bribery and anti-corruption policies. Engage independent third parties, such as auditor teams to verify transactions and review the company's controls and policies. Refer to anti-corruption policies and laws applicable to the organization. Ensure that the signing of blank purchase orders is prohibited and embedded in the smart contracts' codes.
5	Goods received and delivery	Improper claims Imprest funds Variation abuse Contract specification abuse	All transactions should be verified. All validated transactions should be posted on the blockchain ledger. All agreements between involved parties should be approved and embedded in smart contracts. Suppliers and third-party vendors should publish their transactions and accounting information on the blockchain network to facilitate the verification of transactions.
6	The invoice received and approved	False invoices	The auditor should ensure that invoices, requisition orders and purchase orders are properly handled in smart contracts to ensure full compliance.
		Overbilling	Conduct verification of all transactions against discounts and terms of the original purchase order on the network.
		Duplicate or inflated invoice	Verify that canceled purchase orders are reflected on the blockchain network to prevent duplicate payment of the same invoice after delivery.

(continued)

TABLE 4.2 (Continued)
Mapping of Fraud Schemes Associated with Purchase Cycle with Good Practices and Controls

S/N	Stages in the Purchase Cycle	Fraud Scheme	Good Practices and/or Controls
7	Quality control and receiving a report	Product substitute	Documentation of the inspection of goods by the quality control department should be included as criteria for concluding a transaction in applicable smart contracts. Smart contracts should reconcile the purchase order with the merchandise receipt information, and send the order to the account payable team for a final review and approval.
		False statement and claims	Rules related to applicable taxes should be embedded in applicable smart contracts. There should be an acceptance process embedded in smart contracts where the actual cost of an order is checked against the initial amount originally calculated on the purchase order.
8	Payment	Check tampering	All payments should be made through the process embedded in smart contracts; and payments should be authorized by appropriate parties on the blockchain network [8]. A fraud prognosis prototype should be embedded into the smart contract in order to detect fraudulent activities [8].

Source: [34].

PUBLIC AND PRIVATE SECTOR TRIPLE-ENTRY ACCOUNTING CHALLENGES

Table 4.3 lists the major challenges facing a successful implementation of a triple-entry accounting system in organizations. In general, technology adoption aims to improve an enterprise's operations, so that it can increase its effectiveness and efficiency. However, organizational transition efforts typically meet with stakeholder resistance, thereby creating new challenges in implementing modern technologies. Very often, users respond negatively to a proposed technological implementation, due to the need to learn new skills and/or procedures. Additionally, user resistance could also be attributed to a lack of confidence in the technology itself, since blockchain is still relatively new. Challenges related to the auditing of blockchain applications are also another important consideration. Overall, organizational readiness is critical to blockchain adoption within the context of the enterprise [1].

SUMMARY AND CONCLUSIONS

This chapter started out with a brief history of accounting entries from the single-entry accounting system, double-entry accounting system and triple-entry accounting system. Additionally a number of sub-topics related to the advent of triple-entry accounting were discussed; including how various blockchain components work with double-entry accounting; the effect of triple accounting for accountants and auditors; as well as the effects of triple accounting in mitigating financial fraud. The information provided in this chapter is an attempt to help accountants and auditors develop a better understanding of the triple-entry accounting system and its benefits in the accounting/auditing profession.

TABLE 4.3
Impact of Triple-Entry Bookkeeping and Its Challenges

Categories	Impacts	Challenges
Financial	Organizations	Need to invest substantial funds in order to develop and upgrade blockchain technology in the hope that the benefits will outweigh the costs over time. Accounting professional and auditing specialists need to upgrade or expand their current limited knowledge of triple-entry accounting, which translates to additional training and professional development costs.
Stakeholders	Governmental bodies Tax authorities Finance ministry Policymakers Auditors Foreign investors	Introduction of new technologies and the fact that different stakeholders often have different interests or priorities which will make the task of bringing stakeholders together difficult.
Technology	Organizations/stakeholders	It often takes significant time for the new initiative to be fully phased in. It is also possible that the proposed technology will not be able to deliver the intended benefits for its intended use case.
Policy implementations	Triple-entry accounting policies must be updated for relevance.	Due to their enhanced accountability and control, policies may affect some stakeholders negatively.

Source: [25].

CORE CONCEPTS

- This chapter discussed the history of accounting entries from the single-entry accounting system which was started in India by Chanakya to the double-entry accounting system that had its start in the 15th century by the Italian Economist, Luca Pacioli. Triple entry was originally conceived by an accounting scholar named Yuji and also by a financial cryptographer, Ian Grigg. In addition to the debit and credit entries, Yuji recommended the incorporation of the third layer of entries with a new collection of accounts to further illustrate changes in income.
- The implementation of a blockchain-based accounting system where double-entry and distributed records are transformed into decentralized public records could offer high levels of certainty as to the integrity of the audited entity records. Ian Grigg developed a new theory, '*the receipt is the transaction*', in order to prevent transaction fraud and minimize unintended errors in internal documentation. A mutual third entry can interpret a digitally authorized receipt supported by a financial cypher among two parties.
- Triple-entry accounting helps in assuring the authenticity of source documents, potentially increasing future job opportunities for accountants with blockchain skill sets.
- Blockchain technology can help in mitigating ACFE fraud scheme such as financial statement fraud, skimming fraud, cash larceny, register disbursement schemes, check tampering, billing schemes, payroll schemes, expense reimbursement schemes, misuse of inventory and theft of inventory due to its immutability, decentralization, distributed ledgers, consensus mechanism and enhanced security nature of blockchain.
- Due to the traceability characteristic of the blockchain, concerns of missing purchase transactions can be easily addressed; and, the execution of triple-entry accounting on the blockchain network will help to reduce the time used and steps in the auditing and purchase cycle.

ACTIVITY FOR BETTER UNDERSTANDING

Using your web browser, go to YouTube and spend some time watching the videos below to better familiarize yourself with the workings of triple-entry accounting; then try to answer the following questions:

YouTube Video 1- Triple-Entry Accounting: https://www.youtube.com/watch?v=wWXy7wUDEoQ

YouTube Video 2: What Is Triple-Entry Accounting? And Why Is It Changing Our World: https://youtu.be/kxC1SbP7r10

1. How can the implementation of a triple-entry accounting system help improve the accounting system in your organization?
2. Based on the chapter discussion, what specific blockchain-related controls can help better mitigate the risk of fraud in your organization?
3. If your organization were ever the target of a fraud scheme, could that incident possibly have been avoided or detected sooner, if a triple-entry accounting/control system were in place?

REFERENCES

[1] Batubara, R. F., Ubacht, J., & Janssen, M. (2018). Challenges of Blockchain Technology Adoption for e-Government: A Systematic Literature Review. *Proceedings of the 19th Annual International Conference on Digital Government Research*, 1–9. https://doi.org/https://doi.org/10.1145/3209281.3209317

[2] Bentley, H. C. (1929). *Brief Treatise on the History and Development of Accounting* (Vol. 64). Individual and Corporate Publications.

[3] Bible, W., Raphael, J., Riviello, M., & Taylor, P. (2020). Blockchain Technology and Its Potential Impact on the Audit and Assurance Profession. Retrieved 21 February 2021, from https://www.cpacanada.ca/en/business-and-accounting-resources/audit-and-assurance/canadian-auditing-standards-cas/publications/impact-of-blockchain-on-audit.

[4] Bryer, R. (2012). Americanism and Financial Accounting Theory, Part 1: Was America Born Capitalist? *Critical Perspective on Accounting*, 1045–2354, 511–555. https://doi.org/doi:10.1016/j.cpa.2012.09.003

[5] Cai, C. W. (2019). Triple-Entry Accounting with Blockchain: How Far Have We Come? *Accounting & Finance*, 1–21. doi:10.1111/acfi.12556

[6] Carlin, T. (2018). Blockchain and the Journey beyond Double Entry. *Australian Accounting Review*, 29(2), 305–311. doi:10.1111/auar.12273

[7] Chohan, U. W. (2021). The Double Spending Problem and Cryptocurrencies. Critical Blockchain Research Initiative, 1–11.

[8] Dai, J., & Vasarhelyi, M. (2017). Toward Blockchain-Based Accounting and Assurance, *Journal of Information Systems*, 31, 3, 5–21.

[9] Faccia, A., & Mosteanu, N. R. (2019). Accounting and Blockchain Technology: From Double-Entry to Triple-Entry. *Business and Management Review*, 10(2), 108–116.

[10] Faccia, A., Moşteanu, N. R., and Leonardo, L. P. (2020). Blockchain Hash, the Missing Axis of the Accounts to Settle the Triple Entry Bookkeeping System. ICIME 2020: Proceedings of the 2020 12th International Conference on Information Management and Engineering, September, 18–23.

[11] Fanning, K. & David, P. (2016). Blockchain and Its Coming Impact on Financial Services. *Journal of Corporate Accounting & Finance*, 53–57.

[12] Farell, R (2015). An Analysis of the Cryptocurrency Industry. Thesis, Wharton Research Scholars, Pennsylvania.

[13] Fraud Tree. Acfe.com (2016). Retrieved February 23, 2021, from https://www.acfe.com/fraud-tree.aspx.

[14] FreshBooks. (2020, July 20). What Is Single-Entry Bookkeeping? Pros and Cons for Small Business. Retrieved from https://www.freshbooks.com/hub/accounting/single-entry-bookkeeping.

[15] Karame, G. O., Androulaki, E. and Capkun, S. (2012). Double-Spending Fast Payments in Bitcoin, in Proceedings of the 19th ACM Conference on Computer and Communications Security, Raleigh, NC, October, 906–917. Retrieved from http://doi.acm.org/10.1145/2382196.2382292
[16] Gröblacher, M. & Mizdraković, V. (2019, January). Triple-Entry Bookkeeping: History and Benefits of the Concept. In Proceedings of the 6th International Scientific Conference -FINIZ 2019. doi:10.15308/finiz-2019-58-61.
[17] Ibañez, J. I., Bayer, C. N., Tasca, P., & Xu, J. (2020). REA, Triple-Entry Accounting and Blockchain: Converging Paths to Shared Ledger Systems. *SSRN Electronic Journal*. doi:10.2139/ssrn.3602207
[18] Jang, J., & Lee, H. (2019). Profitable Double-Spending Attacks. *IEEE*, 1–13. doi:https://arxiv.org/ftp/arxiv/papers/1903/1903.01711.pdf
[19] CFI Institute (2020, September 28). Double-Spending – Overview, How It Occurs, How to Resolve. Corporate Finance Institute. https://corporatefinanceinstitute.com/resources/knowledge/other/double-spending/.
[20] Kang, K.-Y. (2020). *Cryptocurrency and Double Spending History: Transactions with Zero Confirmation*. Springer, 1–48.
[21] Karajovic, M. K. (2019). Thinking outside the Block: Projected Phase of Blockchain Integration in the Accounting Industry. *Australian Accounting Review*, 1–12.
[22] Llc, L. (2019, December 6). Top 5 Blockchain Security Issues in 2019. LIFARS. https://lifars.com/2019/12/top-5-blockchain-security-issues-in-2019/.
[23] Nakamoto, S. (2008). Bitcoin: A Peer-to-Peer Electronic Cash System. http://bitcoin.org/bitcoin.pdf.
[24] Napier, C. J. (1998). Giving an Account of Accounting History: A Reply to Keenan. *Academic Press*, 104–2354(98), 685–700.
[25] Nath, S. M. (2020, July 15). Triple Entry Accounting: One of the Greatest Inventions in the Last Few Centuries in the World of Accounting. LinkedIn. https://www.linkedin.com/pulse/triple-entry-accounting-one-greatest-inventions-last-few-m-m/.
[26] Okoli, B. E. (2011). Evaluation of the Accounting Systems Used by Small Scale Enterprises in Nigeria: The Case of Enugu, South East Nigeria. *Asian Journal of Business Management*, *4*(20041–8752), 235–240.
[27] Ovunda, A. S. (November 18, 2015). Luca Pacioli's Double-Entry System of Accounting: A Critique. *Research Journal of Finance and Accounting*, 6(2222–1697), 132–139. doi:http://acct.tamu.edu/smith/ethics/pacioli.htm
[28] Muncaster, P. (2017, June). World's Largest Bitcoin Exchange Bitfinex Crippled by DDoS. http://bit.ly/2kqo6HU,
[29] Reiff, N. (2021, February 9). How Does a Blockchain Prevent Double-Spending of Bitcoins? https://www.investopedia.com/ask/answers/061915/how-does-block-chain-prevent-doublespending-bitcoins.asp.
[30] Rembert, L. (2020, November 26). 51% Attack. Retrieved March 1, 2021, from https://privacycanada.net/cryptocurrency/51-attack/
[31] Report to the Nations (RTTN) 2020 Global Study on Occupational Fraud and Abuse. Association of Certified Fraud Examiners. (2021). Retrieved February 21, 2021, from https://acfepublic.s3-us-west-2.amazonaws.com/2020-Report-to-the-Nations.pdf.
[32] Rosenfeld, M. (2014, February 12). Analysis of Hashrate-Based Double-Spending. *ArXiv14022009 Cs*, 2–8. doi:https://arxiv.org/pdf/1402.2009.pdf
[33] Saad, M., Spaulding, J., Njilla, L., Kamhoua, C., Shetty, S., Nyang, D., & Mohaisen, A. (2019, April 6). Exploring the Attack Surface of Blockchain: A Systematic Overview. *Air Force Research*, 1–30. doi:https://arxiv.org/pdf/1904.03487.pdf
[34] Salami, O. (2020). A Proposed Purchase Cycle Audit Approach Using Blockchain Technology to Increase Audit Effectiveness and Reduce Fraud. Concordia University of Edmonton.
[35] Sangster, A. (2016). The Genesis of Double Entry Bookkeeping. *American Accounting Association*, 91(1), 1–28.
[36] Tan, B., & Low, K. (2019). Blockchain as the Database Engine in the Accounting System. *Australian Accounting Review*, 29(2), 312–318.
[37] Yu, T., Lin, Z., & Tang, Q. (2018). Blockchain: The Introduction and Its Application in Financial Accounting. *Journal of Corporate Accounting and Finance*, 29(4), 37–47.

5 Blockchain Use in the Financial Services Sectors

Antonio Ramirez, Bhanu Theja Satyani, Jovid Ismailov and Lovepreet Singh

ABSTRACT

This chapter discusses various use cases of blockchain technology in the banking and financial services sectors in North America. The topics discussed focus on the advantages that blockchain technology offers compared to business-as-usual models. For instance, blockchain technology can streamline and shorten the securities settlement process on exchanges, credit card payment transaction processing, and insurance claim settlements can all be optimized in terms of cost and efficacy. According to *The Blockchain Training Institute*, 90% of banks in the USA and Europe are considering blockchain technology to enhance the efficiency of their operations, while seven out of ten banks are aiming at using blockchain technology within the next three years.

Arguably, the population destined to benefit the most from offering blockchain-based financial and banking services is the millions of inhabitants in parts of the world who do not have access to conventional banking and lending services. Blockchain technology has great potential to become a game-changer for this underserved population. To further explain, anyone with a cell phone and a banking/lending app can effectively conduct financial transactions without using a trusted intermediary, such as a conventional bank. Blockchain technology can also be an effective tool in managing microloans to remote territories worldwide where financial services simply are not available. A blockchain solution can enable financial services to farmers and trading participants to get their business started or expanded, thereby creating a real opportunity for these populations to combat extreme poverty effectively.

BLOCKCHAIN USE IN FINANCIAL SERVICES

The global financial services sector is one of the most complex industries due to the diversity of products, services and classification depending on the market and operations. Market estimates indicated that the size of the global financial services sector was expected to reach $26.5 trillion in 2021, growing at a forecasted rate of 6% compared to the previous year. Asia-Pacific is the largest financial sector, followed by North America. The financial services market represents almost a fifth of the global GDP, estimated at around $149 trillion in 2021 [22].

The North American markets for the financial services and insurance sector reached a market size of $5 trillion in 2021, including a growth rate of 1.7% for 2021 due to the COVID-19 impact. The annual growth rate before the pandemic was about 2.3%, with around 1.2 million businesses and 7.3 million individuals employed in this sector. The key US players in this market include United Health Group, JPMorgan Chase & Co., State Farm Mutual Automobile Insurance Company, Bank of America Corporation and MetLife Inc, to name a few [14].

Blockchain technology can bring increased efficiencies to financial systems compliance, including stricter enforcement of regulations against money laundering. Blockchain decentralized technology enables autonomy and privacy over information and the pillars of the financial industry: funding,

DOI: 10.1201/9781003211723-5

investing, lending and exchanging value. Blockchain efficiencies translate into open platforms to enable individuals to store, transform and move information with a higher degree of security; thereby, enhancing organizational accounting, analysis, auditing and risk management processes through increased traceability and real-time access to information [26].

BLOCKCHAIN IN THE BANKING SECTOR

Blockchain technology is increasingly being adopted across industries. For the banking sector, blockchain is a strategic tool for financial institutions to remain competitive in a highly innovative marketplace. Blockchain can help lower transactional costs and increase efficiency to support complex operations that require a high degree of security. These include larger peer-to-peer payment transfers, cross-border payments, and even high-volume local transactions. Blockchain's consensus verification process and a streamlined access platform have been identified as the key elements to overcome reiterative and duplicated processes with enhanced security and data integrity. The banking industry is taking significant steps to embrace blockchain technology by implementing projects involving different industries where banking services are the key driver for operations at a large scale. One example is a blockchain app created by IBM and sponsored by the global leading forex settlement group, *CLS group holdings A.G.*; this app provides a blockchain structure for banks to standardize global forex markets and thereby reduce transactional costs. Another example is *Batavia*, a pilot blockchain platform developed by a joint effort between *IBM, the Bank of Montreal (BMO), Commerzbank, CaixaBank, Erste Group* and the *Union Bank for Switzerland (UBS)*. Batavia platform is an open system IBM blockchain platform in order to leverage automated processes for arranging and securing international trading transactions to exchange automobiles and textile raw materials. The Batavia blockchain solution also integrates the applicable trade agreements designed to automatically trigger supply chain events and smart contract payments.

Supporting operations at a large scale using blockchain technology brings an additional challenge to manage big data. Blockchain technology requires a robust infrastructure and software to manage today's big data volumes. The banking industry has been exploring and partially adopting blockchain technology as one of the most heavily regulated industries on a global scale. Given the complexity of the financial sector, full adoption of blockchain depends on three key variables: trust, cost-effectiveness and compliance [11].

According to PWC – one of the Big Four professional services firms in the market with specialization in the banking sector – the financial institutions have made considerable investments in development and maintenance costs of their legacy systems. It will be a great challenge for some financial institutions to embrace the digital transformation that new technologies incorporate; however, innovation constitutes the main driver for growth opportunities and competitive advantages. Regulators in the financial sector are leaning in a favorable direction for RegTech and Cybersecurity's recent advances. Regarding blockchain implementations, some financial institutions' position is to wait for industry leaders' success stories before considering blockchain as a valuable solution for their own operations. However, this wait-and-see approach can entail significant competitive advantage risks, since the technology is very rapidly evolving. The current financial sector strategy is focused on shifts toward building partnerships across industries to bring greater efficiencies, cybersecurity and compliance. Blockchain smart contracts functionality can help overcome challenges related to high cost, long delays and other time-consuming aspects of trade finance. For the financial markets, blockchain can bring efficiencies to intermediated functions such as trade clearing and settlement [19].

BANKING PAYMENTS SYSTEMS

The current banking payment system is based on a two-tier structure involving commercial banks in coordination with a central bank. The central bank's main role is to provide 'trust in money'; its underlying principle to support retail and wholesale transactions. Innovations to the current banking payment systems are constant; however, the main role of the central banks in the payments system is even more important amid the current emergent technologies revolution. A payment system is composed of instruments, procedures and rules to exchange funds among its participants. There are two main classifications of the payment system – the retail and wholesale systems. The retail system supports large volumes of transactions involving relatively low values. In contrast, the wholesale payment system supports large-value and low-volume transactions between financial institutions. The payment system is most impacted by technology innovations constantly aiming for faster and safer payments transfers.

The central bank provides commercial banks with the physical (fiat) cash for use by the public. Commercial banks deposit their mandated money reserves with the central bank; as such, the money reserves are used in wholesale transactions between financial institutions. The central bank's physical cash and reserves are considered safe and supported by a system that guarantees liquidity and payment finality. Payment systems are composed of multiple participants apart from the two key role players – central and commercial banks. These include payment service providers (PSPs). PSPs and commercial banks are now introducing digital services such as digital wallets and mobile applications to manage payment services to end-users.

The central bank's role as a guarantor of the payment systems entails adequate responses to various threats and risks. The key risk is systemic in nature in an interconnected system, where the participants could be unable to operate, or the system is unable to achieve finality. The second important risk is fraud, where both wholesale and retail digital payments are a primary target. Additionally, there are also illicit financing and money laundering risks that are constantly increasing in the context of cybercrime.

From a different perspective, the availability of the traditional payment services constitutes a challenge for emerging markets, developing countries and low-income individuals due to a lack of convenient access to bank services and their payment, PSPs and credit platforms.

With the introduction of blockchain technology, there have been attempts to create new decentralized payment systems using cryptocurrencies. Cryptocurrencies are generally considered a threat to jurisdictions' monetary policies, since most cryptocurrencies are not currently perceived with trust by the general public due to a lack of regulations. General acceptance, confidence and certainty of a money system that achieves finality (extinguishes obligations) is what constitutes trust and value of the exchange instrument. Central banks are also considering using their own cryptocurrencies. A set of embedded supervisory principles of blockchain, described below, can help achieve the most important element of a payment system, namely trust [3].

The Embedded Supervision Principles in Blockchains

The embedded supervision principles in blockchain emphasize the importance of using a comprehensive regulatory framework for blockchain implementations in the financial services sector. Embedded supervision principles can be automatically deployed while interacting with the distributed ledger. There are four embedded supervision principles. First, a regulatory framework must be enacted by an effective legal system and promoted by industry-leading institutions to guarantee legitimate operations. Second, embedded supervision can be applied to transactions that are considered final under the economic finality principle (it will not be profitable to revoke a transaction). Third, embedded supervision must be applied upon consensus models where all parties participate and adhere to streamlined supervision. Finally, embedded supervision should promote stability, high quality and verification processes at the lowest possible cost [20].

USE OF BLOCKCHAIN ON EXCHANGES

The stock market enables public companies to raise capital for expansion purposes while providing liquidity and eliminating counter-party risks to investors. These financial exchange activities are performed through institutionalized public and formal marketplaces which operate under strict regulations. There are multiple securities exchange marketplaces around the world. The top ten world stock exchange marketplaces include: the NYSE (USD 25.5 trillion), NASDAQ (USD 11.2 trillion), followed by Japan's Exchange Group (USD 11.2 trillion), Shanghai Stock Exchange (USD 4.6 trillion), Hong Kong Exchanges (USD 4.23 trillion), and the Shenzhen Stock Exchange (USD 3.2 trillion), Euronext (USD 3.6 trillion), the London Stock Exchange (USD$2.9 trillion), Canada's TSX Group (USD 1.75 trillion) and the Bombay Stock Exchange (USD $1.51 trillion) [25].

NASDAQ and NYSE are the biggest stock exchange marketplaces in the world; these exchanges have also been investing in blockchain technology initiatives for several years now. One of those initiatives is *tZero.com* which has introduced its first $134 million private digital security token offering. The tokens will become available in the market for participants who meet the Securities Exchange Commission (SEC) mandates. The tZero initiative was developed with alignment to the SEC rules, and its tokens are considered pioneers in the securities market.

The purpose of the SEC rules related to blockchain initiatives is to mitigate the incremental spread and disruptions to the stock markets. For instance, the *DAO Report* issued by the SEC, is an investigation of a virtual and decentralized autonomous entity. The DAO Report's focus of investigation was an offering of securities with blockchain-based tokens used as cryptocurrency. The report provides guidelines to differentiate digital securities and cryptocurrencies to define compliance with US securities laws. According to the SEC rules, digital securities using tokens as cryptocurrency must operate under a regulated alternative trading system (ATS). Subsequent reports disclosed a cyberattack to the blockchain where the offering was operated. The report concluded that security offerings made using decentralized and autonomous principles should operate under SEC regulations, regardless of the underlying technology [7].

Blockchain-Based Transaction Clearance and Settlements

Traditionally, a series of verification and confirmation processes must occur to clear and settle transactions on market exchanges. With current technology, it typically takes three or more days for the market exchange to achieve finality. In contrast, blockchain technology can streamline the clearing and settlement process in real time by eliminating the need for intermediaries. Blockchain has great potential to improve the securities offerings cycle; from issuance to transferring, settling, and tracking processes, including various compliance-related tasks.

The tZero's initiative is a good example of the efficiencies that blockchain can add to the market exchange. The initiative consists of a public blockchain, and market exchanges operating with tokens using an alternative trading system (ATS). The ATS platform is not based on blockchain technology; but integrated into blockchain-based digital securities as a sidestep. The clearing and settlement processes are supported by an Ethereum blockchain with smart contracts [7].

The Investment Management Sector (Mutual Funds)

Investment management is the general term used for managing financial assets and investments such as securities, bonds, shareholding, mutual funds and other financial instruments. The management of assets includes the decision and strategy to diversify the funds between stocks, bonds, real estate and commodities to maximize the profit generated from these assets. Investment management companies managing assets of at least $25 million or giving advice to companies that offer mutual funds must be registered investment advisors (RIAs). RIAs are required to register with the Securities and Exchange Commission (SEC) and the state securities administrators as registered advisors.

Blockchain and smart contracts could revolutionize the investment management sector through automated financial processes, such as opening new accounts, recordkeeping of documents, automatic payments, and securities transfers. Blockchain-enabled investment management is based on a digital identity linked with a customer's financial records and transaction history. Blockchain technology, in general, has four layers; namely, (from top to bottom) an application, service, network, protocol and an infrastructure layer. The bottom infrastructure layer of blockchain, which includes data storage and network mining, works completely different from standard legacy approaches. Blockchain by design provides encryption and verification to the data established by the consensus across the network. The implementation of a blockchain infrastructure in investment management is more suitable for its value chain functions that include activities, such as multiple parties granting approval, law and compliance enforcement, audits, tracking and reporting, portfolio management and client interactions [6].

Investment Management Use Cases

In 2013, The National Association of Securities and Dealers Automated Quotations (NASDAQ) developed a blockchain network called *Linq* to store and manage privately held companies' shares, known as private securities. NASDAQ believes that the critical factors for the financial market development include a secure and documented chain of custody, and recordkeeping assuring asset liquidity and scalability. The Linq platform currently offers shares trading, regulation enforcement and reporting of small private businesses such as chain.com to minimize the scope of risk while engaging the regulatory bodies in a blockchain-enabled business solution. *The Delaware Blockchain Initiative* (DBI) is another blockchain use case designed to further facilitate legal and operational challenges involving larger commercial transactions. This platform does not provide the digital copy of the paper-based ownership; but instead provides corporations' legal electronic ownership rights. DBI creates an electronic database for the public archive, including birth and death certificates, commercial codes, land titles in the distributed ledger form, and automation of access and control using the smart contract. The DBI is making a legal process that clarifies the digital status of the record on the blockchain by creating a security master file with a digital record that allows for transferring securities through smart contracts. Another component of DBI is to use the smart contract and blockchain to keep records of state corporate fillings with all the details from new records to the notification about expired processes [6].

KNOW YOUR CUSTOMER CONSIDERATIONS

The financial sector is heavily impacted by Anti-Money Laundering (AML) and *Know Your Customer* (KYC) regulations to prevent money laundering, terrorist activity financing and fraud. Financial institutions must undertake preventive measures and controls to mitigate these threats and comply with existing KYC-related regulations. The Know Your Customer concept refers to the client identification, verification and authentication processes. A group of mitigative controls are applied to avoid doing business with individuals or organizations associated with corruption, money laundering or terrorism. The KYC process includes verification to ascertain who prospective clients say they are before giving them access to financial products or services. KYC tactics and AML regulations are crucial and expensive to sustain, as they require dedicated resources and complicated iterative processes. There is also a high degree of duplication of efforts between financial institutions and other third-party service providers. Blockchain technology is poised to make KYC an even more viable utility that can effectively mitigate against money laundering and terrorism risks by providing a more effective way forward based on inherent security and immutability features that provide greater integrity and trust in the data [13].

Streamline Compliance and Process Duplication Avoidance

Regulators and financial institutions can reduce regulatory risks and tighter controls by minimizing human input and promoting standardization. In this regard, directly feeding the financial institutions' profiles from authoritative sources plays a critical role in reducing fraud risk and the scope of human errors instead of relying on customer-inputted data fields or physical documents. As such, banks and other financial services providers can automate the process to allocate AML risk ratings via more objective and effective criteria in order to avoid regulatory fines. Additionally, the KYC process can provide regulators with evidence about transactions, parties engaged and underlying information [5].

Consequently, regulators would be better informed about customers' activities because blockchain technology would help record and track all the financial institutions and customers' activities and regularly monitor data activities and trends. From the customer perspective, financial institutions utilize the KYC utility to help further improve the customer experience by making processes increasingly efficient and timely. Although customers are likely to have privacy concerns regarding sharing their sensitive financial and personal data as required by the KYC regulations, blockchain technology gives customers the power to approve the kind of institutions that can access their information and under what circumstances. Digital integration makes such an approval process easy, such as pushing a consent appeal with the one-time password (OTP) to the client's phone for consent. Some options like smart digital signatures or facial scans can also augment customers' identity autonomy, thus allaying customer concerns about sharing their sensitive data. Finally, where it can be harmonized, KYC workflow routing and controls could be codified into smart contracts and routinely executed operations. Similarly, greater digitization is also likely to contribute to multilingual solutions through smart contracts and translation tools [8].

Blockchain as a Money Laundering Prevention Tool

Blockchain's decentralized, cryptographically secure and immutable properties can detect suspicious transactions instantly and efficiently. The distributed blockchain-based system utilizing smart contracts enables financial institutions to securely distribute data via the blockchain technology's Anti-Money Laundering 'engine', which can spot fraudulent transactions and issue alerts and/or exception reports. In this respect, KYC plays an instrumental role in averting money laundering and terrorism, and blockchain can be designed to act as the engine behind effective and efficient KYC operations [5].

BLOCKCHAIN AND CONSUMER CREDIT

The consumer and commercial credit sector is another suitable candidate for blockchain implementation, especially since data security and privacy are also a vital aspect of lending transactions. The technology facilitates secure, fast payment processing services through encrypted distributed ledgers providing transactions' real-time verification without requiring a third-party intervention. Moreover, the lack of third parties serves to enhance payment transaction speed, as this technology makes it easier to directly manage various credit card payment clearing processes. For instance, MasterCard has filed a blockchain-based payment system patent, enabling the company to deliver instant payments to merchants, secure payments verification and fast-track customers [24]. Similarly, blockchain technology promotes peer-to-peer transfers as it has no geographical limitations. The technology exists almost everywhere globally, making it possible to undertake peer-to-peer transfers worldwide [28].

Cardholder Identity and Verification Processes

Blockchain technology's ability to securely store information in an immutable way has simplified the process of verifying individuals' identities for credit card companies. For instance, thanks to this technology, consumers can utilize their mobile apps to verify their identity for credit card companies' new services rather than provide identification in person or sign additional documents. Moreover, with blockchain in fintech, users have the liberty to decide how they would like to be identified and whomever they would like to share their identity with [28]. Although users must register their identities on the blockchain, they are not required to complete every credit card issuer's initial registration process, provided that other such service providers are also driven by blockchain technology [24].

Blockchain technology can also play an important role offline during periods when the network connection is unavailable. For instance, when the Internet connection is hardly available, cardholders can submit their cards for payment and enter their pin in the merchant payment application to unlock the card's encrypted data. This application will automatically update the vital information between the payment card and the cardholders' bank. In this respect, blockchain technology can mitigate the need for an Internet connection. This aspect gives cardholders extra time to connect to the network and receive updates but still complete payments. However, when connectivity cannot be established for an extended period, offline processing can rely on a current blockchain payment technology referred to as *Zerocash*, which depends on the zero-proof authorization to complete the payment card transaction. In this regard, aspects such as zero-proof authorization, available funds and payment card information are combined to help support offline or delayed payment card transactions [28]. A single mining request obtains the personal identification information and security credential to initiate card system information and cardholder identity verification [24].

Blockchain enhances security and privacy through data encryption to help safeguard data against attempted data theft via scanners. For instance, most credit card companies use public-key cryptology to enhance the identity verification process, guaranteeing security for both the card issuer and their clients. All transactions on the network are signed, and linked cryptographically for immutability, thus enhancing transaction security. Blockchain technology's smart contracts play an instrumental role in executing credit card payments after specific conditions have been met; an aspect that also significantly improves security [28].

Blockchain technology can also reduce transaction fees for credit card companies and their clients. A distributed ledger provides reliable, immutable and shared transactions viewed like traditional fiat transactions between parties, but in an untrusted environment. Blockchain technology is considered one of the main drivers to help financial companies achieve considerable cost-saving, as distributed ledger technology can minimize financial services infrastructure costs. Currently conventional credit card payment processing fees are quite costly, usually ranging from flat to percentage-based fees, or a combination of both [24].

Microloans Management for Users in Developing Countries

Microloans are small loans issued by individuals or benevolent organizations – instead of banks – to help people living in developing countries. In general, microloans are useful in demographics where traditional banking financing is very limited or not available. Access to credit usually comes with several constraints translating to high-interest rates, if such funds are not coming from charitable organizations.

According to the *Human Concern International* (HCI) annual report, on a global scale, one out of nine individuals struggles with malnutrition; a total of around 821 million people in 2017, up from 784 million in 2015. *United Nations Children's Emergency Fund* (UNICEF) indicates that around 385 million children live in extreme poverty. Most of them fall into child labor or are exposed to violence or exploitation. According to one World Bank report, millions of children reach young adulthood without even basic mathematics skills, or the ability to read a bus schedule, or complete a job application. Progress in technology is one tool that can aid sustainable development goals and help to alleviate poverty through a partnership with non-profit organizations [12, 16].

Management of donor funds for the purposes of microloans funding through blockchain can be done more effectively. Smart contracts can release the funds directly to the person in need when certain conditions are met. Most donations are made using cryptocurrencies through a blockchain platform. Benevolent organizations receiving the full amount of donated assets are exempt from capital gains taxation, while donors may enjoy a valuable tax deduction. Blockchains also bring more transparency between donors and recipients. Most of the regulatory paperwork and overhead expenses of such charities can also be greatly reduced. There are various blockchain-based projects for donations; one such project is operated by the *United Nations World Food Program* to deliver food to refugees in Jordan. Refugees can get food without any cash or card just by scanning their irises. The tracking of the donated money using blockchain is much more effective, which will also help reduce corruption at various ends. The lower transaction fees and immediate financial support to victims will help build trust between donors and victims. Some of the existing blockchain projects in charity are *Helperbit*, *Blockchain for Social Impact Coalition* and *Bithope* [9].

A Corda Platform Use Case

Corda is open-source distributed ledger software developed in 2016. Corda's design provides a global logical ledger to support organizations, governments, suppliers, customers, third parties and individuals to safely record and manage their commercial agreements privately and consistently using a common platform. Sometimes, two organizations may have a trusting relationship with each other for trading purposes, but cannot trust other parties involved in a transaction. The Corda platform can eliminate this problem by maintaining records of all trading parties on one shared platform. The vision of the Corda is to create a shared network where all kinds of applications, products and services in such manner as to allow asset gains from any of these services to be redeployed to another trading platform with minimal cost and maximum throughput.

In Corda, the digital document that records the exiting transaction and current agreement between two parties is called a state object. The state object contains the information about the sender, receiver, amount sent, legal agreement and quantity of goods to be traded. Privacy of the network is maintained by allowing only legitimate parties to join the network. Furthermore, the consistency of the global ledger is maintained by verifying identities using cryptography hashes and keeping records of previous versions of the chain. Consensus in Corda is achieved by using two factors: transaction validity and transaction uniqueness. Every updated transaction has an associated contract code which is verified by the signatures of all participating entities. The uniqueness of the transaction is achieved by validating that no other previous transaction corresponds to that consensus with the same state. The uniqueness service of Corda is used to validate the transaction uniqueness; i.e., the new transaction is unique and cannot have taken place earlier. The contract in Corda is run through Java virtual machine, making it easier for banks to apply their existing code inside the contracts. Every individual or organization has a unique identity in the Corda network provided by a legitimate authority. It is represented by a certificate that constitutes the real-world name of that entity. Every identity has a private key associated with signing the transaction digitally.

The global Corda network is established around the world using a different set of nodes. All the nodes are pre-configured with unique identities and default settings to locate and transact directly. The pre-configured settings or standards allow the Corda nodes to transact flawlessly without depending on certain factors such as trading groups, locations or individual entities operating them. A business network in Corda is a group of parties with a similar business model or membership criteria. Every business network has different membership criteria, governance, assets and privacy requirements. Some of the key components of the Corda global network are specific network parameters, identity framework, consensus mode, and governance [4].

BLOCKCHAIN AND TRADE FINANCING

Trade financing is required by trading companies to facilitate the international trade of their goods and services overseas in the forms of trade credit, insurance and guarantees for a short period. Trade financing help to secure the funds needed for importers and exporters to buy goods, thereby facilitating such business transactions. Trade finance is an umbrella term with many parties involved, including banks, importers, exporters, insurance companies and trade finance companies. According to the *World Trade Organization* (WTO), world trade in 2020 plummeted by 9.2 percent, due to the COVID-19 pandemic. Nevertheless, every crisis hides some opportunities; the same is applicable to trade finance. International trade has shifted more rapidly toward digitalization as opposed to labor- and paper-intensive processes.

Distributed ledger technology improves the transparency in trade services by first providing insight into DLT-verified records. For example, whenever goods reach certain points during a trade, an update status is recorded and visible to all parties. Second, keeping track of each product through the blockchain can verify the products' authenticity and decrease counterfeit product threats along the way in a trade. The digitalization of the trade services using DLT is achieved by the digitization of trade documents. Some of the blockchain projects in trade finance are listed below.

INTERNATIONAL TRADE USE CASES

The *Skuchain EC3* (Empowered Collaborative Cloud) blockchain network is used for digitizing financial transactions in trade finance and supply chains. For example, EDIBUS is one of the applications of the skuchain that allows the exchange of documents, Excel spreadsheets and CSV files while maintaining privacy on the network. Organizations use the distributed ledger payment commitment – a global standard for payments on a blockchain network – and the Inventory Control and Finance (ICF) system. Smart contracts in skuchain are called brackets; these are used to send trade finance documents, such as a letter of credit on the blockchain network.

eTradeconnect is a Hong Kong-based blockchain platform for trade finance. The eTradeconnect initiative works to transform the traditional paper-based trade finance process during shipment into a digitized version. This DLT-based platform currently offers member-based products and services that include pre- and post-shipment trade finance purchase orders and invoices. The eTradeconnect platform is supported by various banks, such as *Australia and New Zealand Banking Group Limited, Bank of China Limited, Bank of Asia Limited, Hongkong*, and *Shanghai Banking Corporation Limited.*

The *India Trade Connect Platform*, built by *Infosys Finance*, aims at digitizing the trade finance workflow in distributed, trusted and shared networks. Its main area of focus includes the validation of ownership, certification of documents, and payments, letter of credit, bill collection, consumer-to-consumer (C2C), and business-to-business (B2B) transactions. The India Trade Connect Platform supports all major DLT vendors such as R3 Corda, Hyperledger and Ethereum. The company's investors and shareholders include some leading banks of India such as *Bank of Baroda*, *Federal*

Bank, *ICICI Bank*, *Axis Bank*, *Standard Chartered Bank*, *YES Bank*, *Industrial Bank* and India's central bank.

The *Marco Polo Network* is another network of 30 international banks that focus on capital financing on distributed trade finance platforms. The network includes payment commitments with or without financing, receivable and payable financing. The Marco Polo enterprise resource planning (ERP) is integrated with an application programming interface (API) that allows banks and corporate clients to function on a single platform. Besides financial services tasks, it also provides distributed data storage, bookkeeping, asset verification and identity management functions as well [10].

BLOCKCHAIN IN THE INSURANCE SECTOR

The insurance industry revolves around multiple interactions between insurers, participants and service providers that are all strictly linked to terms established in insurance contracts. As such, all decisions and actions are triggered based on the terms of such contracts. Legacy solutions in this industry involve complex solutions based on multiple applications using different technologies, large amounts of duplicated information and a high degree of system integration. The risk of insurance fraud is high due to the exploitation of various types of insurance policies by policyholders by making false claims in order to obtain financial gains. A blockchain solution can provide the ideal foundation to streamline relationships and automate events to enforce contract rights and obligations using smart contracts. Smart contracts are digitally integrated into a distributed ledger and create a common repository for contract policies and customer claims. At the same time, smart contracts create a common platform to support all transactions between the parties involved.

Blockchain Solutions for Various Insurance Use Cases

Automobile insurance is mandatory for vehicle owners by law, protecting the victim's interests in case of accidents. Insurance companies face a high number of fraudulent activities, where fraudsters intentionally make false claims as a means of earning easy money. In general, there are several fraud types that vehicle insurance is vulnerable to; one of the most common types is double-dipping. Double-dipping is the type of fraud where an individual submits the same accident or vehicle theft claim to more than one insurer in order to be reimbursed more than once for the same submitted claim.

Introducing blockchain technology in vehicle insurance claims processing may help avoid certain types of insurance fraud. A proposed solution – using Ethereum blockchain technology – is an application composed of both front-end and back-end platforms. The front-end application is for use by the insurer's employees in order to interact using a decentralized network, whereas the back-end will support all blockchain processes. To further clarify, when insurance customers want to insure their vehicle(s), required information such as vehicle license plate, vehicle identification number (VIN), the client's age and occupation and an average weekly driving distance will be collected by the insurer and recorded into the system as part of the insurance application process. Once the policyholder accepts the insurance terms and conditions, a smart contract will be activated; and, the information recorded on the blockchain will become available to all insurance companies. Therefore, before finalizing the vehicle insurance policy, the information will be verified to check whether the vehicle to be insured has any active insurance coverage through a different company in order to avoid instances of double-dipping fraud. The vehicle information stored on the blockchain remains immutable; hence every insurance company will be aware of the vehicle's history and claim events using a unique identifier such as the car's license plate or VIN [21].

Smart contract-based car insurance policies can streamline the traditional claims process in automobile insurance; an often cumbersome and costly process. There are several parties involved in the claims process: insurer, policyholder, attorneys and the court system in some cases. Thus, an issue in traditional insurance claim processing and settlement is the inherent distrust between the parties involved.

A prototype solution is the *Car Insurance Policy Framework* (CAIPY); a framework based on the Ethereum blockchain that uses smart contracts to record insurance processes and enables cost savings while providing increased transparency for the policyholders. The primary goal of the CAIPY is to record and store insurance-related events such as car accidents or automobile theft claims. Instead of using a private blockchain, CAIPY makes use of public blockchain for two main reasons. First, Ethereum has its own cryptocurrency (Ether), which can be used to compensate the policyholders. Second, the highly transparent and automated nature of blockchains and smart contracts can help mitigate instances where auto insurers may act maliciously by trying to deny a legitimate claim during the course of manual claim processing. Additionally, in order to help support claim audits by external auditors or insurance regulators, the smart-contract-based policy framework uses tamper-resistant sensors. Tamper-resistant sensors have recently emerged due to the high importance of car telematics, and will be deployed in most automobiles soon. In general, tamper-resistant sensors will automatically detect if the car has had an accident or any major component malfunctions. The data originated from the tamper-resistant sensors are considered reliable; therefore, smart contracts will automatically reimburse the customers in case of any insurance events based on those data.

Essentially, there are three types of smart contracts used in CAIPY: the policy contract, the surveyor contract and a smart contract that involves insurance tokens. The policy contract automatically creates the basis for a policyholder's claiming process based on tamper-resistant insurance coverage data. The surveyor contract is more like an intermediary between the policy contract and the cars. In other words, the surveyor contract stores insurance-related events received from tamper-resistant sensors on the blockchain. It can significantly save time by avoiding manual claim inspections by claim adjusters. The insurance token is an isolated token that uses a smart contract to be used as a payment method (reimbursement) to the customers. A practical solution to mitigate financial risks for both the insurer and the policyholder is use insurance-based tokens due to the cryptocurrency market's volatility.

Considering that the smart-based policy framework is built on a public blockchain, it is crucial to also ensure that the customer's privacy is protected; hence, the information will be stored and encrypted using AES (Advanced Encryption Standard). Both encryption and decryption keys will be distributed to the authorized parties – the customer, the insurer and the court if necessary [2].

Blockchain technology also provides several potential benefits to the health insurance claim processes, such as decentralized management allowing real-time claim processing, an immutable audit trail, data lineage, robustness, availability, security and privacy. However, certain challenges exist in implementing blockchain technology in the health insurance domain. Transparency and confidentiality are two important considerations. Patient information on the blockchain network cannot be transparent. Therefore, since the healthcare data contain highly sensitive information, it is significantly important to address patient privacy issues through the use of a highly effective control access methodology. As such, adopting permissioned-based blockchain networks would mitigate such confidentiality and transparency issues. A permissioned-based blockchain allows only authorized parties to participate in the network. In addition, it is crucial to keep the highly sensitive information encrypted at all times [17].

Health insurers, like any other insurance companies are also vulnerable to fraud. Sensitive information cannot be shared between different insurance providers for security reasons; therefore, malicious policyholders may conduct fraudulent double-dipping activities as well by providing false information to the insurance providers by claiming reimbursement for the same event from multiple insurance companies. Blockchain technology is considered one of the best solutions to address challenges that arise in the claim process. The health care data can be recorded and stored on the blockchain securely, thus reducing fraudulent activities [23].

Blockchain applications in the life insurance sector can vastly improve and simplify the beneficiaries' death registration and claim process. According to the current traditional life insurance claim process, a life insurance beneficiary needs to submit the insured's death certificate to the insurance provider as a first step to start the death benefit payment process. However, the approval may, at times, take up to 6 months or longer, making the claim settlement process long and tedious for the beneficiaries. As such, life insurance companies should consider adopting blockchain technology in the future in order to streamline and expediate the death claim process for the beneficiaries. The primary aspects that insurers are taking into account are data integrity, availability and transparency. The information recorded and stored on the blockchain network must comply with applicable insurance regulations and be accessible solely to authorized parties in order to ensure data security. Parties involved in life insurance claim settlement may also include a hospital or medical center, funeral home and/or a police department in addition to the insurer, and the beneficiary. Implementing a permission-based blockchain network reduces the chances of insurance fraud while making communications easier between the parties involved in settling a death benefit claim [27].

Blockchain and Group Insurance Benefits

According to one of Canada's leading health care and wellness providers – *Alberta Blue Cross* – the group insurance sector thrives on customer experience. Creating user-friendly, streamlined and interactive customer solutions promotes the much-needed customer engagement. The digital experience should provide customers with easy access to bills and payments, their customer profile and all information related to benefits, plans and entitlements. Additionally, online administrative features are needed for plan coordinators with interfaces to be used by various service health benefits providers, such as physio or massage therapists, dentists, dietitians or psychologists, to name a few [1].

Blockchain technology can enhance the group benefits management process. HR teams are usually responsible for storing employees' sensitive information, including at times some basic health care data, making the HR department vulnerable to cybersecurity attacks compromising data privacy. Blockchains enable the protection of sensitive information through a decentralized, permission-based, cryptographically secure algorithm.

As previously mentioned, smart contracts can play an important role in the health insurance industry as they significantly reduce the sometimes lengthy process of claim reimbursements. Moreover, storing the healthcare-related information on a permissioned-based blockchain platform allows the insured individuals to switch from one medical center to another with little hassle. Smart contracts can also improve the way employees access their healthcare and other benefits. When a company decides to hire an individual, certain conditions, such as the employee's start date will be applied to the contract. The human resource team can then deliver the group benefits seamlessly and in a secure manner to its employees. The smart contract will be automatically executed once the conditions are met. Furthermore, with smart contracts, the companies can personalize the benefits for each employee according to their needs, thus improving the employee experience.

Although blockchain technology enables various benefits such as efficiency, cost and time reduction, security and transparency, implementing a new technology in most organizations will prove costly. As such, each organization will have to integrate the technology based on its careful review of specific needs, budget and protocols [15, 18].

SUMMARY AND CONCLUSIONS

The distributed ledger concept has quickly evolved from the original intent of supporting cryptocurrencies to open-source technology. The potential of blockchain has been acknowledged by most leading financial institutions and information systems providers. New business models are offering composite solutions supported with cloud services to maximize the advantages of blockchain technology. As per research conducted on the implementation of emerging technologies, embracing change is critical for a complex and diversified sector like financial services. Comprehensive references to success stories and best practices in the industry can help organizations obtain maximum benefits from this technology. Despite certain current challenges, the future of blockchain within the financial services sector is highly promising as new frameworks and solutions are continually being developed.

CORE CONCEPTS

- Blockchain decentralized technology enables autonomy and privacy over information and the pillars of the financial industry: funding, investing, lending and exchanging value. The use of tokens and digital assets can transform transactions using fiat money into a future token economy with a dynamic use.
- Blockchains can support complex operations with a high degree of security, such as large peer-to-peer transfers, cross-border payments, and even local transactions with a high volume.
- Blockchain can facilitate faster, more secure, and lower cost payment processing services achieved through encrypted distributed ledgers providing transaction real-time verification without requiring a third-party intervention.
- Blockchain technology can support offline or delayed processing when the network connection is offline or the Internet connection is hardly available.
- With current technology, it takes three or more days for the market exchange to achieve finality. In contrast, blockchain technology can streamline the clearing and settlement process in real time by eliminating the need for intermediaries. For investment management, blockchain can link digital identity with customer financial records, transaction history and trigger events automatically using smart contracts.
- Through blockchain technology, the AML risk-rating criteria can be more objective. Additionally, blockchain for the KYC process can provide regulators with evidence about transactions, parties engaged and the underlying information used to verify client identity.
- The risk of insurance fraud is often high in terms of frequency and impact. A blockchain solution can provide the ideal foundation to enhance the claim process and overcome its current challenges. Smart contracts can help by streamlining relationships and automating events to execute contract rights and obligations.
- Customer satisfaction with the help of blockchain technology can be a strategic factor in maintaining a competitive advantage. The financial sector strategy is shifting to building partnerships across industries in order to bring greater trust, cost-effectiveness and compliance with blockchain use.

ACTIVITY FOR BETTER UNDERSTANDING

Readers interested in learning more about the impact of blockchain technology in financial services are encouraged to watch some of the YouTube videos listed below. Additionally, consider the following reflective questions based of your understanding of the various points discussed in this chapter.

1. *Six Ways Blockchain Can Be Used in Financial Services* https://youtu.be/jd03hf65yoE
2. *IBM Blockchain: Transforming Insurance Transactions* https://www.youtube.com/watch?v=3jSC3geKrMk
3. *Emmanuelle Ganne (WTO) on the Transformative Impact of Blockchain for International Trade https://www.youtube.com/watch?v=oazflv1ecHg*

Reflective questions:

1. What role can blockchain play in your organization in terms of mitigating various operational and financial risks?
2. In what way(s) can blockchain curb incidents of fraud in your operations?
3. In your opinion, how can blockchain innovate or revolutionize the way your organization and its competitors raise capital for expansion purposes?
4. Considering trust as the core foundation in financial services, how can blockchain technology help achieve it in your own operations?

REFERENCES

[1] Alberta Blue Cross. (2019). *Annual Report.* Edmonton: ABC Benefits Corporation. Retrieved from https://www.ab.bluecross.ca/pdfs/annual-reports/05-2019-annual-report.pdf

[2] Bader, L. C. B. J., Matzutt, R., & Wehrle, K. (2019). *Smart Contract-Based Car Insurance Policies.* New York: IEEE. Retrieved from https://ieeexplore.ieee.org/document/8644136/authors#authors

[3] Bank for International Settlements. (2020). *Annual Economic Report.* Basel: BIS. Retrieved from www.bis.org/publ/arpdf/ar2020e.htm

[4] Brown, R. G. (2018, May). *The Corda Platform: An Introduction.* Retrieved from https://www.r3.com/wp-content/uploads/2019/06/corda-platform-whitepaper.pdf

[5] Chatain, P. l., Zerzan, A., Noor, W., Dannaoui, N., & Koker, L. d. (2011). *Protecting Mobile Money against Financial Crimes.* Washington, DC: International Bank for Reconstruction and Development, World Bank.

[6] Dannemiller, D. (2017). *Investment Management Firms Getting Started with Blockchain.* New York: Deloitte Development LLC. Retrieved from https://www2.deloitte.com/content/dam/Deloitte/us/Documents/financial-services/us-fsi-im-firms-getting-started-with-blockchain.pdf

[7] Elliot, A. R. (2018, December 24). *The Blockchain Technology Revolution Is about to Remake The Stock Market.* Retrieved from https://www.investors.com/news/technology/blockchain-technology-blockchain-stock-market-revolution/

[8] FATF. (2014). *Virtual Currencies Key Definitions and Potential AML/CFT Risks.* Paris: Financial Action Task Force. Retrieved September 23, 2020, from http://www.fatf-gafi.org/publications/methodsandtrends/documents/virtual-currency-definitions-aml-cft-risk.html

[9] Fontinelle, A. (2021, March 2). *Blockchain's Impact on Charity.* Retrieved from https://www.blocksocial.com/charity/

[10] Ganne, E., & Patel, D. (2020, November). *Blockchain and dlt in Trade: Where Do We Stand?* Retrieved from https://www.wto.org/english/res_e/booksp_e/blockchainanddlt_e.pdf

[11] Huang, X., & Silva, E. (2018). Banking with Blockchain-ed Big Data. *Journal of Management Analytics,* 256–275.

[12] Human Concern International. (2019). *Financial Statements.* Ontario: HCI. Retrieved from https://humanconcern.org/wp-content/uploads/2021/03/20191231-Final-FS-Human-Concern-International_signedRM-1-1.pdf

[13] Hussain, S. H., & Usmani, Z.-u.-h. (2019). Blockchain-Based Decentralized KYC (Know-Your-Customer). In I. 2. (Ed.), *The Fourteenth International Conference on Systems and Networks Communications* (p. 5). Lisbon: IARIA. Retrieved from www.thinkmind.org/articles/icsnc_2019_4_30_28005.pdf

[14] IBISWorld. (2020, November 19). *Finance-Insurance-Industry.* Retrieved from https://www.ibisworld.com/united-states/market-research-reports/finance-insurance-industry/

[15] Kar, A., & Navin, L. (2021, 5). *Diffusion of Blockchain in Insurance Industry: An Analysis through the Review of Academic and Trade Literature.* Austin, TX: Elsevier.

[16] Khan, M. (2019). *HCI Annual Report.* Ontario: Human Concern International. Retrieved from https://humanconcern.org/wp-content/uploads/2020/10/HCI-Annual-Report-2020.pdf

[17] Kuo, T.-T., Kim, H.-E., & Ohno-Machado, L. (2017, 11). *Blockchain Distributed Ledger Technologies for Biomedical and Health Care Applications.* Oxford: Oxford University Press.

[18] Mercer. (2019, August 16). *Blockchain in Healthcare and Wellness Benefits*. Retrieved from https://www.mercer.com/our-thinking/blockchain-for-healthcare.html

[19] PwC's Financial Services Institute. (2017). *Top Financial Services.* New York: PwC. Retrieved from www.pwc.com/fsi

[20] Raphael, A. (2019). *Embedded Supervision: How to Build Regulation into Blockchain Finance.* Basel: Bank for International Settlements. Retrieved September 23, 2020, from https://www.bis.org/publ/work811.htm

[21] Roriz, R., & Pereira, J. (2019, 1). *Avoiding Insurance Fraud: A Blockchain-Based Solution for the Vehicle Sector.* Austin, TX: Elsevier B.V.

[22] Ross, S. (2020, February 6). *Financial Services: Sizing the Sector in the Global Economy.* Retrieved from https://www.investopedia.com/ask/answers/030515/what-percentage-global-economy-comprised-financial-services-sector.asp

[23] Saldamli, G., Reddy, V., Bojja, K. S., Gururaja, M. K., Doddaveerappa, Y., & Tawalbeh, L. (2020). Health Care Insurance Fraud Detection Using Blockchain. *International Conference on Software Defined Systems (SDS)* (pp. 145–152). Paris: IEEE.

[24] Sharma, Y., Sharma, B., & Jain, D. (2019). Blockchain – Creating Positive Vibes in the Card Payment Industry. *Annual Research Journal of SCMS, Pune*, *7*, 1–10.

[25] Statitsta. (2020, March). *Largest Stock Exchange Operators Worldwide as of March 2020.* Retrieved from https://www.statista.com/statistics/270126/largest-stock-exchange-operators-by-market-capitalization-of-listed-companies/#:~:text=The%20New%20York%20Stock%20Exchange,Exchange%2C%20and%20Tokyo%20Stock%20Exchange.

[26] Tapscott, A. (2020). *Financial Services Revolution: How Blockchain Is Transforming Money, Markets, and Banking*. Toronto: Blockchain Research Institute. Retrieved from https://ebookcentral-proquest-com.ezproxy.aec.talonline.ca/lib/concordiaab-ebooks/reader.action?docID=6028914&query=Financial+Services+Revolution#

[27] Voichal Prabhakar, R., Shukla, G., & Ratan, U. (2017). *Blockchain: A Potential Game-Changer for Life Insurance*. Teaneck, NJ: Cognizant.

[28] Welch, D. G., Lagrois, R., Law, J., Anderwald, R. S., & Engels, D. W. (2018). Blockchain in Payment Card Systems. *SMU Data Science Review*, *1*(1).

6 Blockchain and Supply Chain Management

Divya Garg, Siddhanth Karunakar Poojary, Jasjeet Singh Raikhi and Purav Thakkar

ABSTRACT

The use of blockchain technology as an end-to-end quality assurance monitoring tool is bound to become increasingly popular in the years to come, as evidenced by its successful adoption by such major entities as Walmart, Maersk, and DeBeers Jewelers. Blockchain addresses a significant constraint in supply chain management; namely the fact that often organizational information systems in a supply chain may not communicate fully, or at all. This constraint has traditionally limited the monitoring capacities of entities along a supply chain in some significant ways.

This chapter discusses the various types and components of supply chains and identifies their major fraud risks, such as counterfeit goods, fraudulent billing, false claims, and misappropriation of assets. The chapter also provides an overview of some common food fraud schemes, such as product substitution, mislabeling, and adulteration. Most importantly, blockchain implementation benefits within various supply chains are outlined, discussed, and further demonstrated by looking at several case studies. The benefits of using blockchain technology for supply chain monitoring purposes include a higher level of quality assurance through more granular tracking and identity management, more effective and secure communication among various supply chain stakeholders, lower supply-chain related expenses – especially lower external failure costs – and more efficient administration of the entire supply chain.

SUPPLY CHAIN DEFINED

A supply chain can be defined as a network that acts as a framework between a company, its suppliers and the end consumers [13]. Alternatively, a supply chain is also defined as the entire process of manufacturing and marketing commercially valuable goods, including all stages starting from the supply of raw materials and the manufacturing of the goods to their distribution and sale and, in some circumstances, the proper disposal of the purchased good, as is the case with products such as automobile tires or toxic chemicals. Effective supply chain management aims at successfully executing the required tasks associated with each phase of a product supply chain. Ultimately, an efficient infrastructure is required to manage supply chains if an organization wants to maintain its competitive advantages [8].

A supply chain network spans over various activities, entities, stakeholders, sets of data and resources. A supply chain network encompasses all the steps and processes needed to make the commodity available to the end consumer starting from the inception of the commodity. As such, a supply chain network consists of producers, vendors, warehouses, transportation companies, distribution centers and retailers [13].

DOI: 10.1201/9781003211723-6

Different Types of Supply Chain

An *internal supply chain* can be defined as the set of processes and protocols in place that enable an organization to produce in a commercially viable manner [19]. An internal supply chain has a huge impact on the final product and its final acceptance by the consumer. It is important to note that while a manufacturer will not be able to manufacture its product without the raw materials from its suppliers (external), internal activities need to make the product marketable enough so that it gets purchased by consumers. As such, a detailed and effective process of supply chain planning and implementation is of the utmost importance. Accurate market demand research, strictly observed production timelines and effective marketing strategies go a long way in constructing a fail-free and sustainable internal supply chain. An optimized internal supply chain also helps with higher employee satisfaction and the on-time delivery of the final products based on a predefined timeline, in addition to enabling organizations to better deal with unforeseen circumstances [25].

A *reverse supply chain* refers to the transfer of goods from the customer back to the original point in situations such as product returns, refurbishing activities or return of inventory surplus back to suppliers. In other words, a reverse supply chain may be explained as the opposite of the traditional supply chain, resulting in the reversal of the typical progression of the supply chain [2].

Supply Chain Components

Research and development (R&D) in an organization helps an enterprise to maintain competitiveness and develop an edge over its competitors. R&D activities may entail tracking down all developing product or merchandise trends in order to assess what competitors are currently doing. It often entails a detailed economic analysis and sound understanding of present trends and scenarios in a particular niche. An R&D department can help develop and design the future of a business as it can go a long way in strategic development and decision-making [4].

Forecasting customer demand can help in better execution of the entire supply chain. Demand could be forecasted using methods like educated guesses or quantitative prediction models, such as analyzing historical data and the use of advanced statistical techniques to understand the data from various test markets [29]. The use of appropriate forecasting tools could be considered as an important base for efficient supply chain planning [9].

Production is the process of utilizing raw materials and/or parts to develop, form or produce a product. An effective production process can only be initiated and maintained if proper planning is put into place. Supply of goods, maintaining inventory and developing effective and well-run production processes are the key components to a sustainable and efficient production system.

Logistics refers to various distribution processes to deliver raw materials required by an organization to create a product for their customers, as well as the process of getting finished goods to various retail outlets for consumer purchase. As such, transportation methods and their careful planning and monitoring play a significant role in transferring raw materials to the manufacturing hub and/or delivering the end products or goods to the target market [30, 33].

Inventory management focuses on how to maintain undisturbed production, sales and customer service at an optimal cost. Proper inventory planning, the timely receipt of material and closely monitored inventory management are the key parameters which can impact production [29].

Marketing consists of a broad spectrum which includes activities done by a company so as to gauge customers current needs or preferences. Key tasks in marketing consist of actively promoting the organization's good and/or services, as well as the development of a policy to price products or services competitively in order to maintain or gain market share [15].

Sales focuses on generating the revenue a firm needs to strive successfully. Sales functional sectors are bound to vary based on product and market attributes, in addition to the size of the company. As such, sales approaches might differ in structure and approach based on a number of market factors. These include for the use of an inhouse versus a field sales force, or a focus on retail or wholesale markets [29].

Finance and accounting operations consist of financial planning, capital financing and responsibly managing the various assets of a company. Finance managers and management accountants' responsibilities are to anticipate and plan for various short-term, as well as long-term, needs of a business, and to scrutinize the effects of additional borrowing (increased leverage) on the economic welfare of the entity [13]. At times a break off between strategy, finance and operations might occur in companies due to parallel process functionalities, however, proper collaboration and financial mitigation can often lead to continued financial sustainability and/or increased financial efficiency or even innovation [9].

SUPPLY CHAIN FRAUD RISKS

Supply chain risk (SCR) may be defined as potential negative future events that may result in brand damage, significant monetary losses, data and security breaches and ultimately the failure of a business [6]. Fraud is always intentional and cannot be categorized as a mistake, coincidence or misfortune, as fraudulent acts translate to economic benefits for the perpetrators. Since most supply chains are complex, multi-phased, and multi-faceted systems involving a large number of stakeholders, processes and technologies, managing its risks is also complex and challenging. Supply chain threats do not only stem from direct suppliers; various contractors and subcontractors along a supply chain also bring about risks to its processes. According to the Association of Certified Fraud Examiners (ACFE), one out of seven businesses that experience fraud will never recover fully. According to a survey by Deloitte, organizations may not notice that they have been a victim of a supply chain fraud until the organization has sustained significant financial losses. The same survey also indicates that the supply chain forensics and analytics effort has increased from 25% in 2014 to 35% percent in 2017. Deloitte also mentions that the increased use of supply chain forensics and analytics technologies is a clear indicator that organizations are accepting supply chain fraud as one of the most significant threats to businesses [22].

The following are a few fraud schemes that may also be perpetrated in supply chains.

Counterfeit Goods

Counterfeit merchandise is a significant global economic problem that has increased tremendously over the past few years. Very typically, counterfeit goods are made from lower quality parts, raw materials and workmanship with the aim of being able to sell the counterfeit goods at a cheaper price than the original brands. Counterfeiting is considered an illegal act in most parts of the world; however, due to the different laws and enforcement policies varying from country to country, it continues to thrive. As such, in many respects, the sale of counterfeit goods is a less risky criminal endeavor than drug trafficking. Purchasing counterfeit goods in developing countries is as welcomed as in developed countries like the USA, Europe and Australia. In 2013 alone, the US Department of Homeland Security seized counterfeited goods valued at over USD 1.7 billion. In developed countries, consumers are buying counterfeit goods to help create an affluent image within their social class [5].

Fraudulent Billing

Fraudulent billing, also known as improper billing or balance billing, refers to the practice of making fraudulent invoices or overcharging consumers for goods or services. Some major areas affected by fraudulent billing schemes are healthcare sectors, attorney services, and food supply chains. According to a news article published in 2020, the United States Department of Justice obtained more than 3 billion dollars in settlement for fraudulent and false claims made against the government in 2019. Out of that 3-billion-dollar total settlement figure, 2.6 billion was from healthcare fraud committed by various hospitals, pharmacies, medical devices providers, physicians and laboratories. The healthcare frauds involved federal healthcare programs like Medicaid, Medicare or Tricare, consequently increasing the healthcare costs and putting patients' health at risk [11].

Misappropriation of Assets

Using organizations' or customers' assets for personal use or gain can be considered as misappropriation of assets. For instance, taking office supplies to use at home or in a side business would be a misappropriation of assets, as would using a company vehicle for personal transportation purposes or other unauthorized uses. Accounting department personnel are sometimes guilty of asset misappropriation fraud because they commonly have access to the organization's funds and are in a better position to commit financial fraud while being able to cover their tracks – at least for the short run. According to the *Fraud Triangle Theory*, opportunity, incentive and rationalization are the three main factors for committing occupational fraud. To further explain, having access to a large amount of corporate cash on hand or having easy personal access to other company assets can be considered as opportunity. A strained relationship between the organization and the employee or an employee's personal financial issues can lead to incentive. Finally, a process of self-rationalization helps perpetrators justify their fraudulent acts in an attempt to avoid guilt or remorse [16].

FOOD FRAUD

The process of cheating consumers, suppliers, retailers and manufacturers or manipulating ingredients in the food industry can be considered food fraud. Food fraud can stem from manipulating manufacturing or processing standards or required food safety protocols in order to increase profits.

In addition to the previously discussed occupational fraud schemes, the following are some common examples of food fraud schemes:

Food substitution entails changing or reducing the quality or quantity of a food merchandise and is currently one of the biggest issues in the food industry. In one instance, the Food Safety Authority of Ireland (FSIA) had found horse meat being sold under the label of cow meat in 2012. The FSIA took samples from several grocery stores, burger shops and restaurants in Dublin city center and tested the sampled meats in a private laboratory. The tests proved the presence of porcine, bovine and equine DNA which meant that the cow meat was mixed with horse meat [23].

Adulteration or economically motivated adulteration (EMA) has been described as 'The intended substitution or adding of other materials in a product to increase the apparent value of the product or reduce the cost of its production for economic gain'. Using unapproved enhancements, cheaper ingredients, hiding damage or not disclosing proper ingredients is also considered food adulteration. There are several examples of food adulteration, such as diluting milk with water, mixing red chili power with brick powder, adding milk fat to vegetable oil or mixing lead chromate with turmeric powder because it has the same color [18].

False claim(s) can be defined as printing misleading information on food labels. Today, there are millions of products available in grocery stores, food courts, restaurants and many other places where consumers trust various food products based on their labels. There have been numerous incidents reported where an organization has lost the customer's trust by providing false nutritional information on the food labels, such as claims that a food item is preservative-free even though the product contained preservatives or even dangerous chemicals. The US Food and Drug Administration (FDA) sued US ice cream manufacturer, *Ben & Jerry*, in 2012 for putting 'All Natural' on their product label while it contained artificial flavors. In another instance, *Cadbury Schweppes* was also sued by the FDA for claiming the soft drink *7UP* as 'Natural' while the product contained HFCS (high-fructose corn syrup). Another case included baby food manufacturer *Gerber*'s packed fruit being labeled 'with all real fruit and all-natural ingredients', whereas the organization had used corn syrup instead of real sugar [15].

Mislabeling – Some Major Food Fraud Cases

Peanut Corporation of America (PCA) was involved in mislabeling their product where they knowingly released contaminated products into the market. PCA's CEO confirmed that PCA had shipped products to various markets on several occasions after testing positive for *Salmonella* contamination. PCA used to get ingredients like shelled peanuts from the USA, Mexico and China. After receiving and blanching, PCA would grind the shelled peanuts or chop them into different sizes, as needed. As peanuts may contain *Salmonella*, PCA was required to get the peanuts tested in their laboratory to find the percentage of *Salmonella* in peanuts. After getting a result of a higher percentage of *Salmonella*, the PCA didn't stop the use of the same peanuts and didn't recall their products already on store shelves. The mislabeling of PCA caused nine deaths and 714 confirmed cases of illness due to high levels of *Salmonella* in their peanut butter [16].

The E. coli 157 bacteria in Romaine lettuce in Canada and the USA outbreak occurred in December 2017. Canada reported 29 confirmed cases of *E. coli* poisoning. Ten individuals were hospitalized and two suffered from hemolytic-uremic syndrome (HUS). A collaborative investigation between Canada and US public health safety regulators tracked down the information and found that the contaminated Romain lettuce has been harvested in California. The bacteria had originated from the intestines of cattle, poultry and other animals which came into contact with the lettuce via feces or from the soil or water [24].

The E. coli 157 bacteria in spinach (USA) outbreak in 2006 was an earlier food contamination incident caused by the contamination of spinach with *E. coli*. The spinach harvested by *Mission Organics* in California and processed by *Earthbound Farms* in 2006 was contaminated with this bacterium. The spinach was harvested for *Dole Baby Spinach*, and the product was distributed in 26 states of the USA and in Ontario, Canada. The majority of illnesses were reported between August 26 and September 16, 2006. Overall, a total of 205 cases of illness were reported with 31 among them experiencing acute kidney failure. The *Centers for Disease Control and Prevention* estimated 4000 people were sickened due to raw spinach consumption. Since 1996, 12 types of *E. coli* bacteria have been identified, including a few that may cause serious poisoning. Most of the identified bacteria were traced to the Salinas Valley in California, also known as America's Salad Bowl [27].

The Mucci Farms Fraud Case in Ontario was related to tomato mislabeling. After three years of investigation in July of 2014, the *Canadian Food Inspection Agency* (CFIA) filed a case against Mucci Farms for selling tomatoes with a 'Product of Canada' label, when, in reality, the tomatoes were actually imported from Mexico. The organization had sold about $1,000,000 worth of the misbranded tomatoes. The court fined Mucci Farms and its executives over $1,300,000 for this infraction [10].

Currently, seafood mislabeling and fraud is of high concern in North America. According to research, about 41% of the seafood in Canadian markets is mislabeled due to species substitution. Mislabeling violates Canada's *Food and Drug Act Section 5(1)* which clearly states that the incorrect labeling, faulty processing or deceitful advertising of any food product is illegal. In 2017, *Mariner Neptune Fish and Seafood Company Ltd* pleaded guilty to falsely advertising, labeling and selling European Walleye (Zander) as Canadian Walleye (pickerel) and was fined $25,000 [28].

BLOCKCHAIN BENEFITS IN SUPPLY CHAIN MANAGEMENT

Data transparency is one of the many advantages of blockchain in supply chain management. Blockchains allow participants involved in the supply chain process to access relevant and timely information during the various phases of the supply chain. Since all records on the blockchain are transparent, immutable and cryptographically secure, information can be trusted and secured; as such, a greater level of trust is built between all the involved supply chain stakeholders. Although blockchains maintain transparency between all involved stakeholders, their main aim is to provide a greater degree of quality assurance to the end consumers. With the use of blockchain in supply chain management, suppliers are required to closely follow all mandated standards in order to make sure that all merchandise and products meet the required quality levels. Data transparency can also better protect customers from supply chain threats, such as false claims and mislabeling schemes. The use of blockchain technology in supply chain management can give consumers an opportunity to closely verify the various quality-related details associated with purchased products [8].

A more effective *reverse supply chain* management is another major benefit that blockchain offers. This refers to the monitoring of products until the end of their lifespans in order to better manage reverse logistical plans [8]. To further explain, the involved parties can also monitor the supply chain product flows starting from its the disposal stage all the way to the procurement phase. Reverse supply chain monitoring and analysis can help optimize inventory management. Additionally, defective product returns made by customers can also be tracked down more effectively, and product parts in defective units can be refurbished or reused again, if appropriate. This 'parts recycling' process can lead to lower manufacturing expenses and ultimately lower selling prices for products. Furthermore, keeping return policies and procedures transparent to consumers will increase consumer confidence and trust, which will ultimately lead to further business growth.

Blockchains can also help shorten various product distribution channels, as shipments and deliveries can occur with fewer distribution-related middlemen. Having a one-to-one relationship with all suppliers allows companies to sell their products with greater confidence in the market. In food-related SCR, blockchains can be designed to verify that food products originated from approved sources and were kept within stipulated temperatures and travel trajectories in order to avoid spoilage or other food quality degradation [14, 32].

Identity management is another major advantage of blockchain use in SCM. As discussed, blockchains can help track who is performing what actions, at what time, and at what location. These identity management attributes can help establish better product accountability. For organizations where a large number of nodes are involved in delivering the best quality product to the end-users, blockchains can play a crucial role by keeping track of the smallest transactions and/or events occurring within the supply chain for better traceability at an affordable cost [14, 26].

Traceability is yet another essential part of blockchains. Blockchain increases the traceability of the data in the supply chain by making the data permanent and easy to track. Customers can easily find detailed information about the product they are buying, such as its ingredients, packaging date, manufacturing specifications or travel conditions [21]. This leads to complete transparency between the seller and the buyers, thus building a strong relationship between both parties. An example of this is a flexible tag data-logger (FTD), which can be used with wine bottles. FTDs help identify temperatures and humidity levels until the bottled goods reach the supermarket or liquor stores. The stored data in FTDs then can be read using smartphones to help determine the safety and/or quality status of the wine [1, 3].

Digitalization in blockchains allows data to be stored in a tamper-proof digital environment, which in turn can help reduce accounting and paperwork-related costs, thus reducing issues and disputes that could arise due to erroneous information recorded manually [21].

Cost-saving is achieved by eliminating the need for storing a physical copy of all records. An important advantage of blockchain technology in supply chain management is to transform old-style paper-based work into a digital one. Since data are validated by every party involved in the supply chain, updated copies of the data are sent to all stakeholders. Therefore, it can be concluded that blockchain reduces human error and enhances flexibility and efficiency in management [14, 34].

Enhanced security is also achieved in supply chain management. There are numerous stakeholders involved in different parts of the chain, thus making the process cumbersome to connect across the supply network. Blockchain ensures security between the stakeholders in order to exchange information. Without the need for constant human intervention to manage all the nodes that belong to various channels of the process, blockchain automatically ensures that everyone is responsible for their actions and the subsequent outcomes, while making the work transparent to all involved parties, thus greatly reducing the chances of fraudulent activities [21].

WALMART'S SUPPLY CHAIN BLOCKCHAIN

Walmart (WMT) is the largest retailer in the world, with thousands of brick-and-mortar stores and a significant online presence. Walmart's 'everyday low prices' strategy can be credited for attracting consumers with different merchandise needs, including household goods, clothing items, electronics, groceries and more. Retail giant Walmart is currently using blockchain technology to tackle food safety scandals worldwide. In two blockchain pilot programs, Walmart is utilizing IBM's Hyperledger Fabric platform to track pork in China and mangoes in the Americas. WMT's blockchain-based tracking system reduced the time for tracking mango origins from seven days to 2.2 seconds, allowing the chain to provide its customer base with added transparency. IBM refers to it as 'complete end-to-end traceability' [7].

Problems faced in 2011 included China witnessing a massive pork mislabeling scandal and the recall of donkey meat products made from fox meat. In 2013, the EU supply chain was disrupted when bad actors substituted horsemeat for lamb and beef. The disparate record-keeping methods were the reason as to why it was so hard to trace products through the supply chain. The commonly accepted 'one up, one down' approach of knowing only the immediate supplier and customer for a food product is no longer sufficient. China imports nearly half of the world's pork and produces nearly half of it. Purchasing decisions are increasingly based on trust, as food safety and quality have become a high priority. China's government is investing significant funds to further increase food safety by increasing inspection points, closer monitoring of production processes and supply chains, and partnering with retail giants [12].

Walmart's blockchain technology is integrated into plug-and-play capabilities that provide consensus and membership services. On its blockchain platform, permanent e-certificates are created for products including agricultural treatments, identification numbers, manufacturers, known security issues, granted permissions, followed safety protocols, as well as device updates and completed audit procedures.

Walmart is investing in food-related technologies to monitor and detect pathogens in packaged food. Pork is smart tagged with barcodes at the pens, and the process follows the pork from there to the packaging plant. The movement of pigs is also tracked by radio frequency identification and cameras installed in slaughterhouses. Transportation trucks are equipped with temperature sensors and GPS tags, so that meat products can arrive at the retail destination safely. If the conditions in the shipping containers exceed established thresholds, alerts are prompted in order to apply the required, corrective actions. Blockchain technology allows procurement managers to track a large array of information remotely, from expiration dates to warehouse temperatures. The product package can be linked to a QR code that includes all relevant information reviews and details that can be uploaded to an e-certificate. The ensuing result was an increase in speed and accuracy during information exchanges between farms and stores which improved food safety, thus improving the public confidence in the product.

Walmart is also using IBM's Hyperledger Blockchain to trace sliced mangoes that are being shipped across the globe. As mangoes are produced, the tropical fruits can suffer from decay, surface defects, internal degradation, chilling injuries, disorders during ripening and more. Walmart keeps track of quality throughout the supply chain from 'tree to market'.

The future of Walmart with blockchain traceability gives supply chain participants important advantages. The transparent record-keeping allows retailers to enjoy lower recall costs while increasing profits and reducing risk exposures. Walmart's blockchain pilot program will be able to document supply chain inefficiencies and their post-cumulative losses. This digital tracking is bound to significantly increase food safety and enhance supply chain efficiencies by reducing food spoilage. Walmart can use IBM's Hyperledger to improve supply chain traceability alongside existing technologies. Building coalitions within the supply chain ecosystem to explore the use of blockchain more broadly is enabling Walmart to scale, experiment and learn [12].

Maersk

An integrated container logistics company, Moller-Maersk is a part of the *AP Moller Group*. Its company slogan 'All the Way' aims at helping its customers grow and thrive by connecting and simplifying trade. Maersk supports global trade with a dedicated team of over 80,000 people operating in 130 countries. IBM Corporation and global shipping giant Moller-Maersk have launched *TradeLens*, which allows companies to track and better manage their supply chains. The partnership between Maersk and IBM was signed in March 2017. The shipping and IT giants created a consortium to commercialize shipping and container management solutions based on Hyperledger Fabric [7].

TradeLens is an open and neutral platform that aims at eliminating communication barriers between supply chain participants by enabling a more open and automated exchange of information and value. The project's key goals are to create a single ecosystem within which stakeholders interact and exchange information on a shared network where processes are standardized and automated. The TradeLens platform offers a framework for improvement by providing an open API environment, standardizing interactions and opening the platform to third parties [22].

DeBeers Jewelers

The international diamond industry has been working hard to eradicate the production and sales of conflict diamonds and child labor issues. *Fura Gems* and *De Beers* are using blockchain technology in an attempt to find a permanent solution to these issues that have plagued the diamond industries for decades. Conflict diamonds, also known as 'blood diamonds' remain a primary industry concern. Typically, this happens when militias, rebels or government-backed troops use the diamond mines to fund violence against civilian populations [20].

DeBeers adopted the use of blockchain by announcing that they intend to develop the world's first blockchain ledger to track stones from mining to consumers. For many years, consumers and the media have been increasingly concerned about conflict diamonds and are demanding assurance that the diamonds are ethically sourced. Another issue related to the industry is counterfeit precious stones. The diamond industry is troubled with attempts to pass synthetic stones off as actual diamonds, emeralds or rubies. Blockchain-based provenance records help alleviate this problem. Since these blockchains are in the public domain, buyers can see that stones were verified in order to be sure that they are not being manipulated by anyone trying to pass off conflict or fake stones as genuine. The precious stone market is clearly experiencing problems that blockchain seems poised for solving [20].

SUMMARY AND CONCLUSIONS

The current supply chain deals with significant fraud risks every day, which include counterfeit goods, fraudulent billing, misappropriation of assets, food substitution, adulteration, false claim and mislabeling. As discussed in this chapter, the *E. coli* O157 Spinach Bacteria outbreak in 2006, and the Mucci Farms Fraud Case in Ontario are just a few of the many cases of supply chain fraud that have occurred in the past.

Applying blockchain technology increases the potential of the traditional supply chain by several folds, allowing these to transform into truly global supply chain systems.

Inculcation of blockchain in supply chains is advantageous as it improves collaboration between the parties involved. Blockchain use provides transparency at every step of the supply chain and helps develop a more efficient reverse supply chain process. Additionally, blockchains aid in safeguarding that applicable, mandated standards are maintained by ensuring traceability of materials and products while reducing paperwork and supply chain costs.

Big retail and industrial giants like Walmart, MAERSK, and DeBeers Jewelers have already started their respective pilot projects in order to effectively incorporate blockchain technologies within their respective supply chain infrastructure.

CORE CONCEPTS

- A supply chain can be defined as a network that acts as a framework between a company, its suppliers and the end consumer. An integrated supply chain could be defined as an enterprise resource planning approach to supply chain management. Reverse supply chain (when goods are returned to the point of origin) refers to the transfer of goods from the customer back to the original point it was dispatched from.
- A typical supply chain is comprised of the following phases: research & development, forecasting, production, logistics, inventory management, marketing, sales, and finance/accounting.
- Counterfeit goods, misappropriation of assets and fraudulent billing are the major supply chain fraud risks. Selling or manufacturing a cheaper quality of products and/or selling fake products under the name of a reputable brand has been a major issue in various industries. This can lead to the brand's reputational damage and consumer's loss of trust.
- Food fraud is a major crime which can put consumers' health at great risk. In substitution and adulteration food fraud schemes, the producer or manufacturer attempts to manipulate the ingredients by using cheaper quality parts or ingredients. Mislabeling and false claims can misguide the consumer by providing sometimes highly misleading information about the product. Some organizations have attempted these kinds of frauds in order to increase sales while boosting their profit margins.
- Since the supply chain is typically a complicated process, the probability of supply chain fraud remains high. As products go through many stages – from manufacturer to consumer – it is challenging to trace the origins of perpetrated fraudulent acts. Blockchain can help consumers or retailers obtain all the relevant product information with great assurance, since on a blockchain the possibility of altering information is very low (or essentially non-existent).
- Blockchains have several important attributes useful for effective supply chain management. The first one is data transparency. One can investigate the material used in manufacturing a product at various levels of the supply chain to help mitigate quality-related fraud risks. Blockchains can also be used as effective reverse supply chain monitoring tools, as returned or defective merchandise parts and components may be more effectively salvaged and/or refurbished for use in building new products leading to lower product manufacturing costs over the long run.
- By tracking the identity of individuals responsible for performing an action, accountability is ensured. As such, complete transparency between a consumer and a supplier can be achieved as anyone can check the material used, production conditions, and the date of packaging of a product.

- Blockchain traceability gives supply chain participants important advantages. The transparent record-keeping attribute allows retailers to experience lower risk exposures and recall costs while also increasing profits and customer satisfaction.
- Within the jewelry industry, a public and open blockchain enables anyone to see which diamond pieces were officially verified to help ensure that the products were not manipulated by entities or individuals trying to pass off fake diamonds as genuine. The precious stone market is clearly experiencing problems that blockchain seems perfect for solving.

ACTIVITY FOR BETTER UNDERSTANDING

Readers interested in further exploring the merits of blockchain technology as a monitoring tool are also encouraged to watch the following short videos on YouTube as a supplement to the use cases discussed in this chapter.

- *Walmart's food safety solution using IBM Food Trust built on the IBM Blockchain Platform* https://www.youtube.com/watch?v=SV0KXBxSoio
- *TradeLens and Blockchain Technology Supply Chain Demo* https://www.youtube.com/watch?v=0O2E9bCpKDk
- *Securing the diamond trade with blockchain* https://www.youtube.com/watch?v=hHkc-DH0ep4

After studying the chapter and the suggested videos, consider the following questions as a first step to a possible blockchain implementation for your own organization's supply chain:

1. How can blockchain technology better serve your organizations' clients or customers through fast and convenient product tracing?
2. In what ways can the implementation of a blockchain help your organization lower manufacturing and/or distribution expenses?
3. What supply chain risks can be better mitigated in your organization if blockchain was used as a monitoring tool?

REFERENCES

[1] Building a Transparent Supply Chain. (2020, April 14). *Harvard Business Review*. https://hbr.org/2020/05/building-a-transparent-supply-chain

[2] Cognizant. (2016). Reverse logistics management and closed-loop supply chain. *Sustainable Operations and Supply Chain Management*, 162–185. doi:10.1002/9781119383260.ch7

[3] Eljazzar, M. M., Amr, M. A., Kassem, S. S., & Ezzat, M. (2018). Merging supply chain and blockchain technologies. arXiv:1804.04149

[4] Elmelin Industrial Solutions. (2019, October 12). What Is the Role of the R and D Department? Retrieved March 10, 2021, from https://elmelin.com/what-is-the-role-of-the-r-and-d-department/

[5] Elsantil, Y., & Abo Hamza, E. (2021). A Review of Internal and External Factors Underlying the Purchase of Counterfeit Products. *Academy of Strategic Management Journal*, 1–5. https://www.researchgate.net/publication/337167481_Gray_Market_and_Counterfeiting_in_Supply_Chains_A_Review_and_Implications_to_Luxury_Industries

[6] Er Kara, M., & Firat, S. U. (2017). Data Mining Approach in Supply Chain Risk Management. *Supply Chain Risks: Literature Review and a New Categorization*, *10*(1), 31–60, 32–60. https://www.researchgate.net/publication/318041210_SUPPLY_CHAIN_RISKS_LITERATURE_REVIEW_AND_A_NEW_CATEGORIZATION

[7] Hackius, N., & Petersen, M. (2017). Blockchain in Logistics and Supply Chain: Trick or Treat? In *Proceedings of the Hamburg International Conference of Logistics*, pp. 3–18.

[8] IBM. (n.d.). IBM: Digitally Perfecting the Supply Chain: Digital Supply Chain. Retrieved March 1, 2021, from https://www.supplychaindigital.com/supply-chain-2/ibm-digitally-perfecting-supply-chain

[9] Jacobs, T. (2021, February 16). Supply Chain Planning: A Simple Guide to Plan and Forecast. Retrieved February 18, 2021, from https://throughput.world/blog/supply-chain-planning/

[10] Jameson, G. (2016, July 12). *CFIA Gets Serious about Prosecuting Food Fraud*. G.S. Jameson & Company. http://food.gsjameson.com/the-feed/2016/7/5/prosecuting-food-offences-canada-mucci

[11] Justice Department Recovers over $3 Billion from False Claims Act. (2020, January 21). United States Department of Justice. https://www.justice.gov/opa/pr/justice-department-recovers-over-3-billion-false-claims-act-cases-fiscal-year-2019

[12] Kamath, R. (2018). Food Traceability on Blockchain: Walmart's Pork and Mango Pilots with IBM. *Journal of the British Blockchain Association*, *1*(1), 1–12.

[13] Kenton, W. (2020, September 11). How Supply Chains Work. Retrieved March 1, 2021, from https://www.investopedia.com/terms/s/supplychain.asp

[14] Kshetri, N. (2018). 1 Blockchain's Roles in Meeting Key Supply Chain Management Objectives. *International Journal of Information Management*, 39, 80–89.

[15] L. Farris, A. (2010). The "Natural" Aversion: The FDA's Reluctance to Define a Leading Food Industry Marketing Claim, and the Pressing Need for a Workable Rule. *Food & Drug Law Journal*, 1–7. https://nationalaglawcenter.org/publication/farris-the-natural-aversion-the-fdas-reluctance-to-define-a-leading-food-industry-marketing-claim-and-the-pressing-need-for-a-workable-rule-65-food-drug-l-j-403-424/

[16] Leighton, P. (2015). Mass Salmonella Poisoning by the Peanut Corporation of America: State-Corporate Crime Involving Food Safety. *Critical Criminology*, 24(1), 75–91. https://doi.org/10.1007/s10612-015-9284-5

[17] Lumen Learning. (n.d.). Supply Chain Management and Logistics. Retrieved March 16, 2021, from https://courses.lumenlearning.com/wmopen-introbusiness/chapter/supply-chain-management-and-logistics-2/

[18] Manning, L., & Mei Soon, J. (2014, December 1). *Developing Systems to Control Food Adulteration*. ScienceDirect. https://www.sciencedirect.com/science/article/abs/pii/S0306919214000943?casa_token=63Uxhm_0pL4AAAAA:qXj5ro8R_abFg-p9h80ncFwdtUuC9YpSk16b72XyElQwnt2M_vgYYacNuz4-9WrydGZWbtCwTTdm

[19] Manuel, M. (2019, October 11). Internal Supply Chain and Its Importance. Retrieved March 1, 2021, from http://zeritenetwork.com/internal-supply-chain/

[20] Marr, B. (2018, March 20). How Blockchain Could End the Trade in Blood Diamonds: An Incredible Use Case Everyone Should Read. Retrieved from https://www.forbes.com/sites/bernardmarr/2018/03/14/how-blockchain-could-end-the-trade-in-blood-diamonds-an-incredible-use-case-everyone-should-read/?sh=5db4651c387d

[21] McKendrick, J. (2018, March 20). 5 Reasons to Blockchain Your Supply Chain. Forbes. https://www.forbes.com/sites/joemckendrick/2018/03/19/5-reasons-to-blockchain-your-supply-chain/?sh=8312f7a6fe13

[22] Musienko, Y. (2019). Maersk Blockchain Use Case – MEREHEAD 5379. Retrieved February 14, 2021, from https://merehead.com/blog/maersk-blockchain-use-case/

[23] O'Mahony, P. J. (2013, April 25). *Finding Horse Meat in Beef Products: A Global Problem*. Oxford University Press.

[24] Public Health Agency of Canada. (2019, November 1). Public Health Notice – Outbreak of E. Coli Infections Linked to Romaine Lettuce. Government of Canada Public Health. https://www.canada.ca/en/public-health/services/public-health-notices/2018/outbreak-ecoli-infections-linked-romaine-lettuce.html#a2

[25] Radford, R. (2017, June 5). How Important Is the Internal Supply Chain? Retrieved March 1, 2021, from https://www.allthingssupplychain.com/how-important-is-the-internal-supply-chain/

[26] Robinson, A. (2018, August 8). *How the Blockchain Enables Supply Chains to Focus on Customer Experience*. Cerasis. https://cerasis.com/supply-chain-management-with-blockchain/

[27] Seltzer, J. M. (2009). Natural Selection: 2006 E. Coli Recall of Fresh Spinach: A Case Study by the Food Industry Center. https://ageconsearch.umn.edu/record/54784/

[28] Spink, J., Goodridge, L., & Cadieux, B. (2019, August 1). Gap Analysis of the Canadian Food Fraud Regulatory Oversight and Recommendations for Improvement. ScienceDirect. https://www.sciencedirect.com/science/article/abs/pii/S0956713519301112?casa_token=X0U-fJbRxEQAAAAA:5i9aLAwnLyuJeoapbZgUXwdp9gza6Y6_bKA66gJ46Z8X6Y2-W724NyUQX_moQsux24cQ1gGX#bib36

[29] Supply Chain 4 Me. (n.d.). Retrieved February 16, 2021, from https://www.supplychain4me.com/7-areas-of-supply-chain
[30] The Eight Components of Supply Chain Management. (n.d.). Retrieved February 17, 2021, from https://www.iqualifyuk.com/library/business-management-section/the-eight-components-of-supply-chain-management/
[32] Using Blockchain to Drive Supply Chain Transparency. (2020, April 24). https://www2.deloitte.com/us/en/pages/operations/articles/blockchain-supply-chain-innovation.html
[33] Wailgum, B. (2017, August 27). What Is Supply Chain Management (scm)? Mastering Logistics End to End. Retrieved February 20, 2021, from https://www.cio.com/article/2439493/what-is-supply-chain-management-scm-mastering-logistics-end-to-end.html
[34] Zhang, J. (2019, May). Deploying Blockchain Technology in the Supply Chain. Blockchain and Distributed Ledger Technology, https://doi.org/10.5772/intechopen.86530
[35] Zax Shen, Z.-J. (2007). Integrated Supply Chain Design Models: A Survey and Future Research Directions. *Journal of Industrial & Management Optimization*, 3(1), 1–27.

7 Ethereum, Hyperledger and Corda

A Side-by-Side Comparison of Capabilities and Constraints for Developing Various Business Case Uses

Manroop Kaur and Navjot Lnu

ABSTRACT

This chapter provides on an in-depth discussion of three currently popular blockchain frameworks, namely, Ethereum, Hyperledger and Corda.

The Ethereum discussion section starts by presenting the various components of Ethereum, a public blockchain with its own cryptocurrency (Ether). Historically, Ethereum was the first blockchain framework to introduce the concept of smart contracts – a game-changer for blockchain technology. The characteristics and limitations of Ethereum smart contracts and consensus mechanism and a brief discussion of several Ethereum use cases in healthcare and supply chain management are also included in the first section of this chapter.

In the second part of this chapter, several Hyperledger frameworks, such as Hyperledger Burrow, Fabric, Indy, Iroha and Sawtooth are discussed in order to enable readers to become better acquainted with these various private, permissioned Hyperledger 'flavors'. Additionally, several Hyperledger-related tools are also introduced and discussed. These include Hyperledger Caliper, Composer, Cello, Explore and Quilt.

The third part of this chapter focuses on the Corda framework, a popular blockchain framework in the banking and financial services industry. Corda's various components are also introduced and discussed. This section also briefly discusses the characteristics of Corda smart contracts, consensus mechanism as well as several Corda use cases in healthcare, construction and finance. A brief discussion about user privacy in the Corda network and Corda's limitations is also included in the third section of this chapter.

Finally, the chapter concludes with a side-by-side comparison of these three frameworks in an easy-to-reference format, along with a decision tree that can be used as a first step to decide on adopting the most appropriate framework, depending on an organizational contemplated use case.

ETHEREUM

The Ethereum project is an open-source project initially proposed in 2013 by its founder, Russian-Canadian, Vitalik Buterin, before officially launching it in 2015. The Ethereum project was

DOI: 10.1201/9781003211723-7

purposed to provide an alternative protocol specifically designed for people to build decentralized applications. One of the primary aims of this project is to empower consenting individuals to carry out transactions with other consenting individuals in a trust-less environment by defining a state-change system using a loaded and unambiguous language; and, designing a system capable of applying such agreements autonomously. The lack of trust could be due to geographical separation, interfacing challenges, or maybe the incompatibility, inability, reluctance, cost, risk, inconvenience or corruption of existing statutory systems to which Ethereum strives to provide a feasible solution.

Ethereum is a permissionless, open blockchain with an integrated Turing-complete programming language for the creation of smart contracts. Ethereum, being a generalized blockchain with a built-in programming language, allows its users to use the blockchain for any decentralized application that they want to create. The term *Turing-complete programming language*, means that Ethereum can support computations of all types [18]. Ethereum empowers its users to write customized smart contracts and decentralized applications by allowing them to formulate their personalized set of ownership rules, state transition functions and transaction formats. In other words, Ethereum provides a universal programmable blockchain packed into an open client available for everyone to use. For example, Ethereum can build financial applications that are entirely trustworthy and transparent because they run on the blockchain and use smart contracts or transaction types that can be mathematically defined. Ethereum contracts are written in Ethereum Virtual Machine code, or EVM code; a low-level, stack-based bytecode language.

Ethereum also has a virtual currency, *Ether*, also known as ETH that is used to pay for blockchain-related transaction fees. *Wei* is the smallest Ether sub-denomination, in which all currency integer values are counted. One Ether is equivalent to 10^{18} Wei [7, 19].

Ethereum is fundamentally designed to be simple, universal, modular, agile, non-discriminative and non-censored [17]. The Ethereum protocol tends not to include any complex optimization except if it provides a valuable advantage, therefore keeping it simple to implement. Moreover, its built-in Turing-complete scripting language enables users to use smart contracts or transaction types that can be mathematically defined; thereby making Ethereum universally compatible with computations of all types. Furthermore, Ethereum allows users to make minor protocol modifications in one place without hindering the rest of the application's operation since it is designed in parts that are modular and thus separable. Additionally, Ethereum protocols are flexible as they are open for modifications that can considerably enhance the scalability or security of Ethereum. Besides, the Ethereum protocols strive not to limit or restrict explicit divisions of usage actively. The protocol's regulative mechanisms strive not to oppose specific undesired applications but manage the harm directly, thereby avoiding discrimination and censorship [17].

Components of Ethereum

In order to thoroughly understand the working of the Ethereum blockchain, it is imperative to have a basic knowledge of its various components (Figure 7.1). As such, each component of Ethereum will be briefly explained in the following section.

World State is a mapping between 160-bit addresses and account states. The database backend that maintains this mapping is named the *state database*. Each account state has a 20-bytes address and a state transition that directly transfers value and data within accounts. An account state contains four fields as follows:

Nonce: A scalar value that indicates the number of transactions sent from a particular address or the number of contracts formulated by a specific Ethereum account with associated code. It ensures that each transaction gets processed only once.

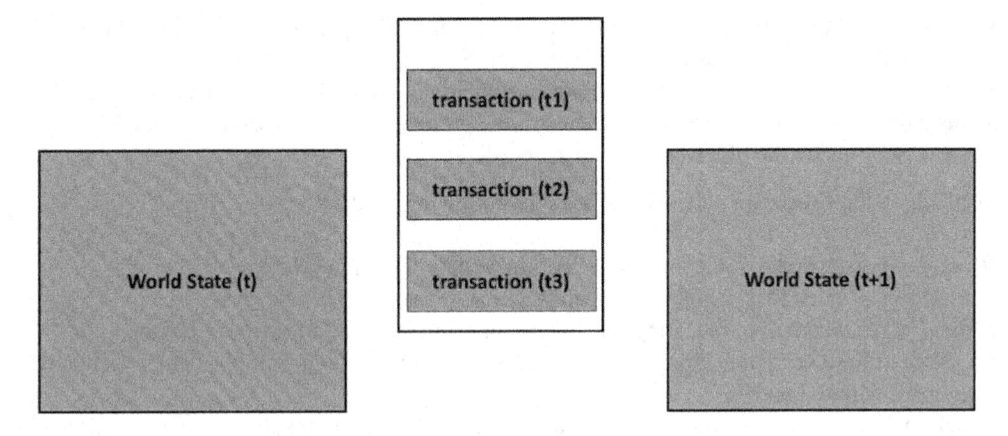

FIGURE 7.1 Blocks and transactions.

Source: [15], recreated with Microsoft PowerPoint.

Balance: A scalar value that denotes the amount of ether owned by the Ethereum account address.

Contract code: The code that gets executed when the respective Ethereum account receives a message call. It also contains the hash of the Ethereum account's EVM code. Because the contract code of an Ethereum account is immutable, it cannot be changed after it is created.

Account's storage: A 256-bit root node hash which encodes Ethereum account's storage content. By default, it is empty.

Externally owned accounts and contract accounts are the two types of accounts in Ethereum. Externally owned Ethereum accounts are those that are controlled by private keys. Contract accounts are Ethereum accounts that are controlled by their contract code.

The *transaction* is a single data package signed cryptographically by an actor external to the Ethereum scope that stores a message to be sent to the actor from an externally owned account. Software tools are utilized for their construction and dissemination. At the same time, the ultimate external actor is expected to be human. There are two kinds of transactions. First are the transactions that result in message calls; and second are transactions that result in creating new accounts with associated code. Both types of transactions contain the following fields:

The message's recipient's 160-bit address. To identify the sender of the transaction, the signature of the transaction is used.

Nonce: The number of transactions sent by the sender as a scalar value.

GasPrice: A scalar value representing the amount of ether paid by the sender per unit of gas for all computation costs incurred during the transaction's execution.

GasLimit: A scalar value that represents the maximum number of computational steps that can be taken when executing a transaction. Before any computation can take place, the amount of gas is paid in advance.

Value: A scalar value that indicates the amount of ether to transmit from the sender to the receiver and as an endowment to the newly established account in the case of contract creation.

An unlimited byte array with the EVM code for initializing the account is also included with the contract creation transactions.

The Blocks: Batches of transactions that contain a hash of the previous block on the chain. The Ethereum network utilizes blocks to assure that all users on the network maintain a synchronized state and consent on the detailed chronicle of transactions. On the Ethereum network, every block contains a reference to its parent block to ensure the preservation of transaction history. Moreover, the transactions inside the block are also strictly ordered.

A block on the Ethereum network contains the following fields:

Block number: A scalar value indicating the blockchain length in blocks.
Timestamp: A value that denotes the inception time of the block.
Difficulty: A scalar value indicating the amount of effort required to mine the block.
MixHash: A hash value that confirms that the block has gone through sufficient computation when coupled with the nonce.
Nonce: A hash value that confirms that the block has gone through sufficient computation when coupled with the MixHash.
Parent Hash: The hash value of the preceding block's header.
State Root: The hash value that denotes the entire state of the system after the execution and finalizations of all transactions.
Transaction List: A list containing the receipts for each transaction on the block.

Each block in the network has a defined size because the network and the minors collectively set a gas limit to evade the block size being arbitrarily large. Because of space and speed constraints, arbitrarily large block sizes can gradually prevent less performing full nodes from keeping up with the network. Any miner mining a new block can change the gas limit by up to 0.1 percent in both directions from the gas limit of the parent block. Because hundreds of transactions may occur, be consented on, and synchronize all at once, Ethereum spaces out the commits, giving all network users ample time to reach consensus. On the Ethereum network, blocks are committed about every 15 seconds, despite the fact that transaction requests come in dozens of times per second. Once a minor has mined a block, it is delivered to the rest of the network, and all the existing nodes add the newly mined block to the end of their blockchain.

Messages are the data structures utilized in the Ethereum network by one contract to send a message to other contracts. The message exists solely in the execution environment of Ethereum. It contains the information about the message's sender, the transaction originator, the recipient of the message, the Ethereum account whose code is to be executed, which is usually the same as the recipient. Moreover, messages contain the following fields:

StartGas: A scalar value that denotes the available gas for the transaction.
GasPrice: A scalar value representing the amount of ether paid per unit of gas by the sender for all computation costs incurred during the transaction's execution.
Value: A scalar value indicating the amount of ether to be transferred from the sender to the recipient in addition to the message.

As discussed so far, typically, a message resembles the transaction; however, it is produced by a contract rather than an external actor after a specific code of the contract is executed by the recipient's account [17, 19].

Smart Contracts

Network participants do not have to come up with a new code to use the EVM for a computation. However, the developers of the applications make use of the EVM storage to upload reusable codes or programs. Therefore, these programs are used for making requests to execute the snippets of the code by using different parameters. These programs uploaded on the network and implemented to fulfil the requests are known as smart contracts.

Simply put, a smart contract is a program that operates on the Ethereum blockchain. It is a set of the different codes and its states located on a particular address on the Ethereum blockchain. This implies that smart contracts have a balance and can submit transactions through the network. Smart contracts are not operated by a single user; but are rather uploaded on a network and run according to a set of instructions.

The Ethereum Virtual Machine

EVM functions as an entity managed by linked computers that help in the smooth operation of an Ethereum client. The environment used by the Ethereum accounts and the smart contract for their operation is the Ethereum protocol; the only reason for its existence is to maintain the constant, uninterrupted and irreversible activity of the EVM protocol. Every existent block in the chain operates as per only one canonical state, and EVM determines the rules to be followed for defining a new state for any block.

Characteristics of Smart Contracts

Smart contracts can be designed and deployed by anyone on the network. All the person needs to know is the coding language of the smart contracts and they must have sufficient ETH required for the deployment of the smart contract. They must, however, be assembled before being implemented, which enables Ethereum's virtual machine to read and store the contract. Smart contracts can be considered as open APIs as they are publicly available on Ethereum. Smart contracts can be vastly extended by calling other smart contracts in one's smart contract. Smart contracts can build other contracts

The developer-friendly languages that can be used for writing smart contracts include the following:

Solidity is based on C++ and JavaScript and is statically typed. It supports inheritance and libraries.

Vyper is Python-based and requires strong typing skills. The compiler code used by Vyper is small and easy to understand. To make contracts more secure and easy to audit, Vyper intentionally lacks certain features in comparison to Solidity. It does not support modifiers, inheritance, inline assembly, overloading the operator, fixed binary points and recursive calling.

Yul is an intermediate language used for coding Ethereum, whereas *Yul+* is the advanced version of Yul. For anyone new to Ethereum, it is advisable to start with Solidity and Vyper. Yul and Yul+ should only be considered by individuals who are deeply aware of the best security practices for smart contracts and know the details of EVM working. Yul+ has extra features as compared to Yul. Yul supports EVM and Ewasm; however, Yul+ was initially used for designing an optimistic rollup contract.

Limitations of Smart Contracts

By design, smart contracts are unable to send HTTP requests; thus, they cannot gather data about real-world events on their own. The sole reliance on information obtained from external sources has a high risk of jeopardizing the consensus. In a worst-case scenario, this can have a significant impact on security and decentralization [20].

Consensus Mechanism

For a blockchain like Ethereum – essentially a distributed system – to work together in a secured manner, the network nodes must reach an agreement on the current state of the system. The agreement, also called *consensus* among the database nodes, is achieved using the consensus mechanism. Until 2021, Ethereum utilized the PoW (proof of work) consensus protocol. The consensus mechanism used in the Ethereum interface was a modified version of the GHOST protocol (Greedy Heaviest Observed Subtree) designed to tackle the issue of stale blocks in the network by incorporating those stale blocks into calculations of the longest chain [17, 18].

In 2021, Ethereum started transitioning from PoW to proof-of-stake (PoS). In contrast to proof-of-work, validators do not require substantial computing power as they are selected randomly and do not compete. Validators do not need to mine blocks; instead, they must create blocks when they are chosen and validate proposed blocks when they are not, referred to as *attesting*. Validators are compensated for suggesting new blocks and attesting to ones they've seen.

Limitations of Ethereum

Ethereum – like other blockchains – also has limitations. It is a public blockchain; hence it inherited all the limitations of a public blockchain. Moreover, Ethereum is relatively slow because it takes 12 seconds for miners to validate and add a block to an Ethereum blockchain network. Furthermore, Ethereum smart contracts use Solidity coding language which has its own limitations; hence it inherited all the limitations of Solidity language.

Ethereum Use Cases

Ethereum is one of the most widely accepted blockchain frameworks on the market for many industries in today's world. The promising characteristics of Ethereum blockchain technology like trustless, decentralized ledgers that support historic immutability, integrity, transparency, confidentiality and auditability offer great potential for achieving security and assurance for many industries. Industry sectors such as banking, cybersecurity, auditing, health care and supply chain management have started to apply the essence of the Ethereum blockchain to their current business processes. Many applications in the marketplace have already utilized the Ethereum blockchain. Examples include *MetaMask*, a web browser plug-in that enables a device for peer-to-peer sharing, token swapping and connection to the Ethereum network; *Civic*, an application for secure identity and data management on the blockchain and *Compound Finance*, an open-source protocol for algorithmic, efficient money markets on the Ethereum network [17].

The next section briefly discusses some real-world use cases from various industries. Chapters 4, 5 and 6 take a closer look at some of these potential use cases.

The healthcare industry is constantly working on developing smart healthcare delivery models that can integrate patients' health data into digital records and share them with care providers for an informed decision-making process. Although electronic medical record (EMR) systems are present, they are heterogeneous and inconsistent in terms of effective implementation of access control as well as security policies. Researchers and developers have put forward some promising proposals that can enable Ethereum-based healthcare solutions to enhance personalized care. One of the solutions proposed is *MedRec*, a unified patient-centered EMR system that utilizes an Ethereum-based decentralized network for managing and integrating the patient's medical records with the patient's care provider. MedRec claims to overcome the present limitations and threats to adequate cross-organizational information-sharing caused by conventional EMR systems or other IT-based healthcare information systems. Moreover, it reckons to ultimately fill the communication gaps between patients and healthcare providers by reducing the involvement of direct third parties. It uses private blockchain and a proof of work consensus model to sustain and secure the network. Access to anonymized and aggregated patient data is granted to the participant who provides proof of work. This assures integrity, immutability, transparency, availability, confidentiality and auditability of patient health data across the network [3].

With the advancements in wireless networking and affordable computer chips, turning a smart device into an *IoT device* is now an achievable task. All IoT devices need to interact and synchronize with each other to provide desired functionalities. A single IoT network can connect to thousands or tens of thousands of IoT devices. Managing such a massive cluster of devices can be challenging. Ethereum blockchain can manage IoT devices by utilizing smart contracts to tailor a Turing-complete code to define IoT devices' behaviors. Moreover, being a general-purpose, open, distributed and permissionless blockchain provides easy synchronization of data without compromising the network's security [7, 17, 19].

As discussed in Chapter 6, *supply chain management* is a complex task that requires transparency and traceability to mitigate the associated risks through closer monitoring. Implementing a blockchain solution can introduce promising improvements for supply chain data flows and liability between buyers and suppliers. The operation of successfully delivering manufactured end products requires an efficiently working supply chain network. Any interruption to this network can cause the whole supply chain to fail. Such failures generally accompany financial, reputational and, in some cases, environmental damage. Current supply chain management systems work on a centralized network that requires a third party for its management. The centralized network raises the risks associated with data security, privacy, technical reliability and interoperability of data storage [7, 17, 20].

HYPERLEDGER

Hyperledger refers to an open source designed for the advancement of blockchain technologies across different industries that aim toward working collectively rather than working individually [2, 16]. This collaboration, introduced by the *Linux Institute*, is prevalent across the world, and includes top finance, banking, Internet of Things, supply chains, manufacturing and technology industries. In 2015, all the firms taking part in this collaboration agreed to pool their resources with the common intention of creating an open source blockchain technology that would help in serving the needs of any of the organizations [2]. Being a part of the Linux Foundation, Hyperledger offers the field of technology a unique and collaborative way to design, legitimize and provide field-specific solutions given by blockchain technology. These solutions can be implemented across the global communities collaboratively [16]. As of August 2015, 230 firms were members of the collaboration of Hyperledger and it involved about 28,000 participants. Hyperledger intends to build an environment that makes it possible to turn the vision of Hyperledger into reality [2].

HYPERLEDGER FRAMEWORKS

The Hyperledger approach facilitates the re-usage of standard building blocks, promotes accelerated component modifications and encourages interoperability among projects. This section provides a brief overview of the various Hyperledger frameworks.

Hyperledger Burrow is a permissioned smart contract guide for blockchain customers designed in part to meet the specifications of the Ethereum Virtual Machine (EVM). Burrow was initially designed and recommended by *Monax*, and was named the fourth highest used distributed ledger platform in Hyperledger in 2017. It offers a highly deterministic blockchain design entirely based on smart contracts. The access-limiting layer of Burrow has proven to be of immense benefit for users. Users can gain access to this layer by using smart contracts and a permission layer based on *Äúsecure* natives.

The consensus engine, Application Blockchain Interface (ABCI), smart contract application engine, and Gateway are all components of Burrow. The consensus engine helps maintain the stack among nodes and ordering transactions of the application engine across the network. The ABCI specifies the interface requirements for establishing a stable connection between the consensus engine and the application engine. The smart contract application engine helps developers successfully operate complicated industrial processes by providing them with a solid deterministic smart contract engine. The gateway improves the system integrations and user interface by providing the users with programmatic interfaces [2].

Hyperledger Fabric can be defined as a manifesto-designed distributed ledger solution capable of high levels of confidentiality, flexibility, resilience and scalability while maintaining a standard architecture. Therefore, any of the solutions put forward by Hyperledger Babric can be implemented by any industry [16].

Hyperledger Indy is a distributed ledger created solely for the purpose of decentralized identity. Indy is a repository for tools, libraries and reusable components that can be used to create and use independent digital identities based on blockchains or other distributed ledgers. Users can use these identities in a variety of administrative domains, applications and organizational departments, implying that associates, opponents and even competitors can all rely on the same source of information. Indy knows the answers to basic questions like 'Who am I dealing with?' and 'How will I test data about the other party in this communication?'.

Hyperledger Iroha is a simple blockchain framework designed to make the incorporation into infrastructure projects requiring distributed ledger technology easy. In October 2016, Iroha, after collaboration with Fabric and Sawtooth, was declared the third distributed ledger framework of Hyperledger. Hyperledger Iroha was initially designed in Japan by Soramitsu [2].

Hyperledger Sawtooth is another scalable Hyperledger flavor developed to design, deploy and run distributed ledgers. Sawtooth functions rely on a different type of consensus, proof of elapsed time (PoET). When compared to proof of work, PoET is fully functional with fewer resources (PoW) [16].

Benefits of Using Hyperledger

The Hyperledger project housed at *The Linux Institute* acts as a 'greenhouse' that enables developers and users from various sectors and market spaces to collaborate at a global level. The benefits associated with using Hyperledger blockchains include ease of use, specialization, greater efficiency, quality content delivery and ease of handling intellectual property.

To further explain, Hyperledger makes it easier for users to keep up with all new developments by reducing the amount of research required by adopting companies thanks to its collaborative environment. This, in turn, streamlines communication by enabling users to access information in an efficient manner.

The greenhouse structure of the Hyperledger also fosters greater specialization. Specialization allows the developers to streamline their energy on a lesser number of tasks in order to gain greater expertise in them. Specialization also helps to boost productivity. Since the participants who have a specialization in the same area do not compete with one another, they are motivated to join any of the fields to help improve and contribute toward Hyperledger's research and development processes.

Collaboration of different parties in the greenhouse organization is highly supported as it helps to reduce the duplicate efforts put in to achieve the same tasks. It also motivates the parties to create recurrent components that could be used to the advantage of the whole community.

Open-source software development projects, such as Hyperledger versions, are often considered to be of excellent quality because the development process involves in-depth code reviews and constant debugging. Hyperledger maintains the quality of the code by reviewing all its projects throughout the various phases of its software development life cycle, performed by its technical governing committee.

Finally, Hyperledger has a straightforward and convenient way of handling intellectual property by making use of a standard approach which eliminates the requirement for compound and costly relationships among users. As all the users clearly define all the expectations and conditions, all the individuals using and designing Hyperledger technologies can engage without overbearing legal restrictions, as it is improbable that users will run into the hidden legal constraints associated with proprietary technologies [2].

The Design Philosophy of Hyperledger

Distributed ledgers may require different parameters for various use cases. For the sake of addressing diversity, all the Hyperledger projects aim to follow a similar design philosophy. The design philosophy of Hyperledger focuses on the following considerations:

A high level of security is one of the critical needed aspects of distributed ledgers. A vast majority of blockchain use cases involve sensitive data, or are based on high-value transactions. Distributed ledgers have turned into a hot target for attackers because they run on large codebases and consist of data flows of value. Since the distributed ledgers should have a wide range of features and functions while still being able to resist persistent adversaries, maintaining the security of blockchain has become an ongoing challenge.

Modular and extensible frameworks aim to achieve modular and extensible frameworks comprised of reusable building blocks. This modularity allows the developers to examine the various components during the development phase and make changes accordingly. The modular strategy allows multiple developers to work on various modules without interrupting each other's work. In this way, the modules can be used more than once for different projects.

Scalablility allows blockchain networks to interact and share information in an attempt to establish more complicated and robust networks. To satisfy this need, Hyperledger smart contracts and applications must be scalable around multiple blockchain networks. This high level of interoperability will make it easier to meet the increased demand for blockchain and distributed ledger technologies.

A well-defined set of APIs is another characteristic of the Hyperledger projects, thereby enhancing interoperability among all the systems. This well-defined set of APIs allows the external clients and applications to interact with the core distributed ledger infrastructure of Hyperledger simply and effortlessly.

Tools of Hyperledger

Hyperledger produces and supports blockchain technology for a variety of businesses which includes providing tools and utility libraries. The Hyperledger approach facilitates the re-usage of standard building blocks, promotes accelerated component modification and encourages interoperability among the projects.

Hyperledger Caliper is a blockchain benchmark tool used to measure any blockchain implementation's performance by comparing them against a collection of predefined use cases. The reports produced by Hyperledger Caliper list resource utilization, transaction latency and transactions per second (TPS) as a few of its numerous performance indicators. Hyperledger Caliper was the first general tool to evaluate the performances of the various blockchain solutions, based on a collection of objective and generally accepted rules. Hyperledger Caliper is not intended to issue benchmark results. Instead, it was designed to be used as an in-house reference tool so that it can help choose the blockchain technology that the organization should implement to fulfill its specific needs.

Hyperledger Composer is an open development toolset that enables the modeling of business networks, including the participants, assets and transactions. This model makes it easy to generate a smart contract and blockchain applications that allow for quick integration within an existing system – speeding up the deployment time to weeks instead of months. Composer also enables the quick modeling of an existent business network and integration of existent data and systems in the blockchain applications. Composer assists the existing infrastructure of the Hyperledger Fabric.

Hyperledger Cello brings an on-demand deployment and management module to the blockchain ecosystem. Cello helps businesses to quickly adopt blockchain technologies by providing automated ways to create, manage and terminate blockchains. Cello provides users a real-time dashboard allowing them to check the status of the blockchain system, view statistics and control blockchains by giving them an option to create, configure or delete them as needed. The major blockchain implementation supported by Cello is Hyperledger Fabric; however, it aims to also support Hyperledger Sawtooth and others in the future.

Hyperledger Explorer provides a dashboard that gives an overview of the blockchain network. It allows the users to generate queries for blocks enabling views of detailed information contained within. It also provides a transparent view of the activities on the network. With Explore, users can identify, troubleshoot and quickly resolve production issues. Users can use Explore in collaboration with any platforms that assure authentication or authorization, either commercial or open source. This collaboration allows the users to use the functions relevant to the privileges of the users.

Hyperledger Composer, Cello and Explore work together to provide an end-to-end toolkit for deploying solutions with Hyperledger Fabric with a long-term goal of supporting all other Hyperledger frameworks.

Hyperledger Quilt is a tool designed to provide a ledger interoperability solution for all Hyperledger projects, thereby enabling the distributed ledger solutions from Hyperledger members, the private ledgers from financial institutions, the wallets from IoT companies, and supply chain systems to successfully connect with each in order to carry out distributed, atomic transactions [2].

Hyperledger Smart Contracts

Smart contract tasks could be as basic as updating data, or as complicated as completing an agreement with several constraints. A smart contract, for example, can be used to update an account balance and can also be used to validate that required funds are present in the account for the withdrawal to be successful. There are two kinds of smart contracts utilized by Hyperledger frameworks: Installed smart contracts and on-chain smart contracts.

Installed smart contracts assist in the successful installation of business logic on network validators prior to network launch. *On-chain smart contracts* use business logic in the form of a transaction that is committed to the blockchain and then called by subsequent transactions. The code defining the business logic is considered part of the ledger when using on-chain smart contracts. Of the five frameworks of Hyperledger, only one framework does not support smart contracts. The four frameworks that support smart contracts are Hyperledger Burrow, Hyperledger Fabric, Hyperledger Iroha and Hyperledger Sawtooth. Across all four frameworks, the transaction requests are processed by the smart contract layer. This layer is also responsible for validating the requests, which is achieved by the execution of business logic [12].

The Architecture of Hyperledger

As previously mentioned, Hyperledger projects are based on a design philosophy that emphasizes a modular approach, interoperability, a prominence on highly secure solutions, a token-agnostic approach consisting of no native cryptocurrency and the development of a rich and easy-to-use application programming interface (API). The components of the architecture of Hyperledger include the following:

The consensus layer has the responsibility of creating an agreement on request verifying the validity of the block of transactions.

The smart contract layer manages the requests for transaction handling and implementing business logic to determine whether transactions are accurate.

The communication layer handles the peer-to-peer transportation of messages amongst the nodes participating in a shared ledger instance.

The data store abstraction allows other modules to access and use different data stores.

Crypto abstraction permits the swapping of different cryptographic algorithms or modules in a way that does not affect the rest of the modules.

Identity services help with the creation of a trusted root during the blockchain instance setup, the admission and registration of identities or system entities during any activity across the network, in addition to managing changes such as drops, additions and terminations. It also offers authentication and authorization.

Policy services are responsible for managing the different policies listed in the system. These policies include the endorsement policy, consensus policy or group management policy. It interacts with and relies on other modules for the successful implementation of such policies.

APIs allow the clients and applications to establish links to the blockchain.

Interoperation allows interoperability between the various instances of the blockchain [14].

Consensus in Hyperledger

Since businesses have different blockchain requirements, the Hyperledger community makes use of four different types of consensus mechanisms in order to ensure modularity. Hyperledger Fabric uses Apache Kafka, Hyperledger Indy makes use of RBFT, Hyperledger Iroha is Sumeragi-based and Hyperledger Sawtooth makes use of PoET. The first three consensus methods follow a similar approach for their operation, based on providing fault tolerance and finality within a matter of seconds. However, PoET makes use of a lottery-based approach to choose the consensus that reduces consensus-related delays while resolving the situation [13].

Limitations of Hyperledger

Even though Hyperledger Fabric is considered the most sophisticated and mature Hyperledger project, the Hyperledger project is still in its infancy. As noted by a Hyperledger developer, there have been several changes, upgrades, project delays and complications associated with the project. However, the Hyperledger project offers tremendous future potential by encouraging innovation and risk-taking [15].

Hyperledger Use Cases

This section discusses several real-world scenarios in which Hyperledger blockchains may have a clear and compelling application. Hyperledger contains effective techniques for each scenario; in some instances, a proof-of-concept has already been created.

When applying for a loan, banks aim to lend money to borrowers who are a reasonable risk. The loan review and approval process requires banks to collect private, personally identifiable information (PII) from applicants, such as first and last name, DOB, government ID number, annual or monthly income and a list of the applicant's assets and liabilities. This type of PII information is therefore used by the bank to access the credit score or credit rating of the applicant. Since banks retain a vast amount of PII information of clients, it makes them vulnerable to a cyber-attack. At the same time, borrowers want to obtain the best interest rates and thus are likely to explore a variety of loan options from various sources. This, however, increases the probability of the PII of the applicant being misused. For this use scenario, Hyperledger Indy provides a transformational identity solution. Indy allows applicants to give only the details that banks require to come to a decision in a way that ensures truthfulness, creates trust in the bank and satisfies regulatory mandates.

Privacy, confidentiality and accountability are the driving forces for blockchain use in the financial services sectors. As such, financial services providers must be able to validate a client's legal identification and permit them to conduct transactions based on regulatory requirements such as Anti-Money Laundering (AML) and/or Know Your Customer (KYC) regulations. Because public blockchains could potentially jeopardize a participant's privacy and confidentiality, these criteria promote the adoption of permissioned and private blockchains. These are the primary reasons why consortium blockchains are getting traction in financial services, along with the enormous numbers of ensuing transactions. Post-trade processing is one of several conceivable use cases for blockchain in this area, particularly in capital markets.

Self-sovereign identification is among the most promising aspects of blockchain applications. This attribute is based on the concept that since individuals own their own identity, they should also be able to control the data associated with them. Hyperledger Indy is a distributed ledger that focuses on self-sovereign identity first and foremost. Traditional enterprise identification solutions, such as 2FA, IDPs, LDAP, OAuth and many more, have various common properties with Indy, including industrial-strength cryptography, rich metadata about identities, sophisticated access control and policy. Indy identities are distributed rather than segregated and federated. Because an Indy identity is transferable, the user can use it at any place that utilizes the distributed ledger. In the case of a hypothetical user, this means that if five systems support Indy IDs, *Lisa M. Private* will not have five different identities. Instead, Lisa's pre-existing identification on the blockchain is accessed by all five systems. Since Lisa is the owner of her identification, an organization can revoke Lisa's access, but not her identity [2].

CORDA

The Corda platform is an open-source software project initially introduced in April 2016 and then officially launched in November 2016 by software firm, *R3*. Corda, since its launch, has undergone a dramatic transformation from Corda 1 to Corda 4, its most recent update at the time of this writing.

Corda's design updated with the collaboration of a diverse global alliance of organizations in numerous industries and regulatory engagement was one of the vital parts of the design process. The financial industry requirements formed the basis for the design of Corda, but experience from the field shows that the Corda's technology is applicable far beyond banking [10]. Unlike 'non-allowing' blockchain platforms, the Corda platform aims to manage real-world transactions with privacy and legal certainty; and, co-exist and operate across the same, open network with multiple groups of participants and related applications in contrast to other 'permitted' blockchain platforms. Corda supports an end-state vision in which real-world entities administer contracts and transfer value without technology constraints or privacy loss [6].

The *Corda Global Network* is defined by Corda, along with a set of standards, network parameters and related governance processes. It allows any organization or individual to transact directly with any other person on the open network. Corda's architecture is fundamentally designed to be scalable, durable, secure, stable and interoperable. The architecture is also designed to legally enforce models and automate real-world transactions across an open network that can run and interoperate among multiple applications while it focuses on identity, the purpose of transactions, privacy and open governance. The governance model of the network is expressly designed to reflect the common interests of the different user groups of the platform. The platform provided by Corda permits transfers and reimbursements of cash liabilities in real-world currencies where rules enable liability settlement from platform-managed contracts. Moreover, the architecture also enables the provision of indigenous assets and platform tokens, aimed at promoting adoption and participation in payment for services. The architecture can be platform-wide or specific to certain business networks on the more extensive Corda network. The architecture defines a new shared platform for all industry sectors allowing new entrants and third parties to provide new innovative products and services [4, 6, 10].

Corda is a platform for *CorDApps* writing and execution; applications that enhance the global database by adding new features. These applications define new data types and new protocol flows between nodes and so-called *smart agreements* that determine changes. Notable among the visions of Corda is the implementation of an extensive range of CorDApps for many companies, suppliers, customers and third parties, through a common, shared, openly governed network, to manage a vast and diverse spectrum of agreements on the same platform. Corda is betting that the savings generated by better data, fewer discrepancies and faster coordination of details between companies will be significant. In addition, deploying this typical business architecture will define a new platform for existing and new suppliers to compete for the needs of their clients. Various Corda software solutions need to share common standards and agree on specific common parameters in contrast to the less useful remote, non-interoperable, independent deployments [6].

Components of the Corda Platform

The Corda platform has various elements discussed below in order to better understand the whole concept behind its working. The following are its main components:

A node is a software part of a Corda system that runs CorDApps and is arranged in an authenticated peer-to-peer network.

The Corda networks are fully connected nodes from peer to peer with connectivity. All communication is direct. As such, lobal broadcast or gossip protocol is not available. The use of advanced messaging queuing protocol (AMQP) through transport layer security (TLS) communicates securely, reliably and asynchronously. Service on the network map lists peers just like a directory service.

A state is an atomic and immutable object representing an event known at a specific point in time by one or more Corda nodes. States may contain arbitrary information which enables nodes to represent any entity. The arbitrary data can be inventories, bonds, loans, KYC data or information on identities, to name a few. States are UTXO (Unspent Transaction Output) and DAG (Directed Acyclic Graph) nodes. Only 'unspent' states determine the current state of the world. The 'spent' states are one of the components of the world state's history. States are one-of-a-kind. Moreover, they can only be consumed or spent once. The distributed ledger state is the aggregate of every state held by all network nodes. Because there is no central ledger and not all nodes are aware of all the states, the overall leader is subjective from the perspective of each participant. At least two different nodes carry the same state in most cases.

A lilateral ledger is a vault – a database that tracks and considers all the current and historical states that it knows, held by every node on the network. In simpler terms, it is the ledger from the point of view of each node and contains the set of all current statements (i.e., not historical) that is known to it.

Transactions define the consumption of existing states and outputs that add new data to existing states to produce a new state; as such, states are nuclear, as they complete or do not affect transactions entirely. Transactions have a life cycle with various phases, but there are no 'in-flight' transactions that are partially finished; although they have a life cycle with multiple phases. A Corda transaction is not like an SQL transaction or Ethereum transaction, where commands and parameters are passed, and the state engine 'decides' what is the new state. Instead, it refers to existing inputs and future outputs. A transaction not yet accepted by the parties is referred to as 'the transaction proposed'. Issuance transactions can add new states to the database without ever using existing states. Moreover, these transactions are not specific (Figure 7.2).

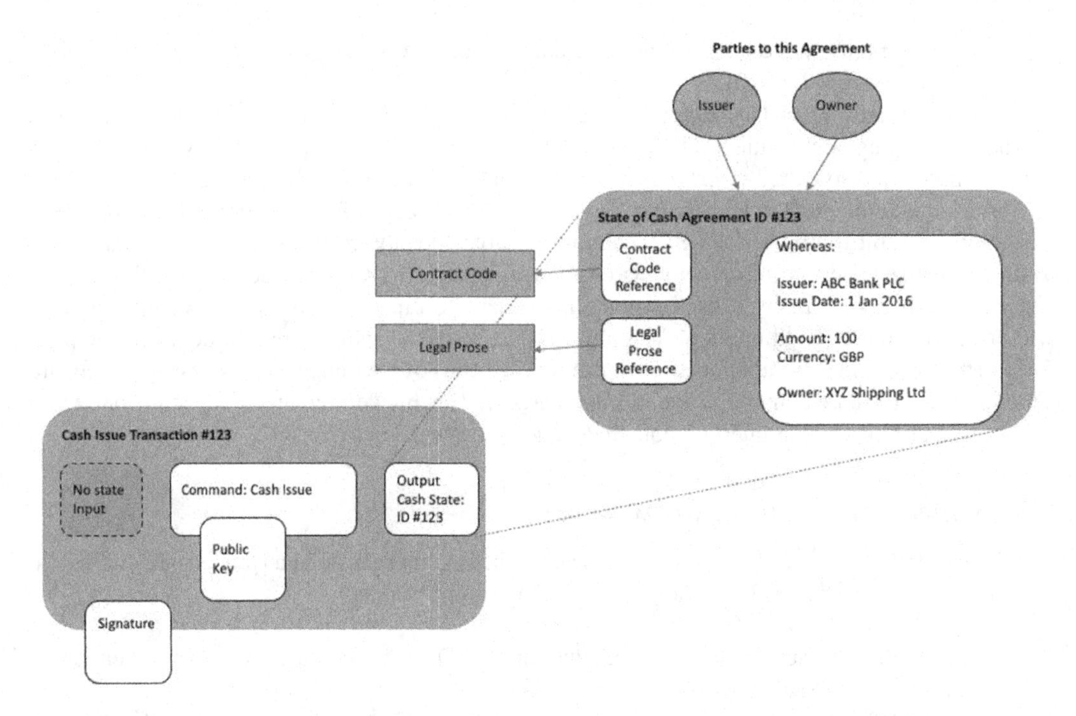

FIGURE 7.2 Cash issuance transaction in Corda.

Source: [6], recreated with Microsoft PowerPoint; © R3 (www.r3.com); https://www.corda.net/content/corda-platform-whitepaper.pdf.

Contracts define the specific properties of states instead of describing the creation of the states. Properties limit permissible state transitions like a checklist. Thus, the proposed state transitions can be validated. In simpler terms, contracts lay down rules on allowable transitions of state and proposed state transactions. A new state of the world is suggested by one or more parties in the transaction, and the agreement accepts this new state. Contracts also refer to legal prose – the concepts that a dispute is presumably acceptable to the courts. States contain a reference to the contract and mention the code which allowed their creation and consumption or prevented them. It is important to note that on Corda platforms that states, transactions and contracts are the fundamental data and transactional elements.

Commands are elements of a transaction that aid in guiding interpretation by providing intent to the transaction. Commands signal the intention of the proposed transitional state to allow the recipient to classify and consider the transactions. Smart contracts can also branch their verification logic on a command basis. In particular, commands are used to parameterize transactions by indicating their purpose, gathering the input and listing the signatories needed to sign using their public keys.

Timestamps are the assertions that claim that, within a given window, something has occurred. The windows can be open or closed. There is no true time in a distributed system; just many desynchronized clocks. Therefore, the timestamp is defined as a window rather than an exact time.

For the implementation of a smart contract code, contract logic shall communicate the transaction's position on the timeline. It is achieved using a timestamp. Other uses for timestamps, such as regulatory reporting or ordering events in a user interface, do not require any obligation. Furthermore, even if they do not precisely match the time observed by other parties, locally observed timestamps may be preferable in some cases. Alternatively, if a particular point on the timeline is needed and several parties must consent, the time window's midpoint may be used by default. While this may not precisely align with any activity, such as a keystroke or verbal consensus, it is often helpful.

Attachments are the .zip and .jar files included in the transaction. An attachment may be composed of legal prose with templates and parameters. It also includes the files of transaction information such as currency definitions, corresponding state definitions, calendars, financial information along with contract code.

Flows are lightweight processes that assist peers in reaching consensus regarding shared states by coordinating multi-step business operations. In other words, flows model the business processes that monitor the development of transaction states. The flow must include the appropriate command and a list of required transactors to authorize the smart contract(s), which must always check for the necessary signatories. Flows can be made to be widely reused or made on the fly. As long as the contracts allow for the planned state changes, a flow tailored to the circumstances can be generated immediately. A flow's complexity is directly proportional to the number of parties involved. More than two parties can be a part of the flow. For example, if a transaction requires regulatory oversight, the flow will indicate an additional step to notify a third party, the regulator.

Consensus is the process through which all parties obtain certainty in shared states. There are two aspects of consensus mechanisms that apply to Corda, namely, validation consensus and uniqueness consensus.

Notary services provide reliable evidence of events and are comparable to conventional notary public services. Simply put, notaries maintain a crucial map of their input states and their transactions. Notary services do not need to know the content of states and transactions. The only thing that a notary service needs to do with a transaction is to identify it reliably. One of two things can take place in the case that a peer sends a transaction to a notary service. If a notary's map contains any of the statements, the notary shall make an exception. If none of the state input is already known to have been used, each information is added to the map by the notary, and the proposed transaction is signed. This process offers consensus on uniqueness. A notary can be made to verify the transactions to protect from a possible DoS attack by which an attacker submits fake transactions. The notary will only take the transaction as a regular node in this case if its smart contracts accept it.

Oracles are trusted sources of information about the outside world. In order to establish a currency, parties may agree to use a third-party quota. This third party is referred to as an oracle. The oracle effectively guarantees the information included in the transaction by means of its signature. It can also provide valid information on request, thereby making it easier for a node to build a transaction approved by the oracle. Moreover, for the purpose of added privacy, only the minimum required information can be transmitted to the oracle using a Merkle tree and Merkle evidence. This is called a *transaction tear-off* [6, 10].

Corda Smart Contracts

As part of CorDApps in the contract code, intelligent contract logic ensures that state transitions are valid under the agreed-upon rules. In Corda, a smart contract is a pure function accepting or rejecting a proposed transaction, and is made up of more specific, reusable functions. One of these functions for each type of state in the proposed transaction shall be executed for each proposed transaction, and all parties are obliged to agree that – given the rules of each state – the proposed transaction is allowable. The transaction is therefore only valid if the contained states agree on all contract codes. A transaction proposer oversees the generation of a transaction that fulfills the constraints of each included state object.

Corda utilizes a programming language called *Kotlin* to write its smart contracts. Kotlin is a general-purpose, statically typed programming language that has cross-platform implementation with type inference. Kotlin is intended to be fully Java-compatible, and the standard Kotlin library version of the Java Class Library is based on JVM. At the same time, type inference allows for more concise syntax. It was designed primarily for JVM, yet JavaScript or native code can be compiled as well. Because of JVM's (JAVA Virtual Machine) wealth of existing libraries and extensive expertise base, and the re-use of industry standards it is possible for financial institutions to re-use their existing code inside contracts. Standardizing bytecodes enables the user to innovate or re-use well-known languages in contract language design depending on their preferences [6, 10].

Corda's Consensus Mechanism

There are two aspects of consensus in Corda as follows:

The transaction validity aspect of consensus ensures that the parties can be certain that a proposed update transaction defining the output states is valid by ensuring that the related contract code, which must be deterministic, is successfully verified by having all of the required signatures.

The transaction uniqueness aspect of consensus ensures that the participants of the transaction can be certain that the transaction under consideration is the sole consumer of all its states that have been inputted in the transaction. Moreover, it also makes sure that no other transaction previously agreed upon exists that used any of the same states.

However, two valid transactions may exist simultaneously, so participants require a way to choose which transactions to consider first. For this purpose, an observer is required among a group of mutually distrustful notary participants. Many people can participate in the same network while having different characteristics and agreements, which is a unique feature of Corda's design. A single service can be composed of many mutually distrustful nodes coordinating through a Byzantine defect tolerant algorithm.

It is important to note that such uniqueness services must only confirm if the previous states were consumed; in such a case, they should not certify the validity of the transaction [4, 6, 10].

Corda Use Cases

Since the blockchain industry is rapidly expanding, businesses have a higher expectation from blockchains when it comes to performance, security and user-friendliness. Corda is one of the platforms that is considered to cater to all these needs. This next section discusses three use cases of Corda. All of these use cases depict Corda's capacity to solve a variety of challenges.

A notable Corda application is in the *healthcare industry*. The healthcare industry recognizes the requirement for a blockchain since even the finest healthcare systems can become fragmented. Increased fraud, delayed claims processing and unsatisfactory patient care are all consequences of fragmentation.

The healthcare system has made using electronic health records (EHRs) a key priority. A Corda blockchain will allow systems to communicate with one another and offer a comprehensive patient health history without any requirement for a third-party validation service. Healthcare can keep every record in a time-stamped, auditable and immutable ledger.

Corda is also often utilized in the *construction business*, as its identification, privacy and workflow qualities make it an ideal choice for this type of operation. Contractors transfer work to subcontractors on a regular basis in the building sector. Corda's capability to fine-tune the data distribution over the network is critical in the construction industry for creating blockchain solutions best suited for already-existent players of the industry. Corda allows real estate developers and contractors to maintain a record of the counterparties with whom they have come to an agreement to do the various jobs and allows such parties to be compensated promptly. Corda also allows project members to see the project's progress and achieved milestones in a uniform and accessible way. Moreover, Corda's state and contract frameworks aid in the definition of shared rules and data structures, as well as how these data structures will grow over time. This will help prevent participants from disagreeing and/or making errors. Even if participants are connected via the same network, the need-to-know data dissemination of Corda prevents other parties from finding specific details of the project [18].

A Corda blockchain's distributed ledger and transparency can also aid *financial firms* in acquiring efficient and reliable access to up-to-date consumer data. As a result, organization performance, non-automated data collection, operational expenses and processing times are decreased, while institutional trust is increased. Transaction finality, scalability, privacy, legally recognized parties, the productivity of the developer and integration of enterprises are some of the advantages offered by Corda to financial institutions.

User Privacy in Corda

Corda uses several approaches to improve user privacy over other distributed ledger systems. In Corda, transactions are not broadcasted globally as is the case with many other blockchain systems. Moreover, Merkle trees are used to structure transactions, and individual subcomponents may be revealed to parties who already know the Merkle root hash. Furthermore, parties may sign the transaction without being able to see the entirety of it [4, 10]. The key administration service on the node generates and uses random keys that are unrelated to an identity. Large transaction graphs involving liquid assets can be 'pruned' by requesting that the asset issuer re-issues the asset with a new reference field onto the ledger. While not an atomic operation, it effectively unlinks the new and old versions of the asset, implying that nodes will not attempt to explore the original dependency graph during verification [6].

Corda's privacy focuses on the encryption and validation of the entire ledger with secure hardware enclaves. It can provide the benefits of homomorphic encrypting and zero-knowledge proofs in case of failure without sacrificing scalability or auditability.

Limitations of Corda

The hefty implementation and operational costs charged by the Corda Network Foundation to enable business networks to enroll are a key impediment to the global Corda network's (the network of networks) natural and accelerated growth. While large financial institutions will not object to these fees, many smaller business networks may refuse to join because the costs would outweigh the benefits of using Corda's DLT technology. As such, the cost of adopting the global Corda network could limit its future growth. Furthermore, these cost barriers threaten to cause an unbridgeable gap between commercial networks that use Corda's open-source version and those that use the enterprise version [9].

A SIDE-BY-SIDE COMPARISON OF ETHEREUM, HYPERLEDGER AND CORDA

Table 7.1 compares three distributed technologies, Ethereum, Hyperledger and Corda, covered in this chapter.

TABLE 7.1
A Comparison of Ethereum, Hyperledger and Corda

Parameters	Ethereum	Hyperledger Frameworks	Corda
General description	A platform for development and deployment of decentralized apps (DApps) using generic blockchain technology.	Modular, scalable and secure platform that facilitates a broad set of industry to accommodate its use cases using plug and play components.	Specialized distributed ledger platform for recording, managing and executing financial agreements.
Governing body	An open-source platform governed by all the developers.	Governed by Linux foundation.	Governed by R3.
Blockchain permission model	Can operate as a permissionless public blockchain or a private blockchain.	Operates as a permissioned private blockchain.	Operates as a permissioned private blockchain.
Consensus model	Uses proof of work consensus model and implements it on the ledger level.	Deploys pluggable consensus models based on nodes' requirement and implements it on transaction level.	Same as Hyperledger but based on notary nodes.
Smart contract	Uses self-executing and strict smart contracts as underlying logic to operate DApps and this contract is applicable for every node on the network.	Supports personalized smart contracts that deploy on only agreed-upon parties and a single node can have multiple smart contracts with different parties.	Same as Fabric, but only applicable for financial agreements with legal prose in it.
Smart contract rectification after deployment	It is nearly impossible to edit or rectify smart contracts once deployed.	Needs an elaborated process to rectify/update it.	Needs node-level administrative operations to edit/update it.
Programming language for smart contract	Solidity	Go, JavaScript (node. js)	Kotlin
Transactions	Used to promote reliability and predictability of a transaction.	Used to promote scalability, trust and transparency of transactions.	Used to maximize integration and interoperability with existing bank systems and latexing bank libraries.
Transaction fees	Includes transaction fees.	No transaction fees.	No transaction fees.
Preferred use cases	Generalized application. Ideal for business-to-consumer applications.	Specifically focused on enterprise application.	Specifically focused on financial service industry.
Built-in currency	Requires built-in currency called Ether.	None	None
Scalability	Ethereum has reached its capacity limits with the increasing number of users. The scalability issue has increased the cost of using the network, necessitating the use of 'scaling solutions'.	No prevalent scalability issues.	Highly scalable.

DECISION TREE TO CHOOSE THE RIGHT FRAMEWORK FOR A BUSINESS USE CASE

Before considering applying blockchain to a business process, it is essential to study its feasibility. R. Lewis has proposed that only uses cases that have concerns regarding trust, consensus, and immutability should consider blockchain as a promising solution. Otherwise, there is no viable reason

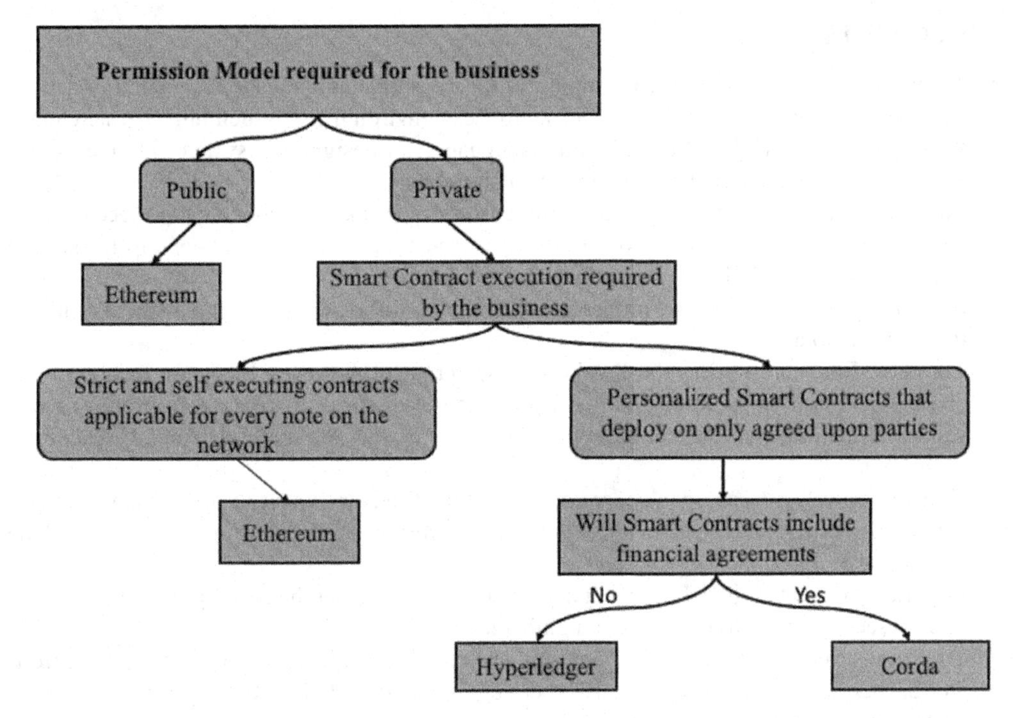

FIGURE 7.3 Decision tree to choose right framework for a business use case.

for the same. After completing a feasibility study that results in a decision to whether implement blockchain technology to the business use case, the decision tree in Figure 7.3 may be referred to.

SUMMARY AND CONCLUSIONS

The idea of implementing blockchain has evolved from the underlying technology of modern digital cryptocurrency to a new class of distributed and decentralized applications that can work effectively in trustless, peer-to-peer environments. The trustless, decentralized ledger's characteristics that support historic immutability have become an increasingly popular idea for achieving security and assurance for many industries.

This chapter intended to briefly introduce three currently popular blockchain frameworks, namely Ethereum, Hyperledger and Corda. These frameworks apply the benefits of blockchain technology to current business processes in banking, cybersecurity, auditing, health care, supply chain management, communication systems and the Internet of Things. If utilized effectively, these frameworks can support existing efforts for securing networks, communications and data in an untrusted environment where the data are distributed and decentralized. To implement these blockchain frameworks to any specific business use case, it is imperative first to understand the underlying and related mechanisms and techniques for implementing a secure blockchain network.

It is worth noting that these frameworks have associated vulnerabilities and limitations along with all their privacy and security features. Before considering applying any of the frameworks discussed in this chapter to a business process, it is essential to study its feasibility. Uses cases with concerns regarding trust, consensus and immutability should explore blockchain technology as a promising solution. Otherwise, other alternatives, such as the use of database technology will often prove more cost-effective.

CORE CONCEPTS

- The primary aim of Ethereum is to empower the consenting individual to carry out transactions with another consenting individual in a trustless environment by defining a state-change system using a loaded and unambiguous language and designing a system that reasonably expects that an agreement will be applied autonomously.
- Ethereum empowers its users to write customized smart contracts and decentralized applications by allowing them to formulate their personalized set of ownership rules, state transition functions and transaction formats.
- Ethereum provides a universal programmable blockchain packed into an open client available for everyone to use.
- Currently, Ethereum utilizes the PoW consensus protocol, however, Ethereum has begun the process of transitioning from proof-of-work to proof-of-stake. In contrast to proof-of-work, validators do not require substantial computing power, as they are selected randomly and do not compete.
- Hyperledger refers to an open source designed for the advancement of blockchain technologies across different industries that aim toward working collectively rather than working individually.
- Hyperledger acts as a greenhouse that helps in bringing together the developers and users from various sectors and market spaces at a global level.
- There are several Hyperledger frameworks, such as Hyperledger Burrow, Fabric, Indy, Iroha and Sawtooth. Additionally, several Hyperledger-related tools include Hyperledger Caliper, Composer, Cello, Explore and Quilt.
- The Corda platform aims to manage real-world transactions with privacy and legal certainty and can operate across the same, open network with multiple groups of participants and related applications in contrast to other 'permitted' blockchain platforms.
- The financial industry requirements formed the basis for the design of Corda, but experience from the field shows that Corda technology seems widely applicable to use cases far beyond the banking sector.

ACTIVITY FOR BETTER UNDERSTANDING

Readers/auditors interested in delving deeper into the above-discussed blockchain technologies are encouraged to further expand their knowledge by studying the following documents listed in this chapter's reference section:

- Buterin, V. (2013). Ethereum [Whitepaper]. Retrieved from ethereum.org: https://ethereum.org/en/whitepaper/
- Hyperledger Architecture, Volume 1: Introduction to Hyperledger Business Blockchain Design Philosophy and Consensus. (n.d.). Retrieved from Hyperledger: https://www.hyperledger.org/wp-content/uploads/2017/08/Hyperledger_Arch_WG_Paper_1_Consensus.pdf
- Hyperledger Architecture, Volume II: Smart Contracts. (n.d.). Retrieved from Hyperledger: https://www.hyperledger.org/wp-content/uploads/2018/04/Hyperledger_Arch_WG_Paper_2_SmartContracts.pdf
- Brown, R. G. (2018, May). The Corda Platform: An Introduction [Whitepaper]. Retrieved from r3: https://www.corda.net/content/corda-platform-whitepaper.pdf

REFLECTIVE QUESTIONS (SUPPLEMENTARY)

1. Which platform discussed in this chapter seems to be the most suited for a use case in your own organization? Use the decision tree provided in the last section of the chapter to start choosing an appropriate platform for your own case use.
2. Based on the chapter discussion, what risks, features and trade-offs do you need to consider when making a choice between the three discussed platforms.
3. In what way(s) will the implementation of blockchain solution help increase your organizational auditing effectiveness?
4. Ethereum encourages transparency of transactions for the sake of mitigating the risk of fraudulent transaction. How can this feature affect the implementation of a blockchain solution for your own use case?

REFERENCES

[1] *A Blockchain Platform for the Enterprise*. (n.d.). Retrieved from https://hyperledger-fabric.readthedocs.io/en/latest/
[2] *An Introduction to Hyperledger*. (2018, August). Retrieved from https://www.hyperledger.org/wp-content/uploads/2018/08/HL_Whitepaper_IntroductiontoHyperledger.pdf
[3] Azaria, A., Ekblaw, A., Vieira, T., & Lippman, A. (2016). Medrec: Using Blockchain for Medical Data Access and Permission Management. *International Conference on Open and Big Data* (pp. 25–30). IEEE.
[4] Benji, M., & Sidhu, M. (2019). A Study on the Corda and Ripple Blockchain Platforms. In *Advances in Big Data and Cloud Computing* (Vol. 750). Singapore: Springer.
[5] Bocek, T., & Stiller, B. (2018). Smart Contracts – Blockchains in the Wings. In *Digital Marketplaces Unleashed* (pp. 169–184). Berlin: Springer.
[6] Brown, R. G. (2018, May). *The Corda Platform: An Introduction*. White paper. Retrieved from https://www.corda.net/content/corda-platform-whitepaper.pdf
[7] Buterin, V. (2013). *Ethereum*. White paper. Retrieved from https://ethereum.org/en/whitepaper/
[8] *Corda: The Blockchain That Doesn't Use Blockchain*. (2020). Retrieved from https://f.hubspotusercontent30.net/hubfs/8639589/Minibook_SettleMint_Corda%20deep%20dive.pdf?utm_medium=email&_hsmi=109470427&_hsenc=p2ANqtz-_DbHQusg0sdkecOrZShm2gIoQ0_LRTLgyb3e4Xp93yKY74rwqII9ZpAMPajr0kRnk7CLZ2WDP6iymbxguqQvGbgYeDmw&utm_content=109470427
[9] Dhillon, V., Metcalf, D., & Hooper, M. (2017). The Hyperledger Project. In *Blockchain Enabled Applications* (pp. 139–149). Berkeley, CA: Apress.
[10] Hearn, M., & Brown, R. G. (2019, August 20). *Corda: A Distributed Ledger*. White paper. Retrieved from https://www.r3.com/wp-content/uploads/2019/08/corda-technical-whitepaper-August-29-2019.pdf
[11] Howard, J. P., & Vachino, M. E. (2019, May 1). Blockchain Compliance with Federal Cryptographic Information Processing Standards.
[12] *Hyperledger Architecture, Volume 1: Introduction to Hyperledger Business Blockchain Design Philosophy and Consensus*. (n.d.). Retrieved from https://www.hyperledger.org/wp-content/uploads/2017/08/Hyperledger_Arch_WG_Paper_1_Consensus.pdf
[13] *Hyperledger Architecture, Volume II: Smart Contracts*. (n.d.). Retrieved from https://www.hyperledger.org/wp-content/uploads/2018/04/Hyperledger_Arch_WG_Paper_2_SmartContracts.pdf
[14] *Hyperledger Gets Cozy with Quilt*. (2017, October 16). Retrieved from https://www.hyperledger.org/blog/2017/10/16/hyperledger-gets-cozy-with-quilt
[15] Mantelli, I. (2018, August 29). *Pros and Cons of Using Hyperledger for Blockchain Applications*. Retrieved from https://www.linkedin.com/: https://www.linkedin.com/pulse/pros-cons-using-hyperledger-blockchain-applications-ivan-mantelli
[16] Muscara, B. (2019, September 19). *Hyperledger Public Sector White Paper*. Retrieved from https://wiki.hyperledger.org/display/PSSIG/Whitepaper
[17] Richards, S., & Cordell, R. (2020, September 24). *Ethereum Development Documentation*. Retrieved from https://ethereum.org/en/developers/docs/

[18] Sharma, T. (2019, November 1). *Top 3 Use Cases for Corda*. Retrieved from https://www.blockchain-council.org/blockchain/top-3-use-cases-for-corda/

[19] Vujičić, D., Jagodić, D., & Ranđić, S. (2018). Blockchain Technology, Bitcoin, and Ethereum: A Brief Overview. *2018 17th International Symposium*, pp. 1–6.

[20] Wackerow, P., Richards, S., & Cordell, R. (n.d.). *Ethereum Development Documentation*. Retrieved from https://ethereum.org/en/developers/docs/

[21] Wood, G. (2020, September 5). *Ethereum: A Secure Decentralised Generalised Transaction Ledger*. Petersburg.

8 Designing a Blockchain Application

Mihir Mashilkar, Kashish Patel, Divy Patel and Het Raval

ABSTRACT

Chapter 8 is the first of three chapters in this book that deal with blockchain application design, development and testing. The chapter begins with an outline of a series of considerations that blockchain application developers must assess during the design process. These include questions such as whether the application needs to be feature-heavy or feature-light; to what extent the application needs to be specialized; whether support for the app needs to be centralized or decentralized; or to what extent the application needs to be scalable. Moreover, this chapter provides an introduction to a fictional case study, comprised of two personas and their user story in order to familiarize readers with the need to create personas, and their user stories as first to determine the functional requirements for an application. Functional requirements are subsequently translated to technical requirements, which lead to the process of creating the specific tasks and steps associated with completing the application design process.

The second part of this chapter briefly discusses some popular application design approaches, such as the Ancile, ontology-driven and model-driven approaches. The chapter discussion concludes with a table outlining eight important blockchain design considerations, including attributes and/or issues.

BLOCKCHAIN DESIGN GUIDING PRINCIPLES AND CONSIDERATIONS

Do We Need a Blockchain Solution in the First Place?

As discussed in previous chapters, a database application is a computer program that stores, manipulates and retrieves data. The data can range from numbers to complex structures [5]. A blockchain consists of a data structure that keeps records of the transactions in the network [3]. It is important to note that while blockchain is a type of database, a database is not necessarily a blockchain [5]. To further explain, there are noteworthy differences between the blockchain and database technologies as follows:

Storage: The major feature that positively distinguishes blockchain from other technologies is its immutability of stored records. To achieve this immutability, it uses cryptographic and consensus mechanisms with the data being stored in distributed nodes. While blockchain is designed to work in a decentralized manner, databases are always centralized, which means that they run on one server, and users cannot download them. Databases are usually secured and protected, since they often contain sensitive user and business data. On the other hand, blockchains do not have a centralized structure. The data are distributed in the whole network where any user of the network can download them [3].

Consensus versus permissions: Another point to be considered is that while blockchains rely on the concept of consensus, databases rely on permissions. A consensus algorithm is a formula that specifies the conditions under which data can be applied to the blockchain. Although

DOI: 10.1201/9781003211723-8

consensus focuses on how new data can be applied to the blockchain and validated, decentralization refers to the idea that the blockchain is not owned by a single entity. This ensures that the database is not hosted on a central server that belongs to a single organization. Any machine that wants to download the blockchain can have a copy of it [5].

Management: Databases contain details regarding the nature of data. On the other hand, blockchains act as a storage medium for chunks of bundled information called *blocks* [2].

Modification: A database can be modified. This means that if a user has the authority, they can record, modify or even delete records from a database that is already backed up. Blockchains contain an added security feature that makes it difficult to hack or tamper with them [18].

Transparency: Another important feature of blockchain is that in the case of an open, public blockchain, users can access the data on the blockchain. Databases, on the other hand – due to their centralized nature – do not allow for the same level of transparency.

Cost: Since database technology has been in existence for a number of decades, a traditional database package typically costs less to deploy than a blockchain solution. As such, conventional database technology may be a more cost-effective alternative for companies to consider, especially if various labor costs are considered in the related cost/benefit analysis.

Talent acquisition: Since blockchain is a relatively new technology, there is currently a small pool of talent available to deal with various blockchain development applications. Furthermore, the acquisition costs related to hiring competent blockchain are currently significant. On the other hand, database-related talent is relatively easy to come by. They are also cost-effective, so even small businesses can afford to hire a database specialist [13].

Append only: Finally, the fact that blockchains are append-only distinguishes them from conventional databases. That is, new data do not replace old data; instead, they are tacked onto the database in such a way that all changes can be traced back over time. In conventional relational databases, new data do not replace old data unless they have been expressly designed to do so. This is because blockchains organize data into blocks, each of which is identified by a timestamp and the hash, or signature, of the previous block of data. This makes it possible to connect data over time and keep track of changes on the blockchain. This functionality can be incorporated in conventional databases (for example, transaction history in banks), but it is not a required feature of it [5].

When it comes to utility, pace and accuracy, often times, database technology is the clear winner. However, when it comes to creativity, authentication and automation, blockchain can offer additional advantages. In a nutshell, for trust, accountability and verification, blockchain is often the way to go. On the other hand, databases are ideal for high-performance applications and services. It is also a great option for apps that need to scale up [13].

Does the Application Need to Be Feature-Heavy or Feature-Light?

When developing the decentralized app (DApp), it is essential to understand users' needs. If the user criteria are not met, it will make the DApp quite complicated and/or frustrating to use. The provision of numerous features for users may seem like a reasonable goal at first, but an overwhelmed user may not opt for an application with extra features [22].

A progressive web app (PWA) is a modern kind of website that provides native app interactions through the use of a browser. A PWA will work even if the computer is not connected to the Internet. To enable rich web-based interactions, PWAs use modern web features such as push notification, cache and stable connections. The PWA feature in the application design requires a longer design time. The design of an application with additional features can create an issue when the application is designed within a short amount of time. Native apps have direct access to the device's hardware. Unlike native apps, the web does not have complete access to the device's hardware capabilities. The hardware's complete functionality can be achieved on a cell phone, although this functionality is not always available on a web browser due to limitations associated with JavaScript. To further explain, JavaScript is a single-threaded program; this means that another operation must be delayed when one operation is being performed. In this critical situation, it will be feasible to use the native code written in *C* or other programming languages that provides the ability to run multiple commands simultaneously [23].

The priority for any user is to utilize an application effectively and efficiently with minimal interruption. The application must be simple, straightforward and easily understood by users. The user interface is the initial step to making the application simple, but this simplicity is not dependent on the user interface alone. Moreover, the application must be aligned to the user's requirements and should be focused on the different functionalities that make the DApp unique. Hence, most attributes of the application's function and design should be feature-light [22].

What's More Important – Collaboration or Security?

Collaboration can result in significant cost and time savings when it comes to setting up and carrying out collaborative procedures, especially when there are trust issues between the third parties involved in the collaboration. Collaboration processes may be carried out by depending on trusted third-party partners [7].

Design for Security and Privacy

Since blockchain applications are linked to a public or private network where central authority is absent, security and privacy are the most essential considerations when it comes to designing blockchain applications. Blockchain applications are usually connected to the Internet like every other network, except for government-classified secret networks. Hence, companies must implement security and privacy appropriately.

Blockchain applications are similar to other Internet-based applications, which is why data exposure is considered one of its major risks. In a blockchain, digital identities are a concept that could help with data privacy. The use of unique digital identities helps in the development of enterprise-grade blockchain security measures. With the use of smart contracts, an extra layer of security can be included in a DApp. To further explain, a smart contract is an algorithm that is responsible for ensuring that if some conditions are fulfilled by the user requests, then a specific function will be implemented – like blockchain transactions, transfer of funds and triggering of a message [10].

Will the App Have Consistency or Specialization?

Consistency refers to the value of always acting or behaving in a consistent fashion, or always executing in a similar manner [26]. Consistency is synonymous with uniformity. This implies that a consistent thing is known to be reliable, well-organized and following certain rules or assumptions [20].

Maintaining consistency is one of the important principles of blockchain technology. By making changes or adding new features, the consistent performance of a blockchain may be affected, thereby resulting in stakeholder dissatisfaction. Therefore, the quality should be maintained at a constant level while designing the blockchain application, especially since visual qualities remain a major part of the customer experience [1].

Moreover, when similar elements have a standardized look and function similarly in a DApp, accessibility and learnability increase. Users can adapt the experience to new contexts and discover new things quickly when the design is consistent. This way, users will concentrate on completing the task rather than understanding how the product's user interface functions every time they switch contexts [17].

As an emerging technology, blockchains maintain user trust by providing consistency; as such, making design decisions according to the client's preferences and dislikes will help create a more consistent application with fewer bugs [1]. For a new user, the interface becomes much easier to learn. Users are less confused, thus fewer user mistakes occur. Consistency decreases design complexity and saves users time and effort. Furthermore, consistency helps to develop a clear picture for any application [26].

There are several types of consistency as follows:

Visual consistency is focused on making similar objects; it is all about colors, shapes, icons, fonts and other essential information. Visual consistency is also important in order to create a strong visual hierarchy and to create relations between various elements [26]. The elements which are included in DApp must be made with great attention to details. Such elements include typography, symbols, shapes, etc. [20].

Functional consistency ensures that identical structures behave in the same manner. Users can predict what will happen by clicking buttons and execute certain functions [26].

Internal consistency binds visual and functional consistency in order to ensure that the application is usable and easily learnable [20]. Internal consistency applies to various aspects of the interface or brand that appear and behave as though they are part of the same system. The new features added in DApp should be fully compatible with those that already exist in a DApp [26].

External consistency ensures that users will have the same experience irrespective of the platforms or systems on which the application is being run. For example, Amazon users use both Amazon's mobile app and web platforms to make purchases, which is why Amazon provides the same user experience on both platforms [20].

Will Support Be Centralized or Decentralized?

A decentralized network has no central authority that controls and handles the entire network, which is why stakeholders need the permission of other network users to make or change any rules. However, major services like Facebook, Uber and Instagram are still centralized, because the user cannot modify the predefined rules. For example, Facebook has full authority to change its rules and its users have to abide by them. Readers are reminded that there is always a single point of failure risk in a centralized system. Blockchain provides a high level of security to the stakeholders as a result of the absence of single points of failure [9].

The Mechanics of Decentralization in DApp

Data are stored across the network in the decentralized system rather than data collection by a central point in order to eliminate the risks of single points of failure. Ad hoc message passing and remote networking can be used to transmit data in the decentralized system. The decentralized system uses the public key cryptography method to provide a reliable authentication mechanism. The public key is accessible to all the users in the system, which helps to encrypt information. Each node or miner in the decentralization system has a duplicate copy of data which provides data equality and also all the users are treated equally in decentralization. The transactions are broadcasted to the network through the DApp. The messages are transmitted in real time within the period specified, which ensures that there will be re-tries to transmit messages, if needed [15].

The factors and characteristics relevant to the blockchain and decentralized application are as follows:

Open access: At some point in its operations, a centralized system will be technically or logically locked to users by the controlling third party. This means that a user will require permission from the central authority to access the centralized application. On the contrary, since there is no central authority due to the decentralized application's open-access capability to control the decentralized blockchain application, a decentralized network is open and available any time to users and developers [9]. The administration of the decentralized application should be independent, and any modification should be managed by consensus of the blockchain's users [15].

Non-hierarchical: With blockchain technology, each user has an equal authority and control based on a non-hierarchical pattern. An example is the proof-of-work mining system where all miners are essentially equal members on the blockchain, but their contribution is directly proportional to the network's computing power.

Transparency: Transparency ensures that each user obtains the same amount and level of information on the blockchain. An exception discussed in Chapter 7 is blockchains such as Hyperledger Fabric, which has the capability to hide the information from participants as dictated by various privacy laws and best practices [9].

Monolithic or Modular?

Modular architecture refers to a method of breaking down a problem's complexity into smaller, more manageable components. The modular architecture consists of some guidelines, principles and patterns, and is a design style that allows developers to see the system as a set of smaller and physical components. Modules aid developers in comprehending, expanding and controlling the system both during the design and runtime stages. Moreover, modules have a clear business structure and communicate their scope through a public interface. The physical composition of the application is the main subject behind modular architecture [19].

Organizations often tend to prefer a modular, versus a tightly coupled system – a monolith. A monolithic application has a single code base that encompasses a number of modules. Modules are categorized into either business or technical aspects. A monolithic approach offers a single build mechanism that includes the entire system, including dependencies.

In general, a modular process can be carried out in the following manner. The first step is to divide up the functionalities of the application. After dividing the functionalities and identifying the respective modules, the next step is to isolate the modules in order to keep them independent. Next, it is important to place the various functionalities as near to the information as possible with the aim of reducing communication between modules after the isolation of the modules. The final step is to eliminate direct communication between the modules [4].

Personas

Personas are fictional characters built based on a user class analysis in order to reflect the various user types that may use a service, product, website or application similarly. Personas are used to construct trustable and accurate images of key user segments for reference. Qualitative and quantitative considerations, as well as web analytics, are assessed in understanding and developing personas. The creation of personas assists with understanding users' expectations, perspectives, behaviors and goals.

Persona Example 1

Professor Shaun is a department head at a university in Canada. Shaun has a responsibility to help manage students' course registration processes and data. As a department head, Shaun does not have time to check and validate all the hard-copy student registration forms manually completed by students. In addition, he has to check all course registration eligibility requirements before approving each student's requests. He feels that this process is extremely time-consuming, and thus highly inefficient.

Persona Example 2

Anika is a doctor at a medical clinic in Alberta, Canada. As a medical professional, she has the responsibility of handling patients' data. With each doctor checking an average of 20 patients every day, doctors have a big problem recognizing patients, especially when a former patient returns for a follow-up appointment, leaving doctors with the task of finding specific documents that detail all the patients' health status.

User Stories

A user story is a casual, general interpretation of an application feature written from the end user's point of view. It aims to explain how an application feature can benefit the user. DApps could allow more than one kind of user. User stories examine a product's capabilities through the eyes of a particular end-user to reflect the interests and desires of a client class.

Prof. Shaun's User Story

Shaun is happy about the new course registration e-form system. He can begin his day by accessing a platform and reviewing all the course registration request e-forms containing all the info he needs to quickly approve/reject a course registration request, such as a student's grade point average, grades, course prerequisites, etc. Shaun feels that this approach is far more time-efficient than the previous paper-based system that needed to be filled out by hand by each student, reviewed and signed by him, and then taken to the course registration department manually.

Dr. Anika's User Story

Anika is happy to find that the new blockchain DApp will help save data for each patient effectively, and with no need for hard-copy patient records. Moreover, the DApp will provide robust security for patient data. Each doctor is validated by the department head and only they (dept. head and the assigned doctor) have access rights to retrieve patient data. Once the doctor approves new patient data, they will be stored as a block on the blockchain-driven platform.

Functional Requirements

Functional specifications define how an application behaves under particular circumstances. This will help determine needed design features and functionalities that DApp developers must incorporate into the solution. Some major blockchain functional requirement decisions include whether a blockchain needs to be private, public, open or closed, in addition to determining the best consensus methodology for validating each block [6].

Technical Requirements and Tasks

Based on the completion of the above-mentioned functional (business) requirements, technical requirements refer to the technical (coding) processes that need to be completed. Each technical requirement entails a series of more granular tasks that must be performed. In other words, the related tasks, in turn, help in fulfilling the technical requirements and defining the estimates. If a task cannot be estimated easily, it should be broken down into subtasks. With each task, the estimated time is calculated to complete the task for a skilled developer [16].

The following is a partial example of the tasks that may be needed for Dr. Shaun's user story, in the form of a checklist:

Student's interface:

- Prerequisite: This interface is only visible to university students who are taking part in the Research Methods course.
- Function: The participant students can register for the Research Method (RM) course by filling out an e-form.
- Tasks/required skills sets:
  1. Place button called 'Register for Research Method' on the interface.
     - 10 minutes – estimated time required for User Interface Designer to complete the task.
  2. Button click should call getRegForm() function
     - 30 minutes – estimated time required for back-end developer to complete the task.
  3. Registration form will open containing the labels and textbox and submit button should be placed on the bottom of the form.
     - 20 minutes – estimated time required for user interface designer to complete the task.
  4. Button click should call the function setRegForm() which will store the request in the database.
     - 30 minutes – estimated time required for back-end developer to complete the task.

Teacher's interface:

- Prerequisite: This interface is only visible to the professor who is registered with the Concordia University of Edmonton and has the required privilege to approve the Research Method request.
- Function: The professor can view, approve or decline the requests for the Research Method.
- Tasks/required skills sets:
  1. Place button called 'View Requests for Research Method' on the interface.
     - 10 minutes – estimated time required for user interface designer to complete the task.
  2. Button click should call getAllRegForm() function
     - 30 minutes – estimated time required for back-end developer to complete the task.
  3. The result from Task 2 will be passed to the show AllRegReq() function.
     - 40 minutes – estimated time required for back-end developer to complete the task.
  4. The result from Task 3 – containing the information of all the students' grades and the e-form for the registration, will be shown in tabular format.
     - 30 minutes – estimated time required for user interface designer to complete the task.
  5. 'Approve' and 'Decline' buttons are placed at the end of each row in the table.
     - 10 minutes – estimated time required for user interface designer to complete the task.
  6. The click of 'Approve' will call function approveReq() and similarly, the function declineReq() will be called on the click of the 'Decline' button.
     - 30 minutes – estimated time required for back-end developer to complete the task.

POPULAR APPLICATION DESIGN APPROACHES

This section introduces and briefly discusses a number of application design principles and approaches that may be suitable to blockchain application development. Depending on applicable application design considerations, blockchain application design stakeholders ought to investigate and choose the design approach that seems most compatible with their intended blockchain use case.

ANCILE

For the privacy-protection approach, Ancile's main aim is to solve data protection and security issues in the management and handling of electronic information. This design approach enables customers and third parties to access and share electronic records in a safe, interoperable and productive manner. The Ancile framework proposes an encryption technique for safe, multi-party access to information on a blockchain by only storing data references (hashes) on the blockchain to achieve this aim.

Ontology-Driven Approach

For the design of blockchain frameworks, an ontology-driven data modeling technique aims to track objects in a supply chain while enabling corporations to be conscious of data that are spread across multiple databases, particularly in cross-organizational processes. Another benefit of the ontology-driven approach is that it introduces systematic verification within organizational operations in a blockchain. The ontology-driven approach accomplishes this by translating data models found in business processes into a given Object-Oriented Programming (OOP) language using a Universal Modeling Language (UML) methodology. Furthermore, the formal ontology derived from OOP is converted into a smart contract code that describes the data produced by the initial business process. ODA is useful for both the design and development phases of a DApp [24].

Software Engineering Strategies for DApps

A software engineering strategy aims to ensure that project agreements between project owners and developers are well-defined and that all stakeholders clearly understand the required specifications. The approach suggests using UML models to describe the requirements and design of a DApp, such as use-case diagrams, class diagrams and operation diagrams.

The Model-Driven Approach

This model is useful for designing the states and behaviors of blockchain applications. The modeling of a blockchain relates to the intersection of traditional software systems and blockchains, which allow the storage of any data generated by a system on a distributed ledger. Purchase orders, service contracts or any other type of data may be included and stored on the distributed ledger. The majority of the time, the production of addresses or transactions originates on the software side. Flow-based visual programming languages are one form of visual design technique [8].

The User Design Approach

As user-related issues may become significant, experts have emphasized the importance of improving users' perception and awareness of blockchain technology. As such, user input and participation is essential for user-centered designs as blockchain technology continues to evolve. It is a difficult task for the majority of users to grasp and comprehend the inherent worth of blockchain technology, and this establishes a substantial entry barrier. While user-centered design takes significant time and expense, it enables efficient development by minimizing the redesign stage and thus improving the developer's understanding of user behavior.

There are several ways for users to contribute to the design process. Users can contribute through interviews, focus groups and on-site observations; criteria can be extracted and efficient task sequences can be planned early on. Furthermore, user satisfaction and usability can be quantified efficiently through usability testing and questionnaires.

Among user-centered design methods, usability testing is recognized as an important method for determining the specific difficulties faced by end-users. As technology has become more complex, usability testing has become a critical component of product and service creation to reduce development costs and efficiently attract customers [14].

The Design Process

The method depicted in Figure 8.1 shows how the current taxonomy could also be used to direct system design at various stages in the design process. The decision of whether to decentralize the

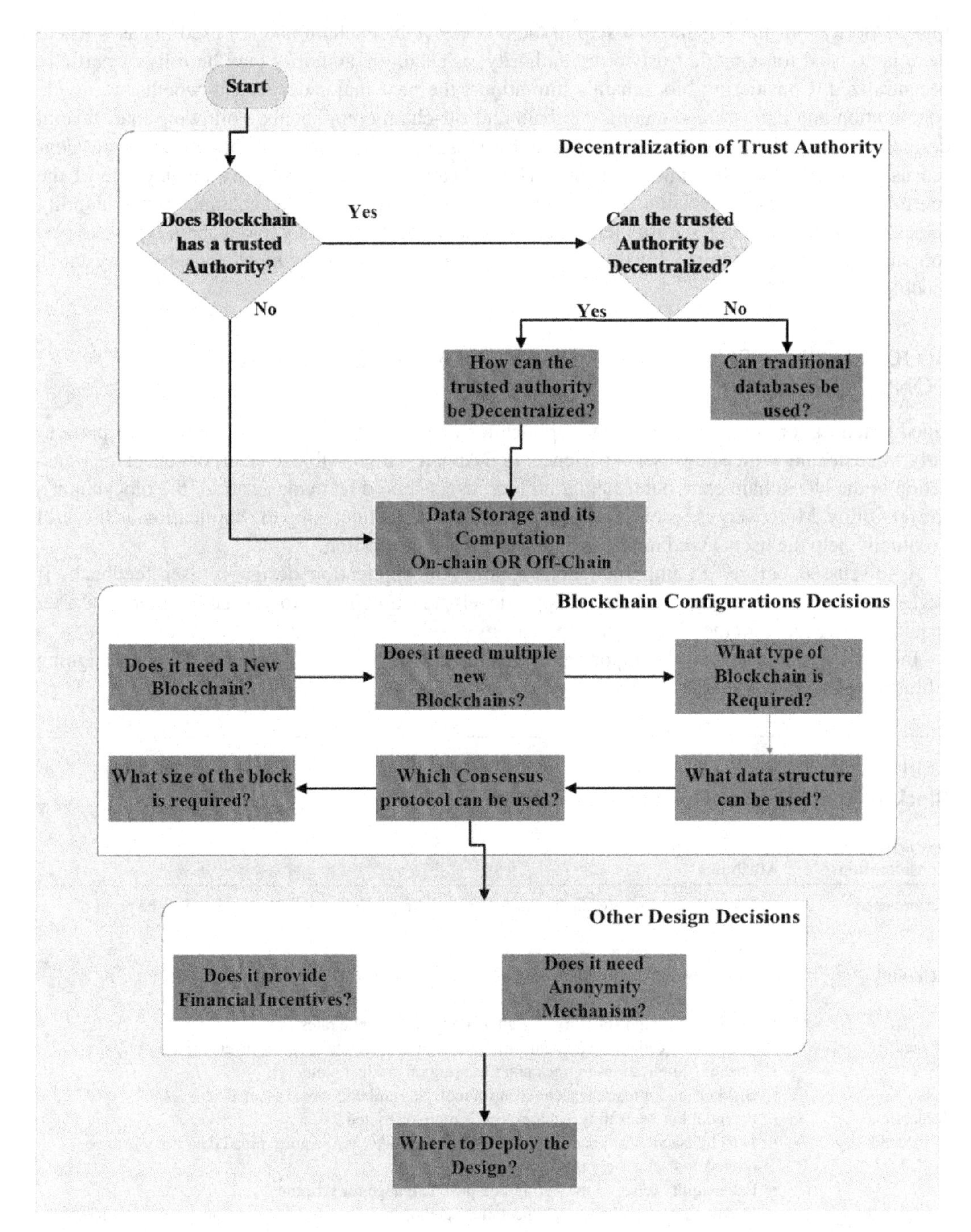

FIGURE 8.1 Design process.

Source: [25].

trust authority – or not – is the first step in the process. A blockchain is often used in cases where there is no need for a single trustworthy authority; as such, the authority may be fully or partially decentralized. Considering blockchain's limitations, the next major decision is whether to divide computation and data storage among on-chain and off-chain components. Following that, several design decisions about blockchain configuration must be taken, including blockchain type, consensus protocol, block size and frequency. The indicators in Figure 8.1 only display one of the several potential design decision stages. Some decisions have a significant impact on scalability (block size and frequency), security (consensus protocol), cost-efficiency (blockchain type) and performance (like data structure). Finally, determining where to deploy the blockchain-based system's modules is key [25].

BLOCKCHAIN APPLICATION DESIGN: GOOD PRACTICES AND CONSIDERATIONS

Good practices for designing blockchain applications begin with maintaining consistency, particularly when dealing with a new user experience. In DApp, it is impossible to revert or cancel the transaction in the blockchain once published; therefore, users should be made aware of the blockchain's irreversibility. Moreover, users must be guided in every step while using the application as this will eventually help the users avoid mistakes and user-related frustration.

As discussed before, an important consideration in application design is user feedback. If neglected during the design process, DApps developers are likely to misjudge important user requirements needed in the blockchain applications [11, 12].

Table 8.1 contains a list of the important design factors that should be considered when designing a blockchain application [21].

TABLE 8.1
Blockchain Application Design Issues and Considerations

Blockchain Design Considerations	Attributes
Performance	• The inefficiency of resources, such as high energy consumption levels related to the proof of work consensus. • Excessive latency for real-time output.
Reliability	• As no single point of failure should occur, high availability. • Reduce the probability of secondary failures. • The inherent high volatility in digital currency exchange rates.
IT security	• User identification and authentication through appropriate asymmetric encryption. • Enabling applications to track and trace (digital product memory). • Blockchain-appropriate access control tools and authorizations, as applicable.
Scalability/ extendibility	• Potential low scalability as block size is often restricted. • Each transaction is broadcasted on all network nodes and requires broad data streams to be filtered and selectively processed. • Extensibility relies on the Turing-complete language for scripting.
Compatibility/ interoperability	• Due to numerous client implementations and open-source code, high compatibility is achievable.
Fault tolerance	• Blockchain transactions remain unchanged, therefore a fault in any node will be transparent to all other nodes.
Quality of service	• Quality of service (QoS) determines the system's ability to offer its services reliably and with a response time and throughput that are satisfactory to the blockchain nodes. • For vital data transactions between nodes, quality of service is a must.
Failure management	• The blockchain must be designed in such a way that it can survive attack-related failures. • There should be systems in place that will identify the cause of a failure and initiate appropriate recovery plans, allowing the blockchain to recover from failures automatically.

SUMMARY AND CONCLUSIONS

This chapter has taken a look at the guiding principles and considerations related to designing a blockchain application, specifically how the application design process should start with the creation of personas. Next, user stories – based on personas -- should be created in order to help define the functional (business) requirements of the application. Based on the needed functional requirements, technical requirements and tasks will in turn be defined. In addition, various frameworks, procedures or approaches currently used to design and/or build decentralized applications (DApps) for blockchain systems are also discussed in this chapter. As such, the primary goal of this chapter is to provide auditors with a basic understanding of relevant factors that should be considered during the design phase of a blockchain application before moving on to the development phase, which is the focus of the next chapter.

CORE CONCEPTS

- There are a number of basic design considerations that must be considered based on user personas and user stories. These include decisions related to whether the application needs to be monolithic or modular, centralized or decentralized; or whether the application should focus on user consistency or specialization.
- Often, maintaining consistency is a good practice when designing blockchain apps. The various types of consistency include visual, functional, internal and external consistency.
- The main aim of software engineering for DApps is to ensure that project agreements are well defined. Numerous approaches can be used to develop decentralized application developments. The Ancile project's main aim is to solve data protection and security issues by enabling the access of electronic records to customers and third parties in a safe and interoperable manner. The ontology-driven approach (ODA) is one of the design approaches which aims to track objects in a supply chain while enabling organizations to understand data that are spread across multiple databases. ODA is useful for both the design and development phases of a DApp.
- The software engineering strategy entails the use of UML models to describe the requirements and design of a DApp. While the model-driven approach is useful for the design stage and behavior of blockchain applications, model-driven-based approaches do not necessarily require the use of smart contracts. User participation is essential for user-centered design which can be achieved in the user design approach. In the user design approach, users can contribute to the design process through interviews, focus groups and on-site observations.
- The irreversibility of the blockchain should be made clear to users. During the application's use, users must be directed through each step. This will assist users in avoiding costly or time-consuming blunders. DApps will often fail if user feedback is ignored during the design phase.
- An application should be easy to understand and use for users. Function and design aspects of the application should often be feature-light, unless there are compelling reasons not to create a feature-light application. The user interface is the first step in making the application as easy to learn and use as possible.

ACTIVITY FOR BETTER UNDERSTANDING

Use the blockchain design knowledge acquired in this chapter to initiate a blockchain application design project in your organization. Consider the following questions:

1. Is there a problem in your organization that can be solved using a blockchain application? Formulate a problem statement by describing the problem concisely and describe how the

problem may be addressed by a blockchain application. Please note that a problem statement need not be described with only one sentence. However, only a short paragraph should suffice to formulate an effective problem statement.

The video below will provide interested readers with additional information regarding problem statement development.

Title: *Identify the Problem*

https://www.youtube.com/watch?v=FPfEciCFtPc&ab_channel=GreggU

2. Go over the chapter section entitled *Blockchain Design Guiding Principles and Considerations* and develop a list of considerations pertinent to the blockchain application design. Which considerations discussed should take priority in your enterprise?
3. Continuing from step a above, develop personas and user stories as the first step in the blockchain application design.
4. Next, try to formulate – in a group setting – the functional and technical requirements for the blockchain application design.
5. Define the tasks that assist in fulfilling the technical requirements.
6. Evaluate the design approaches discussed in this chapter. Which one(s) may be best suited to the identified use case?
7. Refer to the blockchain case study for a better understanding of the need for a blockchain application in the organization.

REFERENCES

[1] Anwar, H. (2019, November 4). Blockchain Principles: Design Your Enterprise Blockchain. Retrieved February 03, 2021, from https://101blockchains.com/blockchain-principles/

[2] Bhardwaj, C. (2020). Blockchain vs Traditional Database: What Should Be Startup's Choice? Appinventiv. https://appinventiv.com/blog/traditional-database-vs-blockchain/

[3] Chowdhury, M. J. M., Colman, A., Kabir, M. A., Han, J., & Sarda, P. (2018, August). Blockchain versus Database: A Critical Analysis. In 2018 17th IEEE International Conference on Trust, Security and Privacy in Computing and Communications/12th IEEE International Conference on Big Data Science and Engineering (pp. 1348–1353). IEEE. https://ieeexplore.ieee.org/abstract/document/8456055

[4] Christian, C. (2020, January 30). From Monolithic to Modular. Retrieved February 20, 2021, from https://www.praqma.com/stories/from-monolithic-to-modular/

[5] Devillez, F. (2020). Blockchains versus Databases – What's the Differences? Espeo Blockchain. https://espeoblockchain.com/blog/blockchain-vs-database-comparison

[6] Dhiman, T., Gulyani, V., & Bhushan, B. (2020). Application, Classification and System Requirements of Blockchain Technology. International Conference on Innovative Computing and Communication. New Delhi.

[7] Dumas, M., Hull, R., Mendling, J., & Ingo Weber, I. (2018, August 17). Blockchain Technology for Collaborative Information Systems. Retrieved March 14, 2021, from https://core.ac.uk/download/pdf/267805876.pdf

[8] Harer, F., & Fill, H. A Comparison of Approaches for Visualizing Blockchains and Smart Contracts. doi:https://zenodo.org/record/2585575/files/Jusletter_IT_Blockchain_Visualization.pdf

[9] Hill, B., Chopra, S., & Valencourt, P. (2018). *Blockchain Quick Reference: A Guide to Exploring Decentralized Blockchain Application Development*. Birmingham: Packt Publishing.

[10] Holbrook, J. (2020). *Architecting Enterprise Blockchain Solutions*. Indianapolis, IN: Sybex.

[11] Hrabovska, M. (2018, August 28). Another Story of Designing for Blockchain. Retrieved February 4, 2021, from https://uxdesign.cc/another-story-of-designing-for-blockchain-b64c34407d2e

[12] HyperLedger. (2019). How Walmart Brought Unprecedented Transparency to the Food Supply Chain with Hyperledger Fabric. Retrieved from https://www.hyperledger.org/wp-content/uploads/2019/02/Hyperledger_CaseStudy_Walmart_Printable_V4.pdf

[13] Iredale, G. (2020, December 30). Blockchain vs Database: Understanding the Difference. https://101blockchains.com/blockchain-vs-database-the-difference/#:~:text=Blockchain%20vs%20Centralized%20Database%3A%20Authority%20and%20Control&text=Blockchain%20is%20designed%20to%20work,the%20next%20generation%20of%20technology

[14] Jang, H., Han S., & Kim, J. (2020). User Perspectives on Blockchain Technology: User-Centered Evaluation and Design Strategies for DApps. User Perspectives on Blockchain Technology: User-Centered Evaluation and Design Strategies for DApps. doi:10.1109/ACCESS.2020.3042822

[15] Kulkarni, K. (2018). Learn Bitcoin and Blockchain: Understanding Blockchain and Bitcoin Architecture to Build Decentralized Applications. In *Learn Bitcoin and Blockchain: Understanding Blockchain and Bitcoin Architecture to Build Decentralized Applications* (pp. 70–73). Birmingham: Packt Publishing

[16] Sheetal, M., & Venkatesh, K.a. (2018, April). Necessary Requirements for Blockchain Technology and Its Applications. *International Journal of Computing Science and Information Technology*. ISSN: 2278-9669.

[17] Nikolov, A. (2020, May). Design Principle: Consistency. Retrieved March 28, 2021, from https://uxdesign.cc/design-principle-consistency-6b0cf7e7339f

[18] Raj, R. (2021). What Is Blockchain Database? Difference Between Blockchain and Relational Database. IntelliPaat. https://intellipaat.com/blog/tutorial/blockchain-tutorial/blockchain-vs-database/

[19] Rengaiah, P. (2014, April 29). On Modular Architectures. Retrieved March 1, 2021, from https://medium.com/on-software-architecture/on-modular-architectures-53ec61f88ff4

[20] Sęk, J. (2021, February 22). Why Is Consistency so Important in Software Design? Retrieved March 25, 2021, from https://www.iteo.com/blog/post/why-is-consistency-so-important-in-software-design

[21] Singh, I., & Lee, S.-W. (2018, 25 January). Comparative Requirements Analysis for the Feasibility of Blockchain for Secure Cloud. In *Communications in Computer and Information Science*. South Korea: ResearchGate.

[22] Solomon, M. (2019). Ethereum for Dummies. Hoboken, NJ: John Wiley. doi:https://www.dummies.com/personal-finance/10-design-principles-for-distributed-blockchain-apps/

[23] Strv. (2018, July 25). 3 Limitations of Progressive Web Apps. Retrieved March 2, 2021, from https://medium.com/@strv/3-limitations-of-progressive-web-apps-98b5afca9338

[24] Udokwu, C., Anyanka, H., & Norta, A. (2020, June). Evaluation of Approaches for Designing and Developing Decentralized Applications. Retrieved from https://www.researchgate.net/publication/342381189_Evaluation_of_Approaches_for_Designing_and_Developing_Decentralized_Applications

[25] Xu, X., Weber, I., Staples, M., Zhu, L., Bosch, J., Bass, L., Pautasso, C., & Rimba, P. (2017). A Taxonomy of Blockchain-Based Systems for Architecture Design. 10.1109/ICSA.2017.33.

[26] Yalanska, M. (2020, December 28). User Experience: Insights into Consistency in Design. Retrieved April 1, 2021, from https://blog.tubikstudio.com/design-consistency/

9 Blockchain Application Development

Gagandeep Singh, Manpreet Kaur, Shival Kashyab and Sunil Kajla

ABSTRACT

Following the previous blockchain application design discussions, Chapter 9 focuses on a discussion of fundamental blockchain application development considerations and presents readers with some basic tools and techniques required during the development phase.

This chapter begins with a discussion of the necessity to get things right the first time, by explaining some reasons that blockchain application developers are well advised to adopt a firmware versus a software approach. Next, a few popular blockchain frameworks are presented and discussed. While an in-depth discussion of the Ethereum, Hyperledger and Corda frameworks was the sole focus of Chapter 7, this chapter also introduces two additional blockchain development frameworks, Quorum and Ripple.

The chapter discussions continue with an explanation of the main layers of a typical blockchain application, namely, the application, service, semantic, network and infrastructure layers. Readers are also provided with an introduction to the various blockchain application tools and techniques, as well as a tool download link table for easy access. Finally, the use of an Integrated Development Environment is introduced and discussed.

The final section of the chapter presents and discusses several blockchain application development good practices. While this chapter is not meant to provide advanced training for blockchain application professionals, novice blockchain enthusiasts and information systems auditors should find its content both interesting and useful.

FUNDAMENTAL DIFFERENCES BETWEEN SOFTWARE AND FIRMWARE

A list of data or machine instructions that tells a computer how to function is referred to as computer software. Software is designed to complete a specific mission such as running, extending or monitoring the hardware. Examples of machine software include various computer operating systems, compilers and interpreters and device drivers and assemblers. Another type of software is application software developed to satisfy the needs of particular users, such as human resource or medical management systems within an organizational setting [44].

Firmware is a form of computer software that provides low-level control over a device's specific hardware. Firmware can perform all monitoring, control and data manipulation functions as a complete operating system for less complex devices, or as a standard operating system for more complex applications. A few examples of firmware-enabled devices are home appliances, embedded software, computers and their accessories or TV remote controls.

Software is generated using both low-level languages and high-level languages such as Java, C++, JavaScript or Python, whereas only low-level languages are used for developing firmware. Also, development of software applications takes longer than creating firmware as software size

DOI: 10.1201/9781003211723-9

TABLE 9.1
Difference in Application Development between Software and Firmware

Differences	Firmware	Software
Function	Controls hardware device [26].	Tells the computer what to do [44].
Updating	Difficult to update [26].	Easy to update [26].
Language	Written in low-level language [18].	Written in high-level and low-level language [18].
Level of Control	Offer low-level control to hardware devices [44].	Controls the applications of the computer [44].
Examples	Consumer appliances, TV remote, computer peripherals and embedded systems [44].	Interpreters, operating system, Microsoft Office applications [44].

very often will exceed several gigabytes (Gb), while firmware size is much smaller as it does not contain the data type needed for graphics and user interfaces [18].

Storage for firmware is non-volatile in nature (ROM, EPROM or flash memory), whereas application software runs from volatile and virtual memory [21]. Applications developed using firmware are difficult to update, as the task necessitates circuitry replacement or reprogramming through a special procedure that cannot be managed at the user end. Software-developed applications, on the other hand, are much simpler and can be changed frequently with the application of update patches. Since firmware is used to monitor hardware, it primarily deals with low-level functions. Software is used to manage hardware as well as meet a variety of user needs as software has to do with both low-level and high-level functionality [26].

Another significant distinction between firmware and device software is the frequency of updates. As mentioned, most devices' firmware is not designed to be modified by the consumer. As such, device manufacturers typically restrict users' access to the firmware entirely. This is a required safety feature because when a consumer tampers with the firmware of these devices, correspondingly there is a probability of ending up with a non-functional system [21]. While a device's firmware may be updated, caution should be exercised, as this task is not as common and easily accomplished as building and applying an update patch to a software application [44].

The key differences between software and hardware application development are described in Table 9.1.

An important application development consideration for organizations, when hiring blockchain application development staff, is to strive to hire application development professionals with experience in both software and firmware development. This is due to the fact that certain aspects of blockchain application development – such as smart contract development – is akin to firmware development. To further explain, and as previously discussed in Chapter 2, once a smart contract is constructed and activated on a blockchain, it cannot simply be updated with a performance or security patch, as is often the case with most software applications. As such, a great level of care and diligence must be exercised during the blockchain application development cycle to help ensure that all components of a blockchain application are working adequately.

POPULAR APPLICATION DEVELOPMENT FRAMEWORKS

Blockchain is a technology still in its infancy, offering tremendous use case potential. Interest in blockchain-based applications is growing across the board; therefore, there is a need to select a blockchain system that meets the desired or mandate requirements [11]. A blockchain application framework is a GUI (graphical user interface) or CLI (command line interface) software that enables easy development and deployment of blockchain applications with less interaction with programming language. All major blockchain application frameworks consist of development libraries, network configurations and consensus protocols. Frameworks can be either permissioned or permissionless based on the subjective evaluation of ease of prototyping. A number of blockchain systems such as Ethereum, Hyperledger, R3, Ripple and EOS could be effectively used by organizations in order to create and host blockchain applications [40, 42].

Blockchain frameworks offer different purposes, features and functionalities. Choosing an appropriate framework is largely based on an organization's objectives and priorities. Some criteria for choosing a framework include the type of blockchain needed (such as public or private), the programming language supported by the framework, desired consensus methods and smart contracts uses [1]. Several popular blockchain frameworks along, with their merits and limitations, are explained further. Readers are also reminded that Chapter 7 provides a granular discussion of three highly popular blockchain frameworks at the time of this writing, namely, Ethereum, Hyperledger technologies and Corda.

Ethereum

Ethereum is a permissionless blockchain platform that is open source. In 2013, the Ethereum framework was proposed by a Russian-Canadian programmer, Vitalik Buterin. The project was subsequently crowdfunded in 2014, with its first version going live in July 2015. Ethereum has its own native cryptocurrency, called Ether. If an organization's solution requires a public platform, Ethereum is an excellent blockchain framework to consider. The Ethereum platform supports the proof of work consensus algorithm at time of this writing, with plans to switch the Ethereum network to a proof of stake consensus algorithm in the near future. The Ethereum Virtual Machine (EVM) is a decentralized virtual machine that can run scripts over an international network of public nodes. Ethereum gives programmers the ability to build and distribute next-generation distributed applications [17].

Hyperledger

The Linux Foundation in San Francisco, California, established the Hyperledger project in December 2015. It started out with 30 companies and has since grown to more than 120 sponsoring entities. Hyperledger is an open-source project dedicated to creating a secure set of protocols, software and libraries for enterprise-grade blockchain deployments. Hyperledger does not support cryptocurrencies, such as Bitcoin, but it does include the infrastructure and standards required for the development of various blockchain-based systems and applications for industrial usage. Hyperledger frameworks offer the following architectural layers and components [13]:

Consensus layer: This layer is responsible for reaching an agreement on the order and verifying the accuracy of the block's transactions.

Smart contract layer: In charge of handling transaction requests and only approving legitimate transactions.

Contact layer: The peer-to-peer message transport is handled by this layer.

Identity management services: These services are needed for maintaining and validating user and system identities, as well as establishing trust on the blockchain.

Application programming interface (API): External applications and clients can interact with the blockchain using the APIs.

Currently, the more widely used Hyperledger frameworks include *Sawtooth*, *Iroha*, *Fabric* and *Burrows* as further explained below [11].

Hyperledger Sawtooth: An Intel-sponsored blockchain suite based on the proof of elapsed time (PoET) consensus algorithm and a modular architecture. For each node in the blockchain network, the algorithm produces a random wait time, meaning that each node must 'go to sleep' or remain inactive for that assigned time length.

Hyperledger Iroha: This flavor of Hyperledger was made possible by the collaboration of several Japanese companies. Iroha is a mobile-friendly blockchain platform that is simple to integrate into an existing solution that has been developed.

Hyperledger Fabric: This is a well-known Hyperledger architecture developed by IBM. Fabric is an excellent choice for building highly scalable applications, considered as the most often used Hyperledger. Some of the advantages of Fabric include its highly modular and thus scalable nature, private channel communication capabilities that allow users to communicate privately, and the support of a committed and supportive community that includes Intel and IBM. Fabric allows developers to build applications efficiently with clean code and a large suite of ready-made solutions and resources.

Corda

R3, the world's largest banking institution, created Corda, an open-source blockchain technology. Corda works as a specialized distributed ledger customized for the financial industry needs. It works in a private blockchain operation mode which means permission is needed to access the network content. Corda provides enhanced privacy and granular access control to digital records. Furthermore, Corda also allows businesses to interact directly via smart contracts, thereby greatly reducing or even eliminating expensive transactional expenses and fees. Unlike Ethereum, and similar to Hyperledger, Corda does not have a built-in cryptocurrency or token. Although primarily built for banking and financial services potential use cases, Corda is now being used in a variety of other industries, including healthcare and supply chains [42].

Ripple

Ripple is an open-source, peer-to-peer framework that works in a decentralized environment to accommodate money transfers in the form of Bitcoin, Euro, USD, etc. Ripple was first developed in 2012 and initially named *Open Coin*. Ripple also offers its own cryptocurrency called XRP. Advantages of the Ripple framework include fast payment settlements taking only minutes versus days, compared some other traditional settlement systems taking days. The Ripple protocol is also automated to select the cheapest bidder for its transactions. Furthermore, since Ripple consumes XRP to send transactions, a spammer will need to spend hefty amounts of XRP to generate spam transactions in bulk [22].

Quorum

An Ethereum-inspired blockchain framework, Quorum is a permissioned-based decentralized platform used to deploy decentralized apps (DApps). Developed by JPMorgan, Quorum provides a blockchain solution choice for the financial sector that offers high speed – currently up to almost 100 transactions per second – and high throughput while maintaining the privacy of the participants [39]. Quorum uses three consensus algorithms: *Raft*, *Istanbul Byzantine Fault Tolerance* and *Quorum Chain*. These three algorithms are lightweight as compared to the Ethereum consensus algorithm, which makes it faster than the Ethereum blockchain [2]

Table 9.2 provides readers with a bird's-eye view of the different features and attributes of the frameworks discussed in this section.

LAYERS OF BLOCKCHAIN APPLICATION

Blockchain platforms contain a layered data-sharing architecture. The architecture is based on functional overviews, design principles and layered collaboration. The design principles include features such as high security, high scalability, decentralization and high availability [40].

TABLE 9.2
Blockchain Platforms Comparison

	Ledger Type	Transaction Speed	Consensus Algorithm	Smart Contract	Industry Focus
Ethereum 2.0 [10]	Permissionless	Up to 10,000 TPS	Proof of stake	Yes	Cross-industry
Hyperledger Fabric [2]	Permissioned	>2000 TPS	Solo, Kafka, Raft	Yes	Cross-industry
Hyperledger Sawtooth [2]	Permissioned/ permissionless	>1000 TPS	PBFT, POET, Raft	Yes	Cross-industry
Hyperledger Iroha [2]	Permissioned	>=1000 TPS	YAC Algorithm	Yes, but pre-defined	Cross-industry
Corda [2]	Permissioned	>170 TPS	Pluggable Consensus	Yes	Financial/ cross-industry
Ripple [2]	Permissioned	>1500 TPS	Probabilistic Voting	Yes	Financial-industry
Quorum [2]	Permissioned	>100 TPS	Raft, Istanbul, BFT	Yes	Cross-industry

Blockchain as a software system is made up of two layers, which include the application layer and the implementation layer. While the application layer oversees the user-facing components, the implementation layer contains the protocols and codes, or basically anything that brings the application to life. These two major layers in a blockchain may contain other minor layers. Layer zero to layer one of the blockchain should contain the physical infrastructure necessary to support cryptocurrencies. On the other hand, layer two and beyond are the second implementation layers that may include the lightning network layer capable of processing a large number of transactions very quickly (with lightning speed) [5].

There are five layers of blockchain technology as follows:

The application layer: This first layer includes DApps (decentralized applications), a DApp browser, a user interface and app hosting. Users can navigate decentralized applications via the DApp browser. As a result, one can see a completely different user interface. This layer's application hosting allows the user to run all the decentralized applications. No DApps can go live on the Internet without this component. The hosting protocol, of course, will be fully decentralized. Furthermore, running these hosting servers entails minimal risk.

The service layer: This layer comes after the application layer in order to provide access to all the resources the user needs to create and run the DApps layer. This layer accommodates various governance implementations, off chain computing, state channels, supporting side chains and data feeds.

The semantic layer: This next layer contains participation requirements, virtual machines and a consensus algorithm, a much needed procedure in which all the nodes agree on the details of the ledger. Participation requirements are the primary rules that assist the network in determining who is eligible to enter the system and who is not. Furthermore, this feature is primarily for private blockchain technologies. Virtual machines have a safe and efficient platform for all network activities.

The network layer: This layer is the next layer after the semantic layer. It includes a trusted execution environment (TEE) in it that helps the architecture in addressing scalability concerns. It not only aids the network in overcoming this issue, but also improves its security. TEE also helps to store data away from the main network. The network layer also contains RLPX; a network suite that facilitates data transfer between two peers. RLPX provides a user interface to aid communication within the blockchain network. Block delivery networks are a type of network that will send web content or pages if a user requests it. In fact, the same can be seen in the typical Internet architecture. Network layer allows the user to follow the *Roll Your Own Mechanism* protocol that can be customized whenever needed to make the protocols more adaptable.

The infrastructure layer: This is the final layer containing features, such as tokens, that help maintain the ecosystem and decentralized storage of the network in a much more secure manner [1].

TOOLS FOR BLOCKCHAIN APPLICATIONS

The primary purpose of blockchain tools is to resolve the numerous requirements that arise during the creation of a blockchain app at various stages with relative ease and efficiency. Such tools also help to improve awareness and hands-on experience with the technology. Blockchain professionals with in-depth knowledge and experience of the following blockchain tools are currently in high demand in the blockchain job market [25]. The following is a brief explanation of some of the major blockchain development tools and techniques:

Solidity: This is an Ethereum-based programming language proposed by British software developer Gavin Wood in 2014. Solidity is a high-level object-oriented language for creating smart contracts. The main purpose behind the development of Solidity was the need to create a programming language that can be easily learned. As such, Solidity is based on common programming concepts such as variables, string manipulation, classes, arithmetic operations and functions. Moreover, it draws from high-level programming languages such as C++, Python, PowerShell and JavaScript. Solidity also has a compiler that breaks down high-level code into simple instructions [6].

Truffle: This tool was created with the objective of providing a streamlined and efficient platform for developing and testing blockchain smart contracts. It uses Ethereum Virtual Machine to create a highly useful development environment and testing framework. The truffle suite is comprised of *Truffle*, *Ganache* and *Drizzle* [37]. Truffle also comes with a large library of custom deployments that aid in the creation of new smart contracts and the resolution of blockchain development issues [25]. Truffle is also well-suited to the development of sophisticated Ethereum decentralized applications [38]. Automated contract checking is another key feature of Truffle as a blockchain tool. For automated contract checking, Truffle may use *Mocha* and *Chai*. Furthermore, Truffle will assist in the creation of smart contracts, as well as their linking, compilation and deployment. Most significantly, Truffle includes a customizable-build pipeline that ensures that specific development methods are followed [25].

Ganache: This is a Truffle suite blockchain tool that allows developers to build their own private Ethereum networks for checking decentralized apps, inspecting states and executing commands while maintaining complete control over the chain's activity [25]. With this tool, a user interacts with a smart contract using both graphical and CLI modes [46]. JavaScript is used in the implementation of the Ethereum blockchains [3]. Ganache can be downloaded from the link provided in Table 9.3. The website automatically detects the installed OS on the machine and redirects the user to the appropriate binary installation [4]. Due to advanced mining controls and its built-in block explorer, Ganache is an often-used blockchain tool by developers. During the creation of smart contracts, blockchain developers also use Ganache to test them [25]. The key advantage of using this platform is that it allows users to monitor mining speed and gas costs to test various smart contract scenarios [38].

ethers.js: The ethers.js library aspires to be a comprehensive and lightweight interface to the Ethereum blockchain and its ecosystem. This is a client-side GUI that lets users build wallets, communicate with smart contracts, and do a lot more [38]. The size of this tool is quite small (~88 kb compressed; 284 kb uncompressed) and is ready with fully TypeScript containing definition files and full TypeScript source. By using ether.js, JSON wallets (*Geth*, *Parity* and *Crowdsale*), BIP 39 mnemonic phrases (12-word backup phrases) and HD wallets (English, Italian, Japanese, Korean, Chinese, etc.) can be imported and exported. *JSON-RPC*, *INFURA*, *Etherscan*, *Alchemy*, *Cloudflare*, or *MetaMask* are used to connect to *Ethereumxdx* nodes. ethers.js performs function such as protecting client private keys. It provides the complete functionality needed for Ethereum and contains extensive user documentation [16].

MetaMask: This is a web browser plugin that allows user account management, blockchain connection and blockchain application. In account management, MetaMask helps the user to store their public and private keys. For blockchain connections, metaMask is used to connect to other Ethereum nodes. The main advantage of MetaMask is that a user can interact with a smart contract without being an Ethereum node. This means that the user need not download the full blockchain data on the local system. For blockchain application MetaMask has injected web3(JavaScript API) that helps the user to handle JSON RPC text (XML formatted). For every transaction, MetaMask displays the transaction content for the user's confirmation [24].

Infura: In order to start working with an Ethereum blockchain, the developer needs to make a system as an Ethereum node. Based on the data size, there are three types of nodes the developer could configure on the machine, namely, a full, light or archive node. Running one's own node is somewhat complicated in terms of data and power management. The alternative is a third-party API provider, such as *Infura* or *Alchemy*. Infura is an application that maintains Ethereum blockchain nodes at their end, not on the client machine. In the case of Geth, the client needs to install an Ethereum blockchain and keep it running on the machine to sync the blockchain ledger. To use Infura, the user needs to sign up on https://infura.io in order to be provided with a link and unique token used to connect to the blockchain [3].

Mist: This is an Ethereum wallet recognized as the official wallet for ETH. Mist is a browser-based application. Mist is developed and managed by the Ethereum team and still in its Beta version as the time of this writing. A client-side script such as JavaScript runs on the browser to fetch various APIs used to communicate with Geth. Mist's main goal is to create a third-generation web (Web 3.0) that eliminates the need for servers by leveraging Ethereum, *Whisper* and *Swarm* as alternatives to centralized servers [34].

Geth: In order to interact with an Ethereum blockchain, a client is needed. Transaction broadcasting, mining and other tasks are handled by the client. There are two popular clients available for Ethereum; namely, *Geth* and *Parity*. Geth, short for *Go Ethereum*, is an Ethereum blockchain client that allows a user to establish a peer-to-peer channel with other clients, mining, exploring, deploying and interacting with smart contracts. Geth is a command-line application written in the go-programming language. The Geth download package supports all the major operating systems [4, 34].

Decentralized applications: DApps are digital applications or systems that operate and run on a blockchain or peer-to-peer network of computers rather than a single independent device or system. A typical web app, such as Uber or Twitter, operates on a computer system that is owned and controlled by a company, giving it complete control over the app's functionality. In other words, there might be several users, but the backend is managed by a single entity. DApps can run on either a peer-to-peer (P2P) or a blockchain network. *BitTorrent*, *Tor* and *Popcorn Time* are software that operate on computers that are part of a peer-to-peer (P2P) network, where multiple users are consuming, or seeding content, or performing both functions at the same time. DApps operate in a public, open, decentralized world with no restrictions when functioning with cryptocurrency. For example, a developer might build a Twitter-like DApp and deploy it on a blockchain, allowing any user to post messages. No one, including the app's developers, can remove the messages after they have been uploaded [14].

IPFS: This blockchain-compatible storage system started as a project by Protocol Labs in 2015 to create a framework that could radically alter the way information is transmitted across the world and pave the way for a distributed, more robust network. IPFS has evolved to serve a variety of use cases and is developing knowledge processing for a variety of industries, ranging from disintermediation in the music industry to agribusiness weather risk security. IPFS is a distributed storage and access mechanism for files, websites, programs and data. It is transport layer agnostic, which means it can communicate using a variety of transport layers, including TCP, UTP, UDT, QUIC, TOR and even Bluetooth. IPFS has rules and protocols that govern the movement of data and content through the network. An example is *Kademlia*, a peer-to-peer distributed hash table (DHT) popularized by its use in the BitTorrent protocol [20].

Remix: An open-source mobile and desktop framework, Remix promotes a quick development cycle that includes many plugins with user-friendly interfaces. Remix is utilized for a contract creation process as well as a learning and teaching environment for Ethereum. The Remix IDE is part of the Remix Project, comprised of a forum for plugin-based development tools. Remix Plugin Engine, Remix Libs, and, of course, Remix-IDE are among the sub-projects. Remix IDE is a popular open-source platform for writing Solidity contracts directly in the browser. It is coded in JavaScript and could be used in the browser or on a desktop. The Remix IDE includes modules for smart contract training, debugging and deployment, among other things [36].

Oracles: These off-chain data repositories act as a link between blockchains and the rest of the world. Off-chain data are inaccessible to blockchains and smart contracts; however, to carry out certain contractual arrangements, it is essential to have relevant details from the outside world. This is where blockchain oracles come in, as they provide connections between off-chain and on-chain information. Oracles are important in the blockchain ecosystem because they expand the range of smart contracts that can be used. Often, smart contracts would be useless without the updated and independent data provided by oracles. A blockchain oracle is the layer that queries, verifies and authenticates external data sources before relaying the information. Price details, the effective completion of a payment or the temperature measured by a sensor are all examples of data transmitted by oracles. The smart contract must be invoked, and network resources must be expended, in order to call data from the outside world. Some oracles may also send information back to external sources, in addition to relaying it to smart contracts.

Tools make the process of developing blockchain apps easier and aid in the creation of software applications. Thus, they contribute to the expansion of our blockchain awareness [30, 38].

TABLE 9.3
Tools Download Source Links

Tool	Source
Solidity	https://docs.soliditylang.org/en/v0.7.4/
Truffle	https://truffleframework.com/
Oracle	https://developer.oracle.com/ca-en/open-source/
Remix	https://remix.ethereum.org
IPFS	https://ipfs.io/
Ganache	https://www.trufflesuite.com/ganache
Mist	https://github.com/ethereum/mist/releases
Geth	https://github.com/ethereum/go-ethereum
MetaMask	https://metamask.io/
Ether.js	https://github.com/ethers-io/ethers.js/
Infura	Https://infura.io

INTEGRATED DEVELOPMENT ENVIRONMENT

According to Lyn, Integrated Development Environment (IDE) is an application software providing application developers a single platform to design, test and debug computer programs. The process of eradicating the errors or bugs is known as debugging and is achieved with the help of debugging tools. Debugging creates an environment which makes it easy to create or develop eminent software applications. Text editor, terminal, debugging tools and code snippets are the tools generally incorporated in the Integrated Development Environment [29].

IDE's may even make use of either compilation or interpreter tools, or both, thus providing an atmosphere that is conducive to software growth. The modern IDEs come with object explorer, class browser and class hierarchy diagram features [9].

Some of the IDE's work on JavaScript or Python platforms [45]. IDE is a single graphical user interface with debugger, local build automation and source code editor as its pillar. For developer's use, local build automation uses simple redundant tasks. With extensive capabilities such as syntax highlighting, the source code aids in the creation of software code [19].

IDE tools increase the productivity and capability to perform tasks such as continuous compilation, automated testing, integrated debugging, refactoring and navigation among classes. Eclipse and visual studio also work with IDE. While using visual basics, IDE developers can set breakpoints or clear breakpoints just next to a line of code. Furthermore, in the immediate window at the bottom of the IDE, the developer can use visual basics statements. Hence, the focus of an IDE is to provide comprehensive facilities to software programmers in software application development [41].

GOOD PRACTICES FOR BLOCKCHAIN DEVELOPMENT

Use caution calling external contracts, favor own: It is strongly advisable to favor in-house contracts over those created by a third party, as much as possible. In other words, all contracts developed in-house have already been subjected to the same level of security, review, testing and acceptance criteria as the rest of the solution, so they will comply with the same quality and safety standards set for the rest of solution components. Adequate audit and testing time and effort in order to carefully review the source code of any external contracts should be allocated. When using a third-party contract, it is essential to ensure there are no inherent security flaws created by the third party.

Handle all errors; catch all exceptions in external calls: In smart contracts, functions must be surrounded by try/catch statements and every exception must be handled explicitly, no matter how unlikely it appears to be. All exceptions that aren't explicitly caught and managed should be thrown, and exception should be raised to its calling function or contract. Contracts should try to gracefully address exceptions and attempt to move forward without user intervention. If this is not possible, an error message should be sent with suggestions regarding how to resolve the error. As mentioned, when dealing with external contracts, there should be no assumption that another developer has taken the same precautions. Users will view an application as having a defect if they receive any unexpected or cryptic errors [8].

Push versus pull payments: A push-based financial transaction happens when a seller uses an automated transaction to electronically remove or add funds from or to a customer's account. Subscription or utilities payment plans that perform automatic monthly or annual debits are an example of a push-based financial transaction. In a blockchain environment, pull payments offer an important benefit. To further explain, a pull payment should divide the payment process into two phases. The first step should be a notification informing the user that there are available funds waiting to be pulled from a smart contract. In the second step the payee should act on the notification in order to pull the available funds out of the payee's correct wallet address. This methodology will help avoid issues with funds being sent to an incorrect wallet

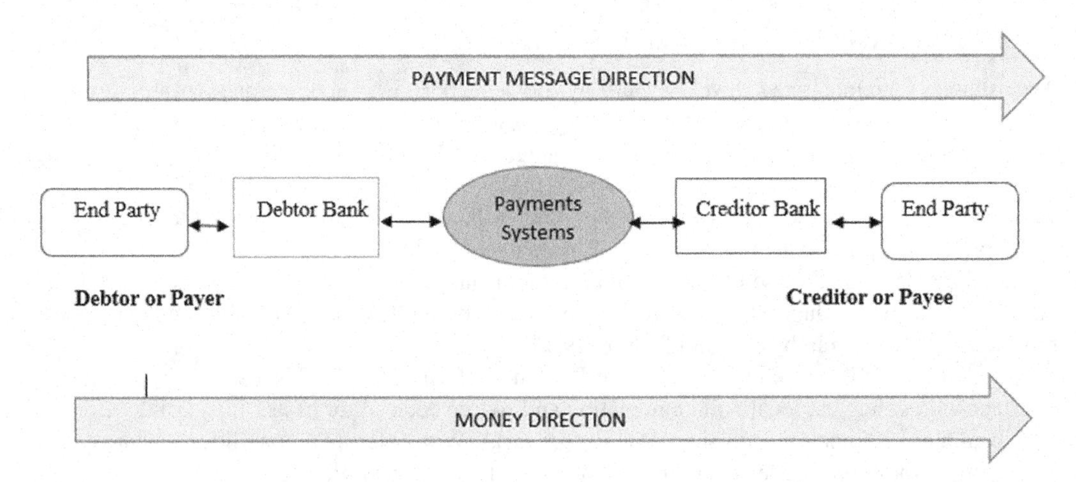

FIGURE 9.1 A push payment transaction.

Source: [32].

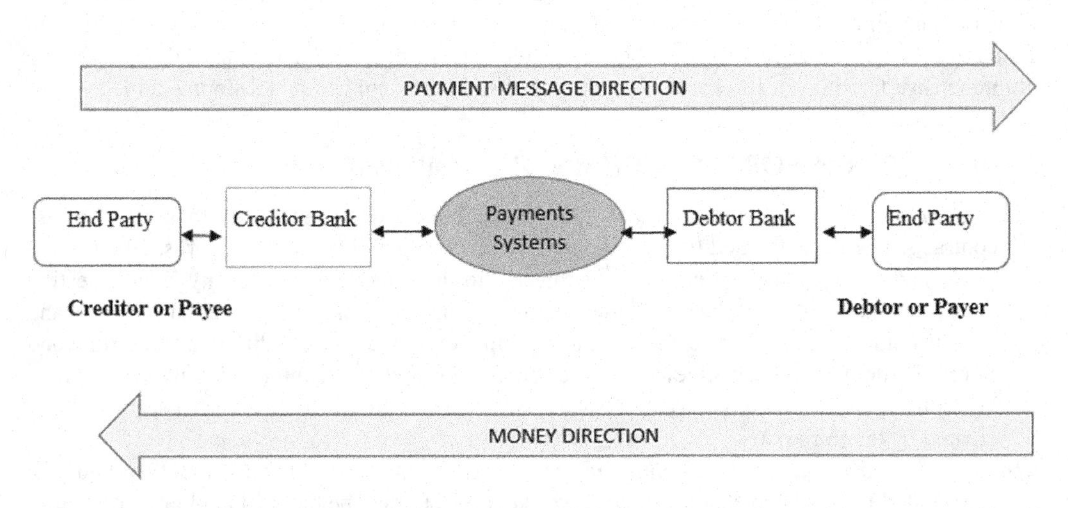

FIGURE 9.2 A pull payment transaction.

Source: [32].

address which cannot be reversed, unlike traditional banking transactions [7, 12, 32, 35, 43] (Figures 9.1 and 9.2).

On-chain public data: A common example of on-chain transactions is cryptocurrency transactions that take place on the blockchain and are validated by the state of the blockchain. These offer high security and transparency. Off-chain public transactions are typically better protected as compared to the on-chain public transactions as more granular transaction information is not exposed publicly. Furthermore, off-chain transactions are typically less expensive and faster. The on-chain public data have high availability, but the loss of privacy and confidentiality are trade-offs, since any node can freely participate in the network at any time. Other disadvantages of using a public blockchain include low transaction throughput and storage scalability, as well as processing fees. In order to better protect on-chain public data, an auxiliary chain could be used within the on-chain configuration to construct a storage hierarchy based on performance, cost and privacy trade-offs [23, 31, 33].

Start with local testing to test networks, then to production: Developers need to make sure to perform extensive tests on a new blockchain in a local development before pushing them to a test or production blockchain. This is especially important when using an Ethereum public blockchain such as Ganache (TestRPC). Due to the nature of the blockchain, all records on the blockchain are permanent (add-only ledger), and smart contracts are permanently stored on the blockchain, so even outdated and deactivated contracts remain on the network. Testing software locally before testing it on external test networks like Rinkeby or Ropsten is a good practice. Once deployed on a production blockchain, costs start escalating quickly, therefore thorough testing is necessary. Embedding blockchains in production requires some real costs, so developers need to be sure to run plenty of tests first to prevent unnecessary expenditures [8]. Chapter 10 discusses testing procedures and methodologies in much more detail.

Agile approach during pre-release and after releasing to production: The agile methodology is often best for small, self-organizing teams that are likely co-located and working on projects with changing requirements. Agile methods are thought to be capable of delivering frequently and quickly, as expected by DApp projects. The question to consider is whether agile approaches can still be used when designing a blockchain application. The nature of smart contracts and the blockchain provides an answer to this question. Since every smart contract creates a permanent record on the blockchain that cannot be deleted once it is activated, it is not suggested to maintain an agile cycle of continuous release and development once a contract is put to production since killing or calling a contract's self-destruct function (if implemented) entails major consequences [7, 8, 28].

Keep complexity in the middle layers: Developers should strive to keep the bottom and top layers of a new blockchain application as basic as feasible. As such, the new blockchain's user interface and the smart contract layer should be kept as simple as possible. Business logic, the more complex part of the application, should be implemented in the middle layers [8, 15].

SUMMARY AND CONCLUSIONS

This chapter discussed a number of blockchain tools and techniques designed to help familiarize information and internal auditors with a new blockchain's application development process. Similar to an onion, blockchain architecture is made up of several layers. Using an Integrated Development Environment (IDE) is beneficial as a user interface for writing code, organizing text groupings and automating programming redundancies is provided by this environment.

A number of recommended application development concepts have been discussed in this chapter. These include favoring pull payments, following a 'sandwich complexity model' during the development process, favoring in-house contracts, to name a few. These concepts will help in creating solutions that are free of unnecessary security vulnerabilities.

CORE CONCEPTS

- Blockchain technology has grown tremendously in popularity in the last few years. Many people find it difficult to differentiate between blockchain and database because of the many striking similarities. Blockchain is a form of database. The opposite is however not the case.
- Blockchain platforms help to develop a blockchain-based application which can be either permissioned or permissionless. Ethereum, Hyperledger, R3 and Ripple are a few of the built frameworks that allow people to develop and host applications on the blockchain.

- The primary aim of blockchain tools is to resolve the numerous requirements that occur during the development of a blockchain app at various stages. The function of blockchain tools is to make the process of designing blockchain software solutions easier.
- An Integrated Development Environment (IDE) is a software application that facilitates software developers by allowing them to design, build, test and debug computer programs within one environment. Some of the tools that may be included in IDE include debugging tools, text editor, terminal and code snippets.
- In the context of developing a blockchain solution, many question whether agile methodologies can still be applied. The answer lies in the fact that blockchain development should primarily be treated the same way as a firmware (versus a software) development.
- Set up a blockchain payment system that works on a pull versus push basis. A pull payment splits the payment process into two halves, giving the contract two places to check that the proper user is withdrawing the funds and allowing the user to give a new wallet address if the previous one is no longer valid or has been compromised. As a result, one will have a more secure environment.

ACTIVITY FOR BETTER UNDERSTANDING

This publication has primarily focused on three currently popular blockchain development frameworks, namely, Ethereum, Hyperledger technologies and Corda. This learning activity provides readers with an opportunity to also learn about two additional blockchain frameworks: Ripple and Quorum. Please refer to the links provided in Table 9.4 to learn more about the Ripple and Quorum blockchains.

Based on the material discussed in this chapter and presented in the links above, consider the following questions:

- Thinking about your organization's scope of operations, what framework seems to be most well suited to your enterprise? Ethereum, Hyperledger or Corda?
- In what way(s) can use of the Ripple or Quorum framework be a better choice for blockchain application developments for your organization?
- What issues and/or considerations make Ripple or Quorum an unsuitable choice for application development, based on your organization's operations and/or long-term objectives?
- In what ways is Ripple superior to traditional payment systems? Do you think it will replace traditional cross-border payments in the future?

Related guides for further reading:

1. Ethereum vs Hyperledger
   https://www.youtube.com/watch?v=tjrvWvX4diA
2. What is RIPPLE? How does it work?
   https://www.youtube.com/watch?v=UmaWDpg4fMU
3. What is Quorum Blockchain?
   https://www.youtube.com/watch?v=X9I9RMhW5H4

TABLE 9.4
Ripple and Quorum Blockchain

Ripple	https://cointelegraph.com/ripple-101/what-is-ripple
Quorum	https://blockgeeks.com/guides/quorum-a-blockchain-platform-for-the-enterprise/

REFERENCES

[1] Anwar, H. (2019). *Blockchain Technology Explained: A Decentralized Ecosystem.* 101 Blockchains. https://101blockchains.com/blockchain-technology-explained/

[2] Anwar, H. (2020). *Best Blockchain Platforms for Enterprise* https://101blockchains.com/best-blockchain-platforms/

[3] Bashir, I. (2018). *Mastering Blockchain – Second Edition.* Packt Publishing.

[4] Bashir, I. (2018). *Mastering Blockchain: Distributed Ledger Technology, Decentralization and Smart Contracts Explained.* Birmingham: Packt Publishing.

[5] Beedham, M. (2019). *Understanding Blockchain Is Way Easier if You Think of It as an Onion.* HardFork. https://thenextweb.com/hardfork/2019/01/24/understanding-blockchain-easier-onion-layer/

[6] Bernstein, C. (2020). *Solidity. Tech Target.* https://whatis.techtarget.com/definition/Solidity

[7] Bitpay (2020). *Push, Don't Pull: How Blockchain Payments Can Reduce Fraud and Chargebacks.* https://landing.bitpay.com/rs/249-OMD-845/images/_Fraud_E-book_BitPay_small.pdf?utm_source=twitter&utm_medium=organic-social&utm_campaign=push

[8] Blockchain Solution Architect Training Book (2019). *Blockchain Training Alliance.* https://blockchaintrainingalliance.com/collections/blockchain-exam-prep-guides/products/cbsa-official-exam-study-guide

[9] Busbee, K. L. (2018). *Integrated Development Environment.* Rebus Community. https://press.rebus.community/programmingfundamentals/chapter/integrated-development-environment/

[10] Council EC (2020). *What Is Ethereum 2.0 and Why Does It Matter?* https://blog.eccouncil.org/what-is-ethereum-2-and-why-does-it-matter/#:~:text=Ethereum%202.0%20will%20have%20shard,only%2030%20transactions%20per%20second.

[11] Daria, R. (2020). *Top 6 Blockchain Frameworks to Build Your App.* Ruby Garage. https://rubygarage.org/blog/best-blockchain-frameworks

[12] Dunn Ed (2018, September 26). What You Need to Know about NLP Blockchain Push Transactions. https://medium.com/swagg-scientific/what-you-need-to-know-about-nlp-blockchain-push-transactions-309e33c5dd63

[13] Frankenfield, J. (2021). *Hyperledger.* Investopedia. https://www.investopedia.com/terms/h/hyperledger.asp

[14] Frankenfield, J. (2021, February 12). Decentralized Applications – DApps. Retrieved May 9, 2021, from https://www.investopedia.com/terms/d/decentralized-applications-dapps.asp

[15] Gitbook (2021, February). *Design rationale: Ethereum Builder's Guide.* Ethereum Builders. https://ethereumbuilders.gitbooks.io/guide/content/en/design_rationale.html

[16] Github (2021). *ether.js*, https://github.com/ethers-io

[17] Guru99.com (2021). Ethereum Tutorial for Beginners: What is, Meaning, History https://www.guru99.com/ethereum-tutorial.html

[18] Hammad, M. (2020). *Difference Between Software and Firmware* https://www.geeksforgeeks.org/difference-between-software-and-firmware/

[19] Hat, R. (2021). *What Is an IDE?* https://www.redhat.com/en/topics/middleware/what-is-ide

[20] IPFS: C. (2020, December 10). An Introduction to IPFS (Interplanetary File System). Retrieved May 9, 2021, from https://blog.infura.io/an-introduction-to-ipfs/

[21] Ivanova, A. (2017). *What Firmware Is and How It Differs from Software.* Make Tech Easier. https://www.maketecheasier.com/firmware-vs-software/

[22] James, F. (2018). *Ripple Quick Start Guide: Get Started with xrp and Develop Applications on Ripple's Blockchain.* Packt Publishing.

[23] Kenton, W. (2021, March 12). On Chain Transactions (Cryptocurrency). https://www.investopedia.com/terms/c/chain-transactions-cryptocurrency.asp

[24] Liao, C., Lin, H., & Yuan, S. (2020). Blockchain-Enabled Integrated Market Platform for Contract Production. *IEEE Access*, *8*, 211007–211027.

[25] Lichtigstein, A. (2020, September 2). List of 10 Best Blockchain Tools, https://101blockchains.com/best-blockchain-tools/

[26] Lithmee (2018). *Difference between Firmware and Software.* PEDIAA. https://pediaa.com/difference-between-firmware-and-software/

[27] Lopez, D. (2020). *A Multi-Layered Blockchain Framework for Smart Mobility Data-Markets.* Science Direct. https://www.sciencedirect.com/science/article/abs/pii/S0968090X19300361

[28] Marchesi, L., Marchesi, M., & Tonelli, R. (2020, December). ABCDE – Agile Block Chain DApp Engineering. https://www.researchgate.net/publication/348140766_ABCDE_-agile_block_chain_DApp_engineering
[29] Muldrow, L. (2020). *What Is an Integrated Development Environment?* Digital Ocean. https://www.digitalocean.com/community/tutorials/what-is-an-integrated-development-environment
[30] Mou, V. (2021, April 29). Blockchain Oracles Explained. Retrieved May 9, 2021, from https://academy.binance.com/en/articles/blockchain-oracles-explained
[31] Paik, H.-Y., Xiwei, X., Hmn Dilum, B., Sung, U. L., & Sin, K. L. (2019, December). Analysis of Data Management in Blockchain-Based Systems: From Architecture to Governance. https://www.researchgate.net/publication/338144930_Analysis_of_Data_Management_in_Blockchain-Based_Systems_From_Architecture_to_Governance
[32] Paul, J. (2017, July 11). Push and Pull Payment Transactions. https://www.paiementor.com/push-and-pull-payment-transactions/
[33] Pinto, R. (2019, September 6). On-Chain versus Off-Chain: The Perpetual Blockchain Governance Debate. https://www.forbes.com/sites/forbestechcouncil/2019/09/06/on-chain-versus-off-chain-the-perpetual-blockchain-governance-debate/?sh=56768d0d1f5e
[34] Prusty, N. (2017). *Building Blockchain Projects: Building Decentralized Blockchain Applications with Ethereum and Solidity*. Birmingham: Packt Publishing.
[35] Pymnts.com. (2019, September 26). Deep Dive: The Benefits and Challenges of Real-Time Push Payments. https://www.pymnts.com/news/faster-payments/2019/benefits-challenges-real-time-push-payments-pull/
[36] Remix: Welcome to Remix's Documentation! (n.d.). Retrieved May 9, 2021, from https://remix-ide.readthedocs.io/en/latest/
[37] Sahu, M. (2020). *What Is Truffle Suite? Features, How to Install, How to Run Smart Contracts*. upGrad Blog. https://www.upgrad.com/blog/what-is-truffle-suite/
[38] Sharma, T. K. (2018). Top 10 Blockchain tools in 2020 to get your hands on, https://www.blockchain-council.org/blockchain/top-10-blockchain-tools-in-2020-to-get-your-hands-on/
[39] Sharma, T. K. (2019, May 15). *Quorum vs Hyperledger: The Ultimate Guide*. Blockchain Council, https://www.blockchain-council.org/hyperledger/quorum-vs-hyperledger-the-ultimate-guide/.
[40] Shen, M., Lieuhuang, Z., & Ke, X. (2020). *Layered Data Sharing Architecture with Blockchain*. Springer. https://link.springer.com/chapter/10.1007/978-981-15-5939-6_3
[41] Shepherd, D. (2015). *Integrated Development Environment*. Science Direct. https://www.sciencedirect.com/topics/computer-science/integrated-development-environment
[42] Takyar, A. (2021). Top Blockchain Platforms of 2021 https://www.leewayhertz.com/blockchain-platforms-for-top-blockchain-companies/
[43] The Finance Buff. (2012, December 5). Transfer Money by ACH Push, Not Pull. https://thefinancebuff.com/transfer-money-by-ach-push-not-pull.html
[44] Vera. (2020). *Firmware VS Software: What's the Difference Between Them?* Mini Tool. https://www.minitool.com/news/firmware-vs-software.html
[45] Walker, A. (2018). *What is an IDE (Integrated Development Environment)?* Learning Hub. https://learn.g2.com/ide
[46] Wu, X., & Sun, W. (2018). *Blockchain Quick Start Guide: A Beginner's Guide to Developing Enterprise-Grade Decentralized Applications*. Birmingham: Packt Publishing.

10 Testing and Auditing Blockchain Applications

Karen Akshatha Franklin, Philip Samuel Panneer Selvam and Samhitha Keshireddy

ABSTRACT

This chapter is the last in the design/develop/test blockchain applications trilogy. Assuming adherence to sound blockchain design and development principles, tools and best practices, it is important to adhere to strict and thorough testing procedures during both the application development and the final testing stages. The discussion in this chapter starts with a discussion of over 10 different blockchain application testing procedures. While it will most likely be very time consuming and expensive to employ all of the discussed testing methodologies, information systems auditors and blockchain testing professionals need to employ a number of testing procedures in order to make sure that the blockchain application is fully functional prior to full deployment.

A discussion of the various types of bugs and a four-step strategy for successful bug management is also provided in this chapter. A number of blockchain testing applications and bug analysis tools are recommended, along with their download links. In addition, the foundation of testing management principles and related concepts are also presented and discussed.

The chapter concludes with a comprehensive blockchain application audit checklist, based on a framework developed by KPMG, India. This checklist should be a valuable tool for information system auditors in planning and conducting blockchain specifics audits.

WHY IT IS CRITICAL TO TEST A BLOCKCHAIN APPLICATION

Blockchain applications do not exclusively exist in the digital domain of cryptocurrency. These applications require to be integrated with other software and systems in order to ensure that the blockchain solution provides access to real-world data and events like smart contracts. The immutability and distributed nature of the blockchain ledger, along with the inclusion of smart contracts, provides an attractive solution for enterprises. Considering how relatively recent the technology is, it does come with its own set of current drawbacks and vulnerabilities. Smart contract-based blockchain platforms have a lot to offer, but lack of testing, best practices and standards is what makes its architecture potentially vulnerable and exploitable by hackers. There are security vulnerabilities present in the technologies that are integrated with blockchain systems. Hence, all blockchain projects should be tested and evaluated for quality, compliance, governance and value at risk.

In 2016, the attack on Ethereum was announced by the Decentralized Autonomous Organization related to a vulnerability on the Ethereum platform being exploited by a hacker. The vulnerability was not present in a smart contract which caused a total loss of 150 million dollars. Such attacks on various blockchains prove what consequences could occur if the codes are not extensively tested for every potential threat [23].

DOI: 10.1201/9781003211723-10

To roll out the blockchain solution successfully, it is very important that trust between participating stakeholders in joining the blockchain ecosystem is present. To ensure this trust, the blockchain components are tested to ensure that they work correctly and also determine if any application integrated with the blockchain framework interacts in a trusted fashion.

Therefore, testing of a blockchain application is critical to provide software quality to all the various components of the blockchain; therefore, testers should employ a best practice-based suite of testing approaches [15].

POPULAR TESTING TOOLS FOR BLOCKCHAIN

Original blockchain developers who have access to the actual system code use a process called *emulation.* Emulation is used to test the overall performance of different private, public and consortium blockchains. There are many emulation tools that are being used to test different platforms. A few are explained below.

Ethereum Tester: An Ethereum-driven open-source library, which helps to evaluate the initial performance for customer-oriented projects. The tool also analyzes the operation of API, web integration, smart contracts and other blockchain components, such as DApps.

Truffle: The first component of the Ethereum-based blockchain tool evaluation known as truffle suite. Deployment of smart contracts is performed along with compilation and testing in the development environment of truffle.

Corda Testing Tool: An open-source environment designed specifically for R3 Corda which provides a wide range of functionality for unit testing and integration testing.

Exonum Test Kit: An open-source framework which uses *Rust* programming language for creating blockchains with emphasis on testing the logic of transactions. The test kit used for testing provides a Java binding tool since it is a part of the software development kit (SDK).

BitcoinJ: An open-source Java-based framework. Designed using Java which implements the Bitcoin protocol (BTC network) to interact with the Bitcoin network and to perform other testing activities. The protocol provides access to maintain a wallet, which allows the user to send/receive the transactions without the need for a local Bitcoin core copy.

Hyperledger Composer: An open-source development toolset and framework provided by the Linux Foundation that supports interactive testing, automated unit testing and system testing. Hyperledger composer uses JavaScript to implement the transactions of blockchain [21].

CHALLENGES IN TESTING BLOCKCHAIN IMPLEMENTATION

Testing as a process is effectively performed in the software development lifecycle, as there are possibilities of bugs passing through the application. To achieve the flow of transaction in a blockchain, all autonomous systems in the distributed network are required to be in collaboration and coordination.

There are several devices with their respective programming languages executing on different operating systems. Often, the number of systems is enormous and the test scenarios for the blockchain's distributed architecture could be a rigorous task. Therefore, there are possibilities for a tester to arrive at different test results that are correct for the same scenario. Considering the wide functionality, scalability and performance of blockchain makes it challenging to have a full test coverage.

In private and public blockchains, the most important aspect to be considered is the network security of the architecture. Though blockchains are often touted to be the most secure platform for transactions, security breaches are still possible. As such, it is important to monitor the nodes' connectivity in such distributed networks [8].

PHASES OF BLOCKCHAIN TESTING

To adhere to an effective test methodology, the following four critical phases are to be kept in consideration in order to achieve quality assurance for a blockchain application:

System initiation phase: Testers are highly advised to be involved at the earliest stage to understand the impacted components/modules in comparison to the modules which work seamlessly. This guides the testers to deploy the various test scenarios for impacted modules, in addition to performing regression testing in later stages.

Test design assurance phase: In regard to blockchain testing, the key design strategy should be considered based on block size, transactions, contracts and the end-point system peers on the network. Tester's will then need to develop and design test strategies based on these application requirements.

Test planning phase: This phase determines the detailed summary of the test cases and scenarios arrived for testing. Additionally, it helps in selecting the required testing tools and methodology for the test modules.

Test execution and result verification: The final phase is to articulate the test execution methods based on the approach outlined in the test plan [24].

TESTING MODELS: FUNCTIONAL TESTING

Functional testing concentrates on the core functionalities of the blockchain. This type of testing covers the entire network architecture and the nodes/peer system in order to acquire the desired validation and verification. The size of the block creates a number of checkpoints for the testers and each block's payload depends on the application.

Each block can be used for numerous transactions, hence it is important to consider the increase in block size and how this affects the transaction rate. In such scenarios, the chain created cannot be limited to a certain length; therefore adequate monitoring is greatly needed.

Transmission of data in P2P network with transparent data availability requires testing of the encryption and decryption processes for transaction integrity. The immutable nature of the blockchain demands the need for testing the transaction legitimacy [15]. The prominent strategies toward functional testing of a blockchain are listed below. Issues identified by the tester in one phase are expected to be corrected before moving to other phases, which also reduces redundancy of effort.

Unit testing: Each module requires to be tested individually based on the functions and structure of the application. The test coverage is to be considered based on the functions involved in the integration of modules.

Integration testing: Performed with the unit-tested modules altogether as a whole system. The next step is to perform the integration testing at which time the testers will be looking for errors with respect to missing functions and/or integration issues regarding module interfaces [13].

Regression testing: Used as a major testing methodology for over 20 years, the testing of distributed systems is enhanced through network emulation testbeds such as *DETERLAB* and *Emulab*. Using these emulation testbeds, distributed systems can deploy and test the programs repeatability [28].

API testing: APIs acting as a middleware platform are responsible for data transaction in and out of the blockchain. The blockchain atmosphere being ad hoc with respect to transactions from internal and external factors requires the APIs to be fully tested. The transactions occurring should be validated against the set of rules, which generates an update to be distributed to the blockchain network. API testing focuses on handling the request and response to transactions occurring on the blockchain. The created rules will trigger from external and internal environments to the middleware application on each transaction occurred [15].

Performance/load testing: It is important to test the performance and latency of a blockchain network for an application to be deployed to production. Key parameters that are involved in testing the performance include the network latency, transaction processing speed, block size and the sequence of transaction nodes. Testing the size of the network and its ability to process the transactions is necessary, as it helps in identifying performance bottlenecks before deployment and also estimate the cost of running the application on the desired environment [17]. Load testing on separate nodes is done to check if a particular node is capable to offload transactions to its peers, versus testing all nodes by overloading the network entirely. By doing so, latency, throughput and capacity of all the combinations of nodes can be evaluated [26].

Smart contract testing: Smart contracts are a set of computer rules or transaction protocols that are automatically executed when certain conditions are met on the blockchain. Hence, it is crucial that smart contracts are rigorously tested to make sure that they function with absolute perfection. Key testing procedures for smart contracts testing include the functional testing of business logic, node testing to ensure consistency of transactions, and validating the effectiveness of the consensus mechanism and the ensuing data immutability. Testing frameworks like Truffle, Mocha and Chai are available to test smart contracts. Smart contract testing involves simulation of all possible expected and unexpected conditions for every contract, testing all combinations of business logic and the proper triggering and correct execution of transactions. Leveraging test-driven development while testing smart contract is a right approach in order to avoid any room for errors or vulnerabilities in smart contracts. Test-driven development involves writing the test case first and then modifying the code until the code passes the written test [23].

Mutation testing: DeMillo, Lipton and Sayward first proposed mutation testing and made it public based on competent programmer and coupling effect hypotheses. The competent programmer hypothesis proposes that errors made by most programmers are very close to being correct after making a few modifications. The coupling effect hypothesis explains that complex defects are coupled with simple defects and test data sets will detect all the faults in the system (i.e., a combination of many simple defects). As such, mutation testing is a process whereby testers alter specific components of an application's source code as to cause errors in the program in order to ensure that a testing software will be able to detect the applied changes [30].

Node testing: The power of the blockchain lies in the shared distributed ledger having the same set of sequence of transactions at every single node of the chain. Therefore, node testing is required to make sure that there is smooth consistency in the transactions and that they are in sync with other validating peers [1]. In order to test the consistency of the transaction, the consensus protocol needs to be tested to ensure that transactions are properly sequenced even when some nodes fail. While testing the nodes, resiliency needs to be tested as well. This can be done by validating that any node restarted or rejoined on the network is in sync with other validating nodes [22]

BLOCKCHAIN BUG MANAGEMENT CONSIDERATIONS

A bug can be classified as any deviation from the requirements that are specified for an application, or any deviations from the expected behavior which can cause potential threats to the environment [11]. Software testing is not just about quality assurance, control and verification of blockchain application, there are other factors to consider. Having an effective quality assurance and bug management approach is essential in making sure that the blockchain application is of high standards before its release [9].

The basic aim of bug management is the identification and detection of bugs in the system, but this is only half the battle. The critical steps in bug management include logging the bugs found, recording and tracking bugs, and organizing and resolving the issues as soon as possible. It is suggested that any security or performance flaw in the system should be identified and corrected in the earlier stages of the software development lifecycle (SDLC) because this would not only provide assurance of quality, but also save significant amounts of money, time and effort. Moreover, finding bugs in the later stages of SDLC can make detection of bugs much harder and more challenging to resolve [11].

A Four-Step Bug Management Strategy

Bug management is to be considered a priority in an enterprise as it not only inculcates within the enterprise's culture but also has an impact on customer satisfaction.

Below is a four-step strategy for identifying gaps and moving closer to a zero-defect goal.

Bug visibility: This part of the bug management revolves around the discovered defects in order to understand what decisions are to be made for an effective product delivery. Criteria on obtaining clear visibility are quality, time, quality and feature.

Bug prioritization: Once clear visibility is confirmed, prioritization is the next step. Quality of data and capturing the right information are essential in prioritizing bugs. The discovered bugs should be prioritized according to their degree of severity measured with such captured information as risk level, the extent of the flaw and its classification.

Bug resolution: After obtaining clear visibility and capturing accurate information, timely and faster decisions can be made earlier in the process of SDLC, which saves time and money. Teams can come up with collaborative steps based on an agreed-upon approach to resolve the defects.

Bug analysis: Once the above steps are completed, an analysis process is carried out where all the inputs of the identified defects are considered. Through this, the teams are closer to reaching that nirvana goal of zero defects.

Through steps 1–3, all critical bug information and corrective action implementation are captured and collected for future bug prevention. These include –but are not limited to – lessons learned and root cause analyses methodologies and results. It is crucial for a system to have the ability to capture defect analysis processes for historical traceability in order to move closer to zero defects [10].

Blockchain Bug Categories

The dependability of blockchain systems is greatly affected by bugs. Therefore, studying and understanding bug characteristics can be helpful in designing efficient tools for preventing, mitigating and detecting bugs on a blockchain system. Categorizing bugs can help in identifying the weaknesses and vulnerabilities present in a blockchain application environment.

The following are the various bugs that are prevalent in blockchain systems and applications:

Concurrency bugs: Concurrency bugs are caused between concurrent threads or processes in programs. Basically, synchronization issues are caused by this type of bug. Such bugs are typically identified in Linux kernels compared to Mozilla and Apache.

Environment and configuration bugs: Environment and configuration bugs are the bugs that hide in the form of third-party libraries in the operating system or in the configuration files. This bug has the second highest rate of occurrence. Typically, it takes several months to fix this type of bug.

Hard fork bugs: This type of bug causes issues when a sudden protocol change causes a previously valid transaction to become an invalid transaction, or vice versa. This type of change in system behavior will also affect the consensus in a blockchain protocol [29].

Performance bugs: Performance bugs are errors that could cause performance degradation, eventually damaging software performance and quality. This could lead to poor user experience, degraded application responsiveness, lower system throughput and wasted computational resources [16]. An example of a performance bug is a high-severity performance bug that was found in Bitcoin. The remote nodes on the Bitcoin blockchain failed to remove invalid transactions from their memory. This led to a denial-of-service vulnerability because of the inability of the remote nodes to clear transactions, thereby over-flooding a victim's node with stale data referred to as 'uncontrolled resource consumption' which could eventually make the node shut down, thereby reducing the performance of the blockchain network [5].

Regression bug: While performing software testing, bugs are identified and subsequently a bug fix is actioned. Following a bug fix, regression testing is carried out to make sure that there is no functionality interruption from the recent bug fix. Bugs that are discovered through this type of testing are known as regression bugs. Regression bugs are hard to deal with, but need to be addressed before the release. These bugs increase project costs, create time complexities and decrease the overall agile velocity of a project [2].

Security bugs: Security bugs are software weaknesses or holes such as design flaws in the application that have the potential to be leveraged by a threat agent in order to cause harm to the entire blockchain architecture by exposing vulnerabilities to attacks [6]. Blockchains may have application-specific security bugs like timing and DDOS attacks, or missing security checks. Security bugs, typically, take the longest time to be addressed.

Semantic bugs: Semantic bugs occur due to a lack of thorough understanding of the system by programmers. These types of bugs are difficult to identify since they are application specific. Semantic bugs are sometimes referred to as dominant running bugs because more than 70% of the bugs are reported as semantic bugs; out of these, 23.9% are caused by missing cases, also known as neglected condition bugs. Other causes of these bugs are incorrect processing procedures, exception handling, missing features, incorrect control flow or input/output issues [29].

User interface (UI) bug: A UI bug causes deviations in the design or flaw to be encountered in the functionality of the application, leading to a problematic user experience. An example of a UI bug is the *Parity Multisig* bug that causes vulnerabilities in the smart contract library code deployed as a shared component of all Parity Multi-sig wallets. In one instance, this bug exposed a vulnerability that allowed a user to make himself an owner of the library contract due to Solidity's default public policy. Subsequently, the user destroyed the entire library using the kill () command, which froze and blocked 587 wallets out of their funds, totaling up to 513,774.16 Ethers (approximately USD 160 million) at the time of the incident [25].

Bugs in Ethereum smart contracts: Ethereum, the largest smart contract blockchain platform, has faced several bug-related smart contract issues in the past. Unfortunately, it is difficult to identify bugs in smart contracts due to a current lack of established frameworks and classifications. Existing tools are just helping to resolve a small portions of bug issues. For new developers, it is challenging to fully understand all the effects that the bugs are causing. Due to this challenge, it is even harder to design new ways of detecting the bugs and resolving them. Classification of bugs may be performed using the applicable NIST framework; however, out of all the identified bugs only a few bugs are classified in the well-defined category, with the remainder classified as 'other category'. Bug detection criteria are a fragile process which makes it difficult for developers to design the tools and algorithms to detect bugs. To address these limitations Ethereum developers are continuing to discover and analyze various bug behaviors.

TABLE 10.1
Tools for Bug Analysis

Tool Name	Characterization	Link
Mythril	An automated security analysis tool used for Ethereum smart contracts. A command line-based tool which uses control flow checking to identify different security vulnerabilities by both the byte code and source code. Detection is based on Laser-Ethereum and highly recommended for smart contract analysis because of its high precision rate.	https://github.com/ConsenSys/mythril
Oyente	An automated security analysis tool used for exposing security vulnerabilities in smart contracts. A command line-based tool which uses the source code for analyzing the bugs.	https://loiluu.com/oyente.html
Securify	An automated formal security analysis tool for Ethereum smart contracts. Uses a UI-based interface tool in combination with byte code and source code to identify different security vulnerabilities, for example, input validation or reentrancy attack (an attack that occurs when a function is created to make an external call to another untrusted contract), etc.	https://github.com/eth-sri/securify2
SmartCheck	SmartDec Security team developed an automated static code analyzer named SmartCheck which analyzes Solidity source code. An UI-based tool that identifies security vulnerabilities leading to possible attacks.	https://github.com/smartdec/smartcheck

At times, when smart contract bugs are collected and analyzed, developers find some bugs that are typically the same but are projected with different names, because there is no bug-naming convention to follow. Hence, the duplicate bugs are merged together based on the nature of consequences caused by the bug.

Two noteworthy smart contract analysis tools include *Slither*, one of the more effective analysis tools used to detect bugs which has an acceptable precision rate. This tool can be easily installed through a Python package called pip3 and can be downloaded at https://github.com/crytic/slither. *Remix* is another analysis tool used for bug detection with a high precision rate. Remix can be installed through the Remix IDE platform and can be accessed at https://remix-wide.readthedocs.io/ [32].

Apart from the above-mentioned tools there are a few other tools that are recommended for both bug analysis and smart contract analysis, as presented in Table 10.1.

Tools for Testing

Table 10.2 provides an overview of the tools discussed earlier, including download links for each tool.

TEST PLAN STRATEGY FOR BLOCKCHAIN APPLICATIONS

An effective test strategy requires sufficient planning. Planning is an essential aspect of a project, and should be considered at an early stage of blockchain development. This approach provides a leap forward in testing environments that result in quality end products and effective testing procedures. In order to implement an efficient test approach, it is important to have detailed information. The information should include details about the test scope, coverage, limitations, durations and tools utilized. This information will play a vital role in testing each phase of the project, and aids the team to structure and proctor testing processes with efficiency. Proper documentation is a part of the test strategy which helps the team to follow order in testing the blockchain application. Depending on the test scenarios, the methodology for testing is designed at the discretion of the organization.

It is also important to analyze and understand the project requirements before beginning a project. Similarly, testing also begins with requirements that are to be analyzed carefully in order to

TABLE 10.2
Blockchain Testing Tools Summary

Tool Name	Type	Purpose	Platform	Links to Access
BitcoinJ	An Open-source Java-based framework.	Designed using Java which implements the Bitcoin protocol (BTC network) to interact with the Bitcoin network and to perform other testing activities. The protocol provides access to maintain a wallet which allows the user to send/receive the transactions without the need for a local Bitcoin core copy.	Bitcoin	https://github.com/bitcoinj/bitcoinj
Corda Testing Tool	An open-source environment	Provides a wide range of functionality for unit testing and integration testing.	R3 corda	https://docs.corda.net/docs/corda-enterprise/4.2/sizing-and-performance.html
Ethereum Tester	–	Evaluates the initial performance for customer-oriented projects. Also analyzes the operations of APIs, web integration and smart contracts.	Ethereum	https://github.com/ethereum/eth-tester
Exonum Test kit	An open-source framework which uses Rust programming language.	Main focus is on testing the logics of the transactions.	Exonum	https://github.com/exonum/exonum
Hyperledger Composer	An open-source development toolset and framework provided by the Linux foundation project.	Supports interactive testing, automated unit testing and system testing. Uses JavaScript to perform transactions on blockchain.	Hyperledger	https://hyperledger.github.io/composer/v0.19/introduction/introduction
Manticore	Accessed through Python API.	Performs security testing on blockchain applications. Can detect crashes and other failure cases in binaries and smart contracts.	Ethereum smart contracts (EVM byte code)	https://github.com/trailofbits/manticore
Populus	Provided by Python testing framework.	Uses blockchain's distributed ledger technology to provide a global trading platform for invoice financing.	Ethereum	https://populus.readthedocs.io/en/latest/testing.html
Truffle	–	Facilitates compilation, testing and deployment of blockchains.	Ethereum	https://www.trufflesuite.com/

Source: [18].

proceed with each test phase. Testers with in-depth knowledge of an application can test the required characteristics of the blockchain application more precisely. As such, testers possessing vast skill sets with respect to the tools and technologies for different blockchain application platforms are valuable organizational assets. In addition, most blockchain applications in current deployment use API protocols for the programming interface which requires testers to also develop adequate API testing skills [27].

Opportunities to Enhance Testing Strategies

Testing is a repetitive process that involves finding flaws and vulnerabilities to the greatest possible extent. Each test will provide a better output, which is then iterated multiple times to arrive at a desired result. Improving the testing strategy depends on considering the various input events causing a failure. The failure caused due to an input event needs to be tracked and monitored for a better test performance. There can be several input events for a given test scenario. These input events are to be considered as key considerations while testing a blockchain application.

Service start-ups: At times, applications fail at the early stage of testing while starting the service. In other words, applications fail to invoke the service to be tested. This is an important input event which requires an adequate test scope, namely, ensuring that applications with long running services do not fail while restarting the services.

Nodes unreachable: This is a common scenario seen on both network and individual stations. Unreachable nodes can cause testing failures. There can be various reasons for this type of failure to occur. Therefore, there should be testing deployed using inputs of the network error that caused the failure. There should also be a keen utilization of adequate tools to arrive at the desired output.

Application configuration changes: Configuration changes are quite frequent in any application based on system requirements. As a matter of fact, advanced systems can automatically detect for configuration changes. This type of testing event requires testing of both valid and invalid configurations in order to end up with a stable and fully functioning application.

Node additions: With blockchain applications, any node addition creates a trigger. As a part of the testing scope, it is critical to test the addition of a node to a network. The test should focus on both the positive and negative test scenarios with respect to the node additions [14, 31].

Regression Testing

Regression testing entails the re-performance of both functional and non-functional tests to help ensure that previously developed software modules continue to work as intended following a change. This requires automated tools and software to run the test suites. The tools are also used to maintain and work on the functionality of applications. The selection of test cases to be executed as a part of a regression suite requires time. Once the test cases are analyzed, the order of test subsets is formed for testing. The purpose of developing regression suites is to optimize the often-used functionalities in the application. The output of regression testing is efficient due to its iterative process. An application can contain a wide range of testing scenarios, but it is important to prioritize the business-focused test scenarios to help achieve the best results through regression testing.

As mentioned earlier, the role of a tester is to find faults to the greatest possible extent. The tester's focus is to keep track and monitor the found faults at an early stage of regression testing. Any application that is to be tested requires keen knowledge over its functionalities. As a result, it is critical for the testers to explore the application functionalities in order to understand the needed test scenario coverage. Reporting faults and errors are essential factors in an effective testing process. Therefore, faults with high severity are to be reported immediately.

Automation Testing

Automation testing ensures a continuous delivery process in blockchain development allowing developers to deploy changes into a production environment. The timeline being the top priority for a majority of projects, automation testing helps the release and rollout of projects into live environments. Considering the time sensitivity of the project, a swift implementation of high-quality test suites enhances the productivity of developers.

Each team member should perform code changes as and when required for various business functionalities. Code changes directly impact the testing process, especially as it pertains to automation testing. Code and version changes need to be automated within an adaptable and often shifting business process. Continuous integration (CI) is vital for all teams in development and testing cycles. The reason for continuous integration is to make sure that both the development and testing teams are aware of the project's pace in order to remain on the same track. Automation is a boon for a development team since the time required for code review decreases significantly. Therefore, more time may be devoted to developing additional features. In short, automation testing helps in creating a strong baseline for both developers and testers to build and implement a robust blockchain application [19, 20].

TEST MANAGEMENT CONSIDERATIONS

Effective test management is necessary in order to create and maintain an organized and controlled testing environment. Managing the test phases enhances the process of reporting, validating and verifying each step while testing the application. Efficient test management is contingent upon a thorough understanding of the testing requirements for the designated application. A testing requirement analysis helps create efficient management strategies to achieve the desired results. A clearly defined test coverage also helps secure the needed resources and budget allocations. Synchronization among the team members and with cross-functional teams (developers and support team) helps in achieving a successful work collaboration environment. There are a variety of tools, platforms and repositories that aid in sharing and recording the progress of a project in a continual manner. Timelines are an important factor in test management, and ought to be fully considered and finalized at the beginning of the project. The implementation of a test strategy and its corresponding duration are to be announced at the earliest stage of the project. Finally, it is also important to define the reporting standards for various testing procedures to help optimize communication [3].

USER ACCEPTANCE TESTING

The final stage before an application goes live should also entail a full set of application testing as per the design requirements. These tests are to be performed considering the user end expectations and comfort levels in using the derived application. As such, the process must involve users with real-time scenarios to test the application. This approach is commonly referred to as *user acceptance testing* or *user interface testing*. Users are involved at this stage to help avoid not only potential financial and reputational losses to the organization pertaining to the development and testing of the application, but to also help ensure end user acceptance. User testing helps the organization with the needed feedback from users, leading to additional improvements of the application. Any flaws remaining from previous testing phases are likely to be found during user acceptance testing and before the live application launch [4].

End users engaged in the application testing should have the right knowledge and skill sets to understand the technology. Engaging the right users in testing helps create efficient comparisons based on initial requirements and the end application. Involving subject matter experts (SMEs) is a highly advised strategy for user acceptance testing (UAT). UAT requires sufficient duration to achieve its intended results. While performing UAT is important, documenting the test results and outcomes is equally crucial for the acceptance testing. Consolidating the entire UAT process requires adequate governance. One of the best models to track and resolve the found faults is RAID (Risk, Action, Issues and Decision) [12].

BLOCKCHAIN APPLICATION AUDITING CONSIDERATIONS

It is also extremely important to consider the need for continually auditing the security of a blockchain application to help provide the needed assurance levels regarding the use of a secure blockchain solution within an enterprise. Therefore, an audit framework needs to be adopted to assist auditors with validating and reviewing the blockchain solutions being implemented.

A FRAMEWORK FOR AUDITING BLOCKCHAIN SOLUTIONS

The audit framework developed by KPMG (India) can help blockchain auditors assess specific risks associated with blockchain platforms. KPMG's blockchain audit framework is comprised of the following audit focus areas:

Key ownership and management: This audit area involves the security, transmission, maintenance and storage of cryptographic private keys, in addition to the management of hash functions.

Interoperability and integration: The focus in this area is on the consistent and effective integration between the various blockchain platforms and the enterprises' legacy systems and current overall IT infrastructure.

Consensus mechanism: During this audit module, the consensus protocol design and performance are assessed. Additional consensus-related considerations, such as how the nodes on the blockchain get validated, along with change control procedures are also looked at by the audit team.

Heterogeneous consensus compliance: This audit area examines the applicable frameworks, standards, procedures and guidelines designed to keep the blockchain platform compliant. Data classifications and sensitivity are also assessed in this audit module.

Access and permission management: Emphasis on monitoring and management of groups and user permissions and the level of access and control provided to various roles are examined in this audit module.

Infrastructure and application management: This audit area looks at the security practices for software development, testing of blockchain applications and code management.

Network and node governance: The focus in this audit module revolves around the governance framework and standards aimed at resolving asset identity disputes and accountability, as well as information protection and transaction validity [7].

A PROPOSED AUDIT CHECKLIST FOR BLOCKCHAIN AUDITS

Table 10.3 represents a proposed blockchain application audit checklist based on KPMG's (India) audit model.

SUMMARY AND CONCLUSIONS

Blockchain technology is heading toward an exponential growth in the global market. Being the ground-breaking and disruptive technology of this era, organizations must not underestimate the importance of adequate testing and the application of effective auditing methodologies to a blockchain environment. With business relying on constant transformation and customer satisfaction, implementing blockchain technology without proper considerations into an enterprise's architecture is likely to have adverse impacts. Therefore, effective testing and bug management principles should be enforced during the development lifecycle of a blockchain application for a smooth integration.

TABLE 10.3
Proposed Blockchain Audit Checklist

Audit Area	Auditor's checklist	Yes	No	Auditor's Notes/ Explanations
Key ownership and management	Is there a cryptographic key policy and procedure document? Does the cryptographic policy/procedure document(s) address the following areas? 1. Key generation 2. Key maintenance and storage 3. Key decommissioning 4. Key management procedures 5. Key traceability and version control 6. Hash algorithm management.			
Interoperability/ Integration	Are the interface/API policies, procedures and interface requirements properly documented? If yes? Is the document regularly reviewed? How often? Are there well-defined data mapping and integration diagrams? Are the data validation and rules check followed? Are the described protocols in place between the intermediary platforms for the integration of blockchain platform and enterprise systems? Is the interoperability performance functioning as desired? Are the interoperability connectors and plugin updated?			
Consensus mechanism	Is there a consensus protocol design with established rules? Are the consensus rules regularly reviewed? How often? Are proper consensus change control procedures in place? Are the audit trails and logs of blockchain transactions present and well maintained? Are the consensus override procedures and consensus hijack risks properly addressed and monitored?			
Regulatory Compliance	Is the platform compliant with the applicable frameworks, standards, procedures and guidelines? Are transactions involved throughout and across the industry compliant with regulatory authorities requirements? Is the sensitivity of data transactions regulated based on read and permissions? Are data classified and compliant based on the framework established? Is the system compliant with applicable jurisdictional laws specific to cross-border data flows?			
Access Control/ID Management	Are user permissions properly established, recorded and monitored? Are system access levels role-based? Has segregation of duties reduced conflict/misuse of authority? Are there strict enrollment and termination procedures for the network nodes?			
Infrastructure and Application Management	Does the implemented SDLC (software development lifecycle) allow an effective and efficient transition from one phase to the other? Is coding based on consistent and well-managed secure coding principles and development practices?			

TABLE 10.3 (Continued)
Proposed Blockchain Audit Checklist

Audit Area	Auditor's checklist	Yes	No	Auditor's Notes/ Explanations
	Does the testing framework include effective phases for proper bug tracking and patch management?			
	Is there an adequate amount of resources allocated for cybersecurity testing?			
	Are platform and application documentations adequate and securely maintained?			
Blockchain Governance	Does the platform have an effective governance framework to resolve disputes over assets, identities and transactions?			
	Is the network compliant with respect to the organization's current operations?			
	Does the blockchain have adequate and effective standards for the accountability, information protection and transaction validity?			
	Do nodes have proper reputation validation procedures?			
	Are there sufficient tools and resources deployed to monitor and analyze the network?			
	Are there effective mechanisms in place to detect data leakages from the network?			
	Does the network have a methodology to identify and address possible points of failure?			

Source: [7].

The chapter begins with a discussion of the testing procedures and ensuing challenges that are to be considered during the phases of testing. Testing – as a process – is all about finding defects/bugs. In the later sections of the chapter, various categories of bugs and recommended testing tools were presented. Finally, a KPMG-proposed audit framework was presented and discussed. Developing an effective audit framework and test strategy using the proposed audit checklist will enable secure blockchain applications.

CORE CONCEPTS

- To produce a positive contribution to the shift toward digitalization, blockchain technology must be tested and integrated successfully with other software and technologies. Thus, the criticality of testing a blockchain application during the phases of the development lifecycle should be considered as it exposes the vulnerabilities that a new blockchain application may possess.
- In order to support the efficient and high-standard release of a blockchain application, quality assurance and bug management strategies should be implemented along with the effective and comprehensive testing methodologies. The discussed four-step bug management strategy can help identify the corrective actions that should be taken for bug elimination.
- Testing blockchain applications plays a crucial role because once a blockchain transaction is processed, it cannot be reversed. When bugs are identified using the tools discussed in this chapter, they need to be properly classified and documented. This document can further help new developers to further understand bugs' behaviors and their occurrence levels. Open-source tools for bug analysis and smart contract analysis are available and recommended. Different tools use different bug detection techniques. For example, Mythril uses control flow

checking, Oyente uses source code to analyze bugs, Securify uses a UI-based interface and SmartCheck analyzes Solidity source codes.

- While performing tests using proven testing methodologies, there are a number of factors that are to be considered, such as block size, chain size, load, security, transmission of data, block additions and cryptographically secure data. Depending on the components of the blockchain application, different software testing methodologies are performed, such as functional testing, API testing, integration testing, performance/load testing and security testing, to name a few.
- There are different frameworks and tools for testing blockchains. Developers and testers should choose the tools and frameworks for testing blockchain according to their needs and objectives.
- Truffle is a popular Ethereum development framework. Testers who prefer a command line interface can use Hyperledger Composer. Rust Programmers can consider the Exonum test kit to perform testing on transaction logics. All tools that are presented in the chapter are available on Github.
- The world is evolving into a phase where day-to-day activities are moving into an automated environment. The need for manual intervention is decreased in blockchain applications by implementing testing models known as automation testing. There can be business scenarios that require continual testing at regular time intervals. These scenarios are automated and subjected to regression testing.
- To track and monitor the testing process there has to be an efficient test management strategy in place. The test management includes the tracking and maintenance of entire testing processes for a blockchain application. It is quite important to test the finished application with respect to the end users' perspectives in order to derive the necessary feedback. Based on the feedback received, adequate time needs to be allocated to address all the issues before deploying the blockchain application.

ACTIVITY FOR BETTER UNDERSTANDING

The following reading should prove useful for audit professionals who are planning to develop an internal audit strategy for blockchain applications.

Guide to Auditing Blockchain Environments

Access Link: *https://www2.deloitte.com/content/dam/Deloitte/us/Documents/risk/internal-auditors-guide-to-blockchain.pdf*

REFLECTIVE QUESTIONS

- A balance between development and bug fixing should be established during the SDLC. How can a categorization/prioritization system for bugs help testers contribute to this balance?
- How can the proposed KPMG audit framework and the ensuing blockchain audit checklist coupled with the reading suggested above (*Guide to Auditing Blockchain Environments*) be of benefit to auditors interested in a better understanding of the key risk areas surrounding the implementation of a blockchain solution? What additional enhancements would you and your audit team incorporate into the audit checklist presented in this chapter?

REFERENCES

[1] Azevedo, R. (2020). *Testing Blockchain Applications.* Retrieved from azevedorafaela: https://azevedorafaela.com/2020/12/29/testing-blockchain-applications/

Chowdhury, A. R. (2019, January 21). *Why Understanding Regression Defects Is Important For Your Next Release.* Retrieved from LambdaTest: https://www.lambdatest.com/blog/why-understanding-regression-defects-is-important-for-your-next-release/

[2] Davis, C. (2018, September 10). *Test Management Best Practices: How to Improve Your Testing Efforts.* Retrieved from https://www.ibm.com/blogs/internet-of-things/iot-test-management-best-practices/

[3] Elazar, E. (2021, February 10). *What Is UAT Testing?* Retrieved from https://www.panaya.com/blog/testing/what-is-uat-testing/

[4] Foxley, W. (2020, September 9). *High Severity Bug in Bitcoin Software Revealed 2 Years after Fix.* Retrieved from https://www.coindesk.com/high-severity-bug-in-bitcoin-software-revealed-2-years-after-fix

[5] Handova, D. (2019, August 28). *What Are the Different Types of Security Vulnerabilities?* Retrieved from https://www.synopsys.com/blogs/software-security/types-of-security-vulnerabilities/

[6] India, K. (2018, October). Auditing Blockchain Solutions. Retrieved from https://assets.kpmg/content/dam/kpmg/in/pdf/2018/10/Auditing_Blockchain_Solutions.pdf

[7] Khoul, R. (2018, April). Blockchain Oriented Software Testing – Challenges and Approaches. *IEEE.* Retrieved from https://ieeexplore.ieee.org/document/8529728

[8] Kualitee. (2019, April 24). *Blockchain Application Testing.* Retrieved from https://www.kualitee.com/blockchain-application-testing/blockchain-application-testing-methods-step-testing-game/

[9] Kualitee. (2019, December 3). *Bug Management: A 4-Step Strategy for Better Products and Processes.* Retrieved from https://www.kualitee.com/bug-management/bug-management-a-4-step-strategy-for-better-products-and-processes/

[10] Kualitee. (2020, June 30). *The Anatomy of Bugs Life Cycle and What Measures Can Be Taken to Catch Them Early.* Retrieved from https://www.kualitee.com/bug-management/the-anatomy-of-bugs-life-cycle-and-what-measures-can-be-taken-to-catch-them-early/

[11] Kurkowski, P. (2005, June). *Top 5 Tips for User Acceptance Testing.* Retrieved from Alight: https://www.ngahr.com/blog/top-5-tips-for-user-acceptance-testing

[12] Leung, H., & White, L. (2002). A Study of Integration Testing and Software Regression at the Integration Level. *IEEE.* Retrieved from https://ieeexplore.ieee.org/document/131377

[13] Mikucki, M. (2019, September 30). *Design + Blockchain.* Retrieved from https://www.linkedin.com/pulse/design-blockchain-miki-mikucki

[14] Neotys. (2018, May). *Blockchain Best Practices.* Retrieved from https://www.neotys.com/wpcontent/uploads/2018/04/WP_Blockchain_Best_Practices_A4_EN.pdf

[15] Nistor, A., Jiang, T., & Tan, L. (2013, October 19). Discovering, Reporting, and Fixing Performance Bugs. *IEEE.* Retrieved from https://ieeexplore.ieee.org/document/6624035

[16] Parasu, J., & Reddy, V. (2019). *Blockchain Testing.* Retrieved from https://www.qteksystems.com/wp-content/uploads/2019/11/Block_Chain_Testing_Final.pdf

[17] Parizi, R. M., Dehghantanha, A., Choo, K. K., & Singh, A. (2018). Empirical Vulnerability Analysis of Automated Smart Contracts Security Testing on Blockchains. *28th Annual International Conference on Computer Science and Software Engineering (CASCON18)*, 103–113.

[18] Rothermel, G., Untch, R., Chu, C., & Harrold, M. (2001, October). Prioritizing Test Cases for Regression Testing. *IEEE*, 929–948. Retrieved from https://ieeexplore.ieee.org/document/962562

[19] Serek, M. (2019, April 5). *Automation Testing Blockchain for Improved Speed and Quality.* Retrieved from https://dzone.com/articles/how-introducing-automation-testing-for-our-blockch

[20] Smetanin, S., Ometov, A., Komarov, M., Masek, P., & Koucheryavy, Y. (2020). Blockchain Evaluation Approaches: State-of-the-Art and Future Perspective. *Sensors*, *20*(12).

[21] Sucharita, S. (2017). *Specialized Testing Requirements for the Blockchain.* Retrieved from https://www.infosysblogs.com/blockchain/2017/01/blockchain_specialized_testing_requirement.html

[22] Sukhwani, A. (2018, December). *Blockchain Testing.* Retrieved from https://ca.nttdata.com/en/-/media/assets/white-paper/404734-blockchain-testing-white-paper.pdf

[23] Sundarraman, A. (2018). *Assuriing Success in Blockchain implementations by Engineering Quality in Validation.* Retrieved from https://www.infosys.com/services/it-services/validation-solution/white-papers/documents/blockchain-implementations-quality-validation.pdf

[24] Technology, P. (2017, November 15). *A Postmortem on the Parity Multi-Sig Library Self-Destruct.* Retrieved from https://www.parity.io/a-postmortem-on-the-parity-multi-sig-library-self-destruct/

[25] Toulme, A. (2020). *Testing Blockchain and Distributed Decentralized Systems.* Retrieved from Neotys: https://www.neotys.com/blog/neotyspac-testing-blockchains-distributed-decentralized-systems/

[26] Turpitka, D. (2020, March 12). *How to Create a Well-Planned Testing Strategy for Blockchain-Based Solutions.* Retrieved from https://www.forbes.com/sites/forbestechcouncil/2020/03/12/how-to-create-a-well-planned-testing-strategy-for-blockchain-based-solutions/?sh=62dedd536feb

[27] Walker, M. A., Dubey, A., Laszka, A., & Schmidt, D. C. (2017, December). PlaTIBART: A Platform for Transactive IoT Blockchain Applications with Repeatable Testing. *In Proceedings of the 4th Workshop on Middleware and Applications for the Internet of Things*, 17–22. Retrieved from https://arxiv.org/pdf/1709.09612.pdf

[28] Wan, Z., Lo, D., Xia, X., & Cai, L. (2017). Bug Characteristics in Blockchain Systems: A Large-Scale Empirical Study. *In 2017 IEEE/ACM 14th International Conference on Mining Software Repositories*, 413–424. doi:10.1109/MSR.2017.59.

[29] Wu, H., Wang, X., Xu, J., Zou, W., Zhang, L., & Chen, Z. (2019). Mutation Testing for Ethereum Smart Contract. Retrieved from https://arxiv.org/pdf/1908.03707.pdf

[30] Yuan, D., Luo, Y., Zhuang, X., Rodrigues, G., Zhao, X., Zhang, Y.,Stumm, M. (2014). Simple Testing Can Prevent Most Critical Failures: An Analysis of Production Failures in Distributed Data-Intensive Systems. *11th USENIX Conference on Operating Systems Design and Implementation*, pp. 249–265.

[31] Zhang, P., Xiao, F., & Luo, X. (2020, September). Framework and DataSet for Bugs in Ethereum Smart Contracts. *In 2020 IEEE International Conference on Software Maintenance and Evolution*, 139–150.

11 Blockchain System Implementation

Bharghava Sai Nakkina, Deepthi Gudapati, Naga Venkat Palaparthy and Sai Sreenath Sadupally

ABSTRACT

This chapter focuses on the implementation aspects of blockchain technology. The chapter starts with a brief discussion of some potential organizational use cases such as the use of blockchain technology in disaster recovery, contract management, product monetization, supply chain monitoring, asset management and compliance management. Next, using the COBIT 2019 implementation approach as the framework, the chapter presents and discusses 10 general IT implementation challenges; then the reader is taken through the seven phases of the COBIT 2019 implementation by outlining roles and responsibilities, as well as each phase's description, objectives, inputs, and outputs. Readers interested in using the COBIT 2019 implementation approach are highly encouraged to review several COBIT 2019-related publications, including *COBIT 2019 Implementation Guide*, *COBIT 2019 Design Guide* and the *COBIT 2019 Governance and Management Objectives* guide available on the ISACA website (ISACA.org), as this chapter aims to provide a high-level discussion of blockchain implementation considerations.

INTRODUCTION

Before looking into a discussion of blockchain implementation considerations, in the following sections we will present some internal implementations use cases, namely values of blockchain as an addition or supplement to an enterprise's disaster recovery, product tokenization, supply chain monitoring and asset management. The next section will present some blockchain-company-based use case considerations, before delving into the implementation aspects of a blockchain platform.

DISASTER RECOVERY

Blockchain technology with its cutting-edge technological applications can be integrated into IT disaster recovery (DR) plans and can find effective uses in emergencies [1]. Disaster recovery refers to the process of documenting, collecting, preserving and disseminating information to the public and government in order to make prompt and efficient decisions during emergencies. Existing mechanisms for addressing emergencies have attained certain levels of automation in the process but are hindered by the lack of efficient sharing, consistency and integrity of information maintained across agencies working independently [2]. Some of the challenges faced during disasters are inefficiencies across various workflows, which are due to the low visibility of activities, lack of transparency and trust, and the improper utilization of resources [3].

Disaster recovery planning and business continuity planning are processes that help prepare organizations for when disasters strike. Disasters can be triggered due to natural hazards such as earthquakes, tsunamis and floods, or man-caused events like security breaches, denial of service attacks and power outages due to acts of sabotage, to name a few. In today's technological era, IT

DOI: 10.1201/9781003211723-11

systems play a major role in an organization's business operations, hence the need to ensure that these systems continue to work effectively and without interruption. Continuity of operations is top of the list of important attributes for quality business, as unplanned outages or sudden disruptions of services can have a huge financial and reputational impact on the organization [4].

At times, business executives assume the safety of their businesses based on geographic locations since certain regions are less prone to natural disasters – earthquakes, floods, storms, etc. than others. Unfortunately, not all disasters come in the form of natural disasters. A disaster may involve pipe breakage and water shutdowns, cyber-attacks, power outages or fires. In fact, according to a report from the World Economic Forum (WEF), the threat of cyber-attacks is ranked third, only superseded by extreme weather events and natural disasters. The WEF report highlights ransomware in particular as a cyber-threat and says that 64% of all malicious phishing emails contain file-encrypting malware [32]. Businesses face major setbacks due to security concerns also. Hackers attack by stealing information and/or deploying malicious programs such as ransomware, backdoors or trojans. Data compromised during these activities are held against companies for monetary gains. These disasters can cause enough damage to either halt day-to-day operations or destroy a business. While a disaster recovery plan deals with the continuity of IT systems, a business continuity plan guarantees the overall continuity of business operations in times of crisis or disaster [4].

According to the University of Texas, '94% of companies suffering from a catastrophic data loss do not survive – 43% never reopen and 51% close within two years'. Many disasters are unavoidable; however, the risks of certain disasters can be mitigated. The plan for mitigating these risks is to reduce the effects of substantial infrastructure disruptions to the barest minimum. A lack of proper planning will substantially increase the number of resources spent on recovery efforts [5].

A disaster recovery plan (DRP) is an essential business document that describes how an organization can quickly resume work after an unplanned incident. The disaster recovery plan is an essential part of a business continuity plan, designed to assist the company in the recovery of system functionality and prevention of data loss to ensure a smooth resumption of operations in the aftermath of an incident. The primary objective is to balance the cost of downtime along with the cost of continuity of business for the critical systems. For business enterprises, disaster recovery is a top priority [4].

Elements of an Effective Disaster Recovery Plan

Disasters can occur anywhere, taking a toll on the human population in that area. The impact of downtime on businesses is often even more significant. As such, organizations should create a DRP that can address all kinds of disasters. Typical elements of an effective DRP should consist of the following:

Disaster recovery team: Organizations should form a team to create a disaster recovery plan. The team will be responsible for creating, implementing and maintaining the DRP. A DRP should recognize the members of the team, describe the roles of each member and include their contact information. In the case of a disaster or emergency, the DRP should also specify who should be contacted. When a disaster occurs, all employees should be duly informed and be able to understand the DRP mandates, along with their own responsibilities [6].

Comprehensive inventory: A full inventory count of all organizational hardware, applications and software presently in use within the organization needs to be performed. Next, systems should be prioritized according to their criticality to operations and the order in which the systems should be restored in the event of an interruption. Information on each piece of equipment – like serial numbers, information on technical support, and contact details – should also be included in the DRP. A list containing access codes and passwords needed to maintain access to cloud-based programs, CRM systems or data backups should also be documented [7].

Identification and assessment of disaster risks: Risks to the organization should be identified and assessed by the disaster recovery team. These identification and assessment processes should cover natural disasters, man-made emergencies and incidents related to technologies. This will assist the team in evaluating what an acceptable recovery point objective (RPO) and recovery time objective (RTO) are for each set of applications. The recovery plans and resources needed to ensure recovery from disasters within a predetermined and appropriate timeline need to be determined [8].

Communication plan: Communication is critical when responding to and recovering from any emergency, crisis event or disaster. Regular means of communication are often ineffective during a disaster. To contact employees, the communication team will need to develop and implement proper protocols and business processes, along with contingency plans in case email platforms, cell coverage or telephone lines being down. Having a written reference framework in place ensures successful post-disaster responses and collaborations between organizations, employees and stakeholders [7].

Backups and off-site storage procedures: The DRP protocols should determine what to back up, how to carry out the backup, the backup venue and how often backups should occur. It is essential to backup all sensitive applications, facilities and records. The latest financial statements, tax returns, a current list of employees, and their contact details, in addition to inventory, customer and vendor lists are obvious examples of data that require protection through regular systems backups. Critical supplies needed for daily operations – such as company checks for purchasing or payroll purposes – should be kept at an off-site location in addition to a hardprint copy of the DRP [6].

SLAs and sensitive information: Organizations should maintain service level agreements (SLAs) with the third-party firms to which they have outsourced projects or technology. The SLA should also define the expected level of service in the case of emergencies or disasters. Protocols related to when the DRP is activated with operational and technical procedures to ensure the protection of data should also be documented and saved properly [8].

Testing and maintenance of the DRP: Disaster recovery planning is an evolving process, as threats and emergencies are continuously changing. Once a disaster recovery plan is developed, annual DRP reviews and updates should also become part of the DRP maintenance process. It is recommended that the DRP be regularly tested by the organization to assess the procedures documented in the emergency plan [6].

In summary, a disaster recovery plan should be clear and concise, focusing on key activities/functions required to restore services. The DRP should be tested, reviewed and updated regularly. The DRP team should always remember that 'when it comes to disaster recovery, the plan is only as good as its last test' [8].

The Use of Blockchain Technology in Disaster Recovery

Blockchain technology can be an effective tool in organizational disaster recovery. The main purpose of any recovery tool is to verify the integrity of the compromised system and the availability of resources. For the most part, in the event of a disaster, a blockchain's role can be attributed to integrity. The following sections describe how blockchains can help an enterprise during a disaster.

Verification of Data Integrity

When an organization is struck down with a disaster, there is a very high possibility of the organization's data being lost. File systems may get corrupted, certain servers may be destroyed or computer systems may perish. These occurrences can lead to the destruction of large chunks of data. The currently compromised system can no longer be relied on to function properly post-incident. In this case, it is important to verify the integrity of the current system against a backed-up system that is usually housed at a disaster recovery site [9].

Instead of writing large blocks of data into the blockchain, the mathematical hash of the whole system can be computed and stored on the blockchain. This mathematical hash can be compared against the compromised system's mathematical hash. If the hash matches, there is no need to restore the system from the backup. This saves time and effort and the recovery operations can be diverted to other efforts such as the prevention of disasters [9].

Availability

The availability of needed resources in an organization is important in ensuring the continuity of business operations daily. When a disaster strikes, often the needed resources are unavailable; this affects IT operations as well. For instance, if a denial-of-service (DoS) attack occurs on a company's server, any requests made to the server will be denied to users trying to access the server. In such cases, a back-up server will be needed to process those requests while the attacked server is taken off the grid. This ensures the availability of resources and the continuation of business operations [9].

Using a blockchain helps with the availability of data. Due to blockchain's status as a distributed ledger system, several copies of each storage block (containing resources) can be stored in various locations. If one block fails or is altered, the others can be used to continue operations [9].

Memory Correction

Memory correction is a common memory repair technique. In most cases, error correcting code (ECC) is used to rectify corrupted memory. The recalibration of the system is important in a post-incident environment; this includes checking if the computer's data storage is n-bit corrupted. Data corruption occurs due to unintentional single-bit errors caused during reading/writing data from the memory. An error correction code can be generated and stored on a blockchain. On a blockchain, this code can be used to detect and correct errors. If data are corrupted, blockchains can be used for data correction [9].

Cloud Storage

In a traditional setting for cloud services, confidential data are stored over a centralized database. In the case of a successful cyber-attack, a centralized database offers the hackers a perfect nest of information in a single location. Without a distributed system, the hacker requires less effort to steal data as all the data are lodged in a single location, resulting in a huge compromise to the data's confidentiality. A blockchain cloud-storage solution provides a venue to store data in a distributed and decentralized manner. *Salesforce* has started a similar concept called *Salesforce Blockchain* built on its CRM software using smart contracts and blockchain-based data-sharing [10].

CONTRACT MANAGEMENT

Contract management – sometimes referred to as procurement administration – is a vital process in organizations that entails terms and conditions related to negotiated contracts. In essence, the contract management process allows several involved parties to work toward a common objective. Contract management also holds each party accountable if any party is unable to hold up their side of the bargain. Misinterpretation and deception involving terms and conditions are the major causes of contract fraud.

Fraud in Contract Management

A procurement system is a collection of processes, procedures and entities involved in the purchasing of goods and services by public or private entities. The primary objective of an effective procurement policy is the achievement of the most value for money. It is important for procurement processes to be void of fraud. There are several ways in which procurement fraud can occur in the different contract management phases. According to the *Association of Certified Fraud Examiners* (ACFE), there are four predominant phases in which fraud occurs [11]:

Pre-solicitation phase: The procuring entity identifies its needs and develops bid specifications. These specifications in turn help to understand and determine who the bid should be awarded to. Fraud may occur in various ways during this phase. Examples of such fraud instances include bid tailoring, bid splitting or unjustified methods of procurement. To further explain, bid tailoring results in non-competitive bids that are favorable to a specific contractor. Bid tailoring happens when the contracting party develops the bid specifications favoring a certain contractor. This results in a non-competitive bid, as one bidding contractor will have an edge over the others. Finally, bid splitting results in subcontracting a contract and/or subletting it without following the proper protocols.

Solicitation phase: In this phase, the procuring entity prepares a bid solicitation document and issues notices to the bidders. The bidders can then submit their bids for evaluation. Fraud occurs in this phase through bid manipulation, leaking of bid data submitted by other bidders or collusion among bidding contractors.

Bid evaluation and award phase: In this phase, the submitted bids are evaluated through discussions and/or further negotiations with bidders. Bidders are given a chance to re-evaluate their submitted bids before the decision is made. Fraud can occur in this phase by bid manipulation (low balling) and/or leaking of submitted bid data by other bidders to give some bidders an edge over others.

Post-award and administration phase: During this final bidding phase, contracts are awarded to the bidder(s) who submitted the most competitive bid(s). Contract modifications sometimes occur in this phase. Fraud can occur in this phase through cost mischarging or change order abuse. Cost mischarging refers to a situation where the contractor charges the contracting party highly inflated material and/or labor costs. Change order abuse is referred to a situation where the contractor applies fraudulent changes to the awarded contract. These fraudulent changes are amendments made to the contract which affect the contract's scope.

The previous sections help demonstrate how fraud can be committed in the various phases of contract management. As such, the need for a fair bidding process and efficient contract management can be addressed through the implementation of blockchain technology, providing an effective way to help mitigate these challenges.

Blockchain as a Tool in Contract Management

Blockchain technology can change the anatomy of a traditional contract by using smart contracts which possess self-executing capabilities. Smart contracts can track performance and require little or no interference from external parties. In addition to fraud prevention, blockchain's self-reliant attribute can lead to a reduction in operational costs, as well as easier and closer cooperation amongst participating entities. A smart contract also eliminates middlemen and third parties which are required for contract enforcement [11].

The code for such smart contracts is written in such a way that if one condition is met (or fails) a subsequent step will be executed. As with all aspects of contract law, participating parties must agree to certain terms and conditions. The smart contract is deployed on a blockchain and can be executed by multiple parties, which when combined with smart contracts' automated nature also offers an audit trail. The audit trail contains draft versions (for every revision made to the contract) which are sequential versions of contract documents and repetitive transactions. This audit trail is helpful for keeping track of amendments [12].

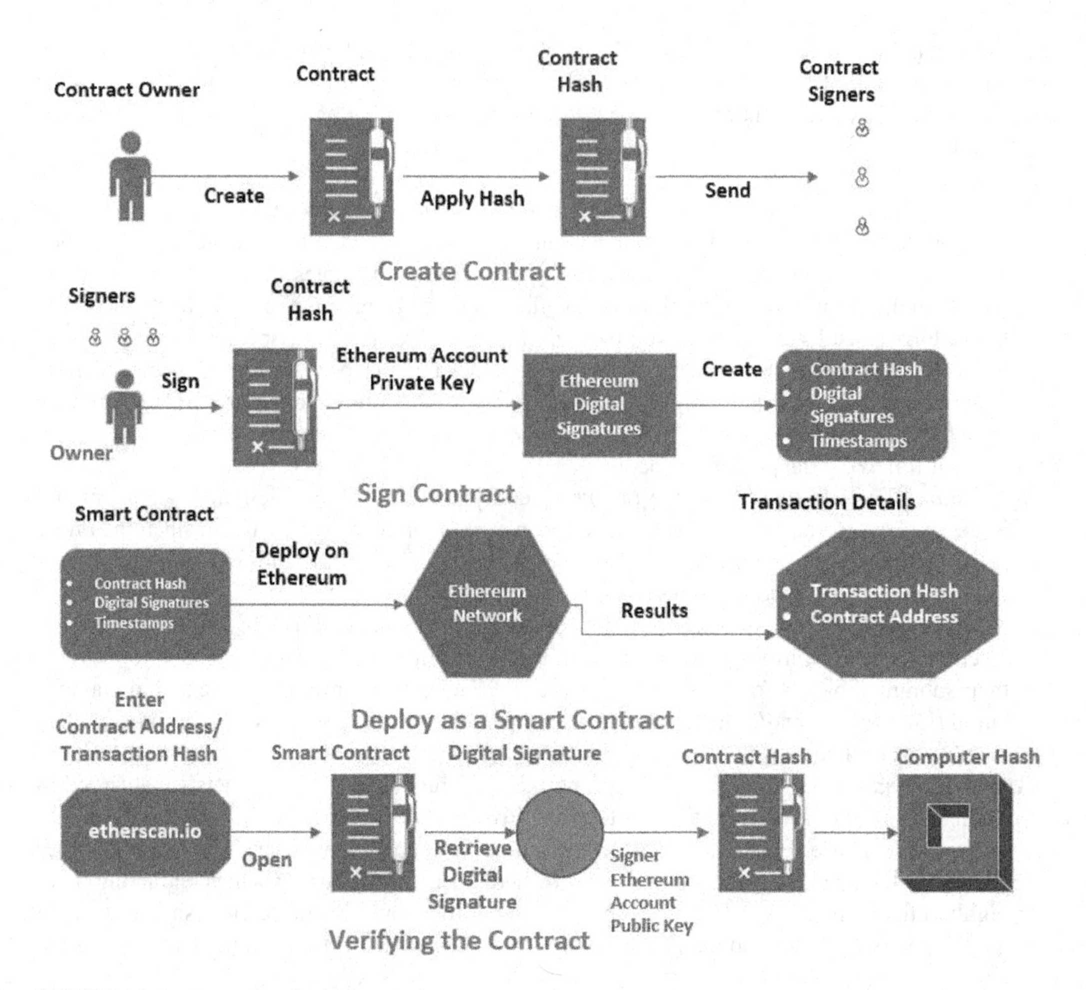

FIGURE 11.1 Example of a self-executing contract.

Source: [13].

Figure 11.1 depicts how contract management may be administered on an Ethereum blockchain platform:

As seen in Figure 11.1, the smart contract phases are as follows:

The smart contract creation phase: In this phase, a contract owner creates a smart contract and applies a hash function producing a contract hash.

The smart contract signing phase: In this phase, the participating signers on the contract along with the smart contract owner/creator sign the contract hash using their Ethereum account's private key. This results in the creation of respective Ethereum digital signatures of all the participants. At this stage, the three main elements of a smart contract are created, namely: (a) the contract hash, (b) participants' digital signatures and (c) timestamps.

The smart contract deployment phase: In this phase, the resultant smart contract is then deployed onto the Ethereum blockchain network [13].

Figure 11.2 illustrates how a typical contract is signed as part of the contract management process on an Ethereum blockchain. A cryptographic hash is applied to the digital contract after

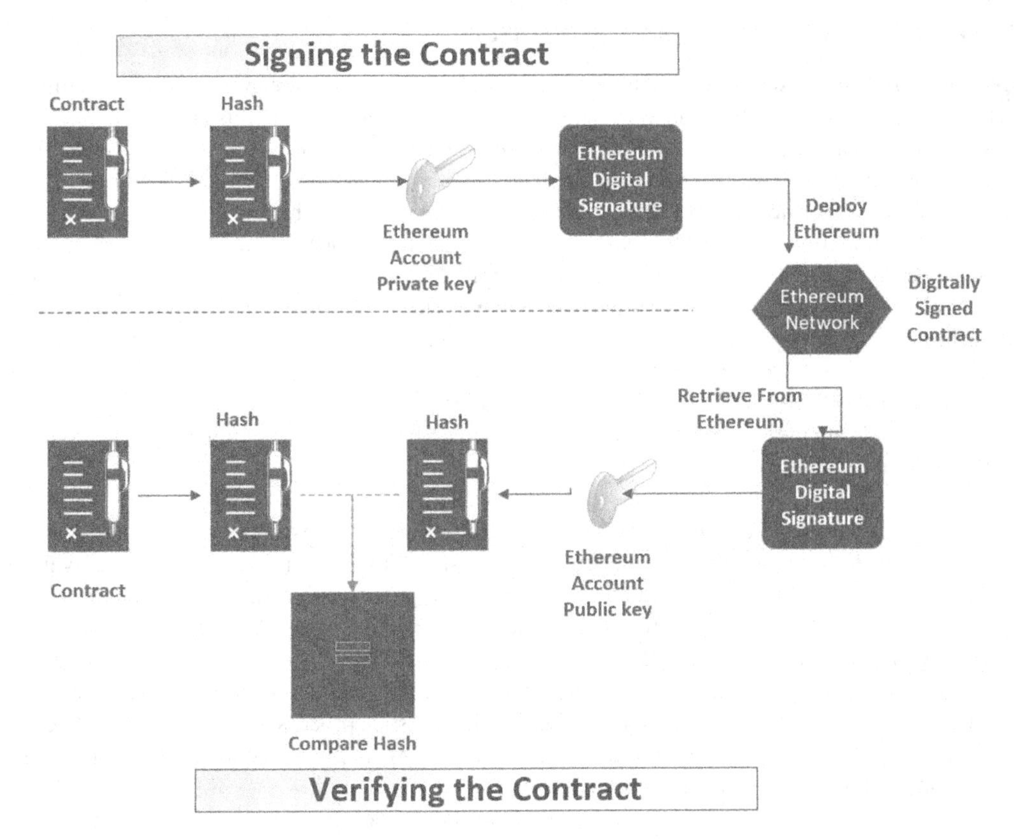

FIGURE 11.2 Signing and verifying a contract.

Source: [13].

which an Ethereum account's private key is used to produce a digital signature. The contract is then deployed into the Ethereum network to be acknowledged by all the nodes in the network, and is thus witnessed, published and verified. To verify a contract, its digital signature is obtained from the network. The Ethereum account's public key is applied to the signature and a hash is obtained. The obtained hash is compared with the original hash and if both hashes are the same, the contract is verified [13].

In summary, the use of smart contract technology in contract management will result in the elimination of third-party transaction fees, as smart contracts are self-executing entities with the capacity to track transactions and resolve contract disputes based on contract code (also known as code law). Currently, several organizations – including *Accenture*, *Icertis*, *Mindtree* and *Konfidio* – use smart contracts for contract management purposes. For example, Accenture uses R3 Corda technology and Microsoft Azure to provide blockchains for contracts [30]. Icertis' Blockchain Framework deploys contracts on permissioned as well as permissionless blockchains [14].

As mentioned, a key feature of blockchains is the elimination of third-party transaction fees. With blockchain implementation, there is often no need for intermediaries and thus no intermediary-related fees or markups. Also, since the blockchain is digital and distributed, there is a lifetime of data retrieval. Since smart contracts are self-executing, they are capable of tracking obligations and thus replace a manual review. Finally, most disputes can automatically be resolved based on contract rules [13].

PRODUCT DISTRIBUTION/MONETIZATION

The Fourth Industrial Revolution – also known as Industry 4.0 – is the new period of data digitalization that led to the growth of data-driven business models and the use of its applicable technologies. The distribution industry – both wholesale and retail – is defined as mortar-and-brick stores or online warehouses that serve as an intermediary between producers and end customers. In the process of transition, the industry faced multiple challenges such as shifting customer preferences, intensified competition, disruption of digital/e-commerce and an increase in real-estate prices [15].

In this current era of digitalization, new marketplaces and technological innovations in supply chains help businesses better manage logistical challenges to stay competitive. To gain consumer trust and achieve value proposition, manufacturers and merchandisers seem to welcome a decentralized, distributed ledger that can be used by both wholesale and retail entities to gain optimal efficiency, provenance, transparency and ease of doing business [15].

The following are the key areas where blockchain could have a significant impact on the products' sale and distribution process:

Customer expenses: Using the conventional banking system means making/receiving recurring payments that involve intermediary fees such as transaction fees and/or commissions. By using a blockchain system, remittance-related costs can be greatly reduced. For instance, the remittance fee for a $100 transfer from Canada to Asia may cost up to 4$, versus just a few cents using blockchain technology [15].

Minimizing counterfeits and bringing transparency: Blockchain can be used to maintain a trail of custody logs, tracking each step involved in the logistics process and ensuring supply chain integrity. This could be a boon to many enterprises, such as pharmaceutical companies where counterfeit drugs are often a major problem [15].

Reduced cost of business: Blockchain technology helps reduce administrative paperwork and operational costs throughout a product's supply chain by providing all the required information with a single scan [15]. Firms may be able to minimize the need for manual involvement in aggregating, amending and exchanging data, thus making regulatory reporting and audit records easier to process, requiring less manual intervention. Employees will be able to concentrate solely on value-added tasks as a result [31].

Compliance and regulatory guidelines: The implementation of blockchain can help trace and monitor mass product recalls, or defective products. Blockchain technology can also help reduce products' external failure costs, defined as the costs of product failures, and recalls to a business [15].

The practical implementation of blockchain technology in various sectors has increased over the years due to the potential benefits it can provide. For example, during the COVID-19 pandemic, one of the largest music events in Thailand, known as *Mystic Valley*, used tokens for enabling cashless payments for a three-day global event in November 2020. The idea behind the adoption of cashless payments using blockchain was to limit customer contact due to imposed social distancing rules [16].

SUPPLY CHAIN MANAGEMENT

The significance of quality management within the supply chain industry is an important consideration due to the increasing number of counterfeit products flooding some markets, causing ensuing product quality scandals. The root cause of this issue seems to stem from the fact that the conventional centralized trust-based systems do not seem to be able to fully address three main challenges namely: (a) the self-interests of members of the supply chain; (b) the asymmetry of details in the manufacturing process and (c) the quality testing costs and technological limitations [17].

As also discussed in Chapter 6, the introduction of blockchain technology has brought some innovative capabilities to supply chain quality management. To begin with, the adoption of blockchain technology to solve problems related to a lack of trust in supply chain management has become a much more manageable issue. By using automated quality management smart contracts, it is possible to develop an automated intelligent system as blockchain technology adopts the enterprise's governance model into a decentralized one [17]. To date, several large enterprises such as Walmart, Maersk and IBM have implemented blockchain technology into their respective supply chains. For example, Maersk and IBM are working on cross-border, cross-party transactions that aim to increase process efficiency by using blockchain technology [18]. Walmart Canada's DL Freight blockchain manages freight invoices and payments to over 60 transportation carriers. Walmart has also been using blockchain technology to better manage food freshness through better blockchain tracking abilities [19].

Blockchain's Value in Today's Supply Chains

Today's supply chains run at-scale without blockchain technology in most instances. Even so, the supply chain sectors were fascinated by the technology. This technology has also inspired numerous papers and encouraged both existing IT entities and start-ups to launch promising pilot projects like the authentication of transactions and the accuracy and efficacy of record-keeping. Walmart tested an application which was designed to trace pork shipments to China from the US and also to adopt blockchain technology into food tracking. Provenance – a UK startup – recently raised $800,000 to incorporate blockchain technology into the tracing of food. It recently pioneered the tracing of tuna in the Southeast Asian supply chain [18].

ASSET MANAGEMENT USING BLOCKCHAIN

Asset management refers to the process of managing various types of assets on behalf of others. In a nutshell, the asset management process has two main objectives: the appreciation of client assets over the long haul, and the effective mitigation of clients' portfolio risk. Asset management enterprises include but are not limited to banks, investment management companies, such as mutual funds or investment advisory services, to name a few. The assets being managed may be financial, public, information systems-based or even equipment. According to the Corporate Finance Institute (CFI), all enterprises need to keep a close tab on all their assets, whether liquid or capital in nature. The enterprise needs to maintain an accurate and updated asset inventory that includes the location, condition, acquisition costs, tax bases and purpose of each asset to effectively manage these assets. Accurate asset management helps enterprises to identify and manage asset risks more effectively, while making sure that asset amortization rates are properly computed and that lost, damaged or stolen assets are removed from a company's financial records in an accurate and timely manner [20]. Chapter 5 discusses the financial services use cases of blockchain in more detail.

Blockchain as a Solution in Asset Management

The distributed ledger technology of blockchain has various solutions that can accommodate rapid and radical changes to the asset management industry, as follows:

Enabling open collaboration: Blockchain allows for the creation of a system involving third parties – such as service or solution providers – thereby further facilitating open collaboration activities for asset management. In short, blockchain technology makes the task of adding new business partners easier.

Transaction transparency and immutability: Since blockchain transactions are immutable and tamper-proof, asset management data can be shared securely and transparently while also maintaining data integrity.

Increased consistency through consensus: The distributed ledger of blockchain ensures that there are no inconsistent records since all data blocks must be verified by all the processing nodes based on an agreed-upon consensus methodology [21].

Increased operational efficiency: A significant increase in efficiency for B2B transactions within the asset management industry is offered by the distributed structure of blockchain technology. The main factor considered for increased operational efficiency is the blockchain infrastructure itself. Rather than organizations sending information to each other continually, blockchain participants need to simply update their current blockchain state for the latest blockchain transaction updates [22].

Client onboarding and Know Your Customer (KYC) activities for financial services providers: Client identification checks and onboarding activities are arguably time-consuming tasks in the highly regulated asset management industry. It can take days or even weeks to validate customer identification, asset ownership, sources of funds, citizenship and risk profile, and investment suitability information from new clients. Blockchains can also facilitate quicker information sharing. As blockchain nodes are spread across different organizations, KYC activities are bound to become as common as current consumer or business credit checks [22]. Other blockchain benefits for financial services provider organizations may include faster trade clearance and settling of securities trades, more effective broker/dealer compliance and significantly less expensive transaction fees for both customers and the asset management entity using smart contracts. Chapter 5 discusses these issues in more detail.

USE OF BLOCKCHAIN FOR DATA CONTROL, SECURITY, LEGAL COMPLIANCE AND ASSURANCE

Apart from the various benefits that blockchains offer in various fields, there are other advantages also worth discussing. Some key features that blockchains offer are data control, security, legal compliance and assurance (auditing). These benefits are discussed below.

DATA CONTROL

Enterprises deal with vast amounts of data such as customer records, including personally identifiable information (PII) or the company's internal and/or proprietary information that are typically prime targets in various information leaks. Information can 'seep through crevices' when moving through multiple organizational and systemic layers of an organization. Centralized organizations such as Facebook, Google or Amazon have a considerable level of control over consumer data. This control does not lie with the actual user but with the organization and can be utilized by the company as they wish. Privacy, security and transparency issues increase when third-party vendors create services that have indirect access to such sensitive information. These companies do not transparently operate with their affiliates, which keeps the customers in the dark. Blockchains offer a decentralized solution that verifies the identity of a third-party cryptographically. This ultimately places the control back in the user's hand, as it eliminates the need for an intermediary to authorize or authenticate a certain subject [23].

SECURITY

Blockchain can help fix some security gaps, thereby addressing the CIA triad of information security by improving resilience, encryption, auditing and transparency. The full encryption of a blockchain ensures that data are not accessible to unauthorized parties; a perfect solution for man-in-the-middle attacks. Blockchains can also be used as an alternative to end-to-end encryption in messaging apps and social media platforms where data security will be achieved by securing the user's metadata. This metadata is randomly distributed across the network, avoiding the accumulation of all data at one point. The security revolving around secure messaging is still being explored and will likely make great advances in the future [24].

Legal Compliance

Smart contracts can be written with certain conditions which upon meeting their predefined criteria can be self-executed. This eliminates the need for intermediaries to review and execute contracts. Blockchains can be used in land registry proceedings where the land deed information can be accessed on the ledger where all parties are pre-authenticated. Public records such as census reports, death records or criminal activity can also be maintained with blockchains.

Blockchains offers a secure and tamper-resistant database. This facilitates a reduction of crimes such as fraud and money laundering which are vital to KYC and Anti-Money Laundering (AML). Smart contracts can automate the AML monitoring process and give real-time updates [25].

Assurance

A blockchain can help provide an assurance baseline that can greatly reduce traditional auditing tasks. The entire history of transactions or the exchanging of digital assets is updated in real time and available transparently to all parties, contributing to an organized auditable record. Since the transactions are interlinked with the previous block's hash, it is virtually impossible to tamper with. The immutability feature of blockchains ensures fraud and money laundering activities are curtailed [26].

BLOCKCHAIN IMPLEMENTATION CHALLENGES

After examining some ways through which blockchains can contribute to everyday life, it is only natural to ask the question: 'How do we implement blockchain technology?'. The answer, however, is not that simple. Blockchains have become a center of attention, especially with the current popularity of cryptocurrencies like Bitcoin and Ethereum. Be that as it may, blockchain technology is still in its infancy. As such, blockchains are yet to be widely adopted across the IT industry. Some of the reasons why this is the case are discussed below:

Cost of implementation: To migrate to a new technology solution in any organization, the first thing to consider is the cost of implementation. Do the expenses associated with a new blockchain solution interfere with the budget allocated for other IT projects within the organization? Does it hinder any ongoing projects? Are there any hidden costs during implementation? Securing sufficient funds is essential when transitioning to a blockchain solution. The selection, implementation and optimization of a suitable blockchain solution should be carefully planned to avoid excessive implementation costs.

Resistance to change within the organization: Upper management approval and stakeholder acceptance for a new IT project do not come easily, hence the need for all stakeholders to be on-board with the plan to implement a blockchain solution. Stakeholders may oppose the idea of implementing a blockchain technology due to doubts surrounding the realization of blockchain's benefits upon implementation. If the idea is accepted, another factor to be considered is the response of the organization's end-users toward the transition from 'the way we do things now' to an enhanced 'blockchain way' of doing things.

Training considerations: Before deploying a blockchain solution, the staff must be adequately trained to use the appropriate blockchain framework. Knowledge transfer from the implementation team to the staff should be prioritized by the management. This ensures that there are no major issues once a new workflow is put in place. At times, training costs and processes do not receive adequate consideration which, in turn, can slow down adaptation to the change [27].

Lack of usability: If the proposed blockchain solution is not adequate and compatible with the overall, current workflow, it will likely create usability gaps that often cause new challenges for the end-users. As mentioned before, such issues can result in resistance to change or a severe under-usability of the new technology. The best advice here is to ensure that end-users concerns, feedback and suggestions are considered carefully at every stage of the implementation.

Data privacy: It is a major responsibility during and after implementation to make sure that data are not exposed to potential leakage. Stakeholders may voice concerns such as the increased possibility of cybersecurity threats as a result of the implementation of a new IT solution such as blockchains. Such concerns need to be adequately discussed and addressed during the planning and implementation phase.

Data migration: When adopting a new solution such as blockchains, a major concern is successful data migration. Data migration is a comprehensive, yet meticulous task involving the movement of massive amounts of data (including paper-based) from a legacy system or infrastructure to a blockchain-based one. As such, the data migration phase requires careful planning and a prudent approach.

Limited technical resources: Technical resource limitations are a function of the IT budget and size of the organization. A lack of technical resources implies that the implementation of a blockchain-based technology may be constrained. These technical resources are comprised of teams responsible for blockchain's implementation as well as implementation tools. Since blockchain is a recent technology, not all organizations have an in-house team that can implement a blockchain successfully. Mostly, this task usually involves outsourcing of the implementation process to third-party vendors.

Interoperability: Interoperability issues should be carefully considered during the implementation process of a blockchain solution, as the process involves information flows across multiple platforms in conjunction with other tools and technologies. A blockchain solution should find it easy to accommodate these changes to ensure everyday businesses operate without flaws [27].

When migrating existing, traditional systems to a blockchain-based infrastructure, a comprehensive, systematic approach should be developed through proper planning to ensure a smooth transition. This is the reason why we recommend the use of a well-accepted implementation approach, such as the one outlined in the next section. Transparency and communication between all participating parties are also important. This ensures a full project commitment from all participating parties.

USING A COBIT 2019 APPROACH FOR BLOCKCHAIN IMPLEMENTATION

It is essential to set and maintain an appropriate environment for the Enterprise Governance of Information and Technology (EGIT) when implementing a solution within an enterprise in line with industry standards. The EGIT implementation needs to be well-governed and managed with the right direction, support and oversight by management. Improper and inadequate support from stakeholders may derail EGIT initiatives due to policies and procedures that lack accountability, responsibility and ownership. As such, the ensuing improvements may not become part of established business practices due to some of the previously discussed factors in the previous section. These improvements will, therefore, lack management structure parameters such as roles and responsibilities, continued operations and conformance to monitoring [28].

Need for COBIT 2019 implementation: COBIT 2019 Implementation Guide establishes a need for creating an appropriate environment to facilitate EGIT implementations as part of the overall governance approach in the enterprise. An appropriate governance structure should be driven by the executive management. Regardless of the nature of initiatives – large or small scale – structures, procedures, processes and practices must be in line with good governance principles to assist the executive management in ensuring better decision-making and authority models leading to informed decisions.

A COBIT 2019 approach for the first-time implementation of a blockchain-based infrastructure in an enterprise is discussed below.

A Three-Layer Implementation Approach

The COBIT 2019 lifecycle has three components: *program management*, *change enablement* and *continual improvement*. These three components are not isolated, but interdependent and concurrent. Each of these components contributes to the creation of the appropriate environment for the aforementioned EGIT initiatives. The cycle can be completed or retired once the project or improvement initiative successfully merges with the daily operations of the enterprise.

Program management consists of tasks that pertain to the implementation of the blockchain. It includes tasks such as defining challenges and opportunities, defining implementation roadmaps, developing and executing a blockchain program and realizing and reviewing its effectiveness.

Change enablement refers to a holistic and methodical process to ensure that stakeholders are geared up and committed to the changes that will move the current state to the desired state in the future. It contains activities such as: (a) assessing the impact of the proposed change on the enterprise's environment including the employees and stakeholders; (b) defining a vision of the future state; (c) building change response plans; and (d) measuring the progress of the change due to improvement initiatives.

A continual improvement approach consists of ongoing activities that put effort into improving the enterprises' products, services and processes. These activities are continuous and include tasks such as recognizing the need for a change, assessment of current and target states, building improvements, monitoring, and evaluating the improvements throughout the implementation cycle [28] (Figure 11.3).

The COBIT 2019 Implementation Approach

As previously mentioned, there are seven phases spread across the three components of the COBIT 2019 implementation cycle, namely, continual improvement (CI), change enablement (CE) and program management (PM). Each of the aforementioned seven phases is described in the subsequent sections with the related tasks in continual improvement, change enablement and program management (Figure 11.4).

Phase 1: What Are the Drivers?

Change drivers are identified in this phase. Change drivers can be defined as considerations that imply a need for a change. Examples of change drivers are industry trends, performance shortfalls, organization goals or software implementations.

The purpose of Phase 1 is to understand the scope of the proposed improvement, i.e., implementation of blockchain, and to understand how the blockchain system affects various stakeholders,

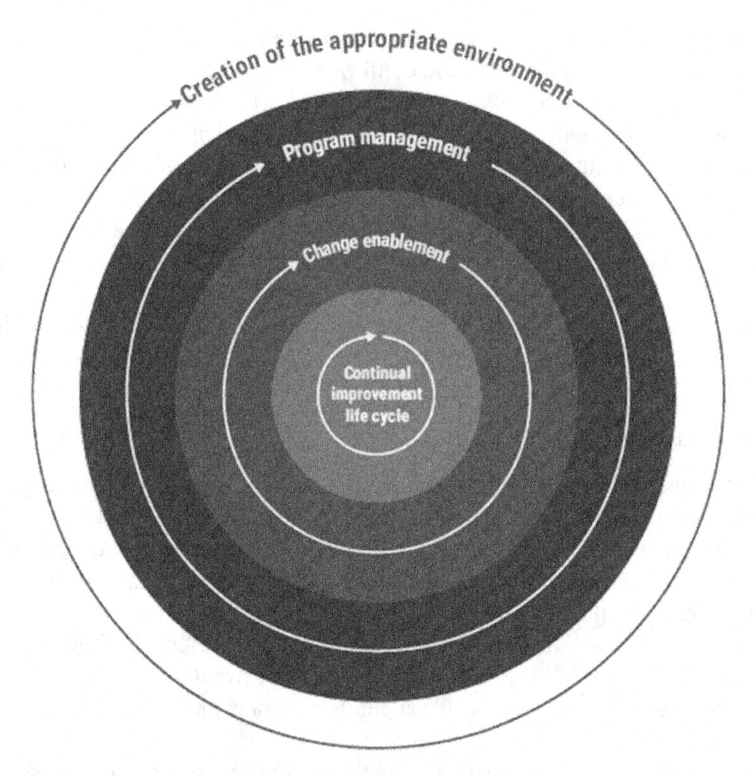

FIGURE 11.3 COBIT 2019 continual implementation lifecycle.

Source: [28]; COBIT 2019 Implementation Guide, ISACA (2019), all rights reserved.

as well as the readiness level to adopt the technology. Tables 11.1 and 11.2 describe the various stakeholders and their roles and responsibilities in this phase.

Phase 2: Where Are We Now?

Phase 2 aligns blockchain objectives with the enterprises' goals and strategies. It also prioritizes the ongoing projects to better align with governance and management objectives. Based on the alignment goals, governance and management objectives should be determined to be of the appropriate capability to produce the desired outcomes. The management should identify any deficiencies by performing process capability assessments on the current system.

An effective blockchain implementation team should be formed with the right skillsets. The team members should have adequate expertise and experience and be equipped with the right mix of internal and external required resources. The blockchain team should define and commit to a clear vision for the project. Also, problems that might cause friction when implementing the blockchain project should be identified and addressed proactively (Tables 11.3 and 11.4).

Phase 3: Where Do We Want to Be?

This phase assesses the current enterprise environment as a step to implement a blockchain solution. It sets a target for improvement and executes a gap analysis to identify the value of the potential blockchain solution. A high-level change enablement plan suitable for the blockchain program should be developed in this phase. An effective communication strategy should also be created to address the various groups with the vision, benefits, impact, purpose and involvement of stakeholders.

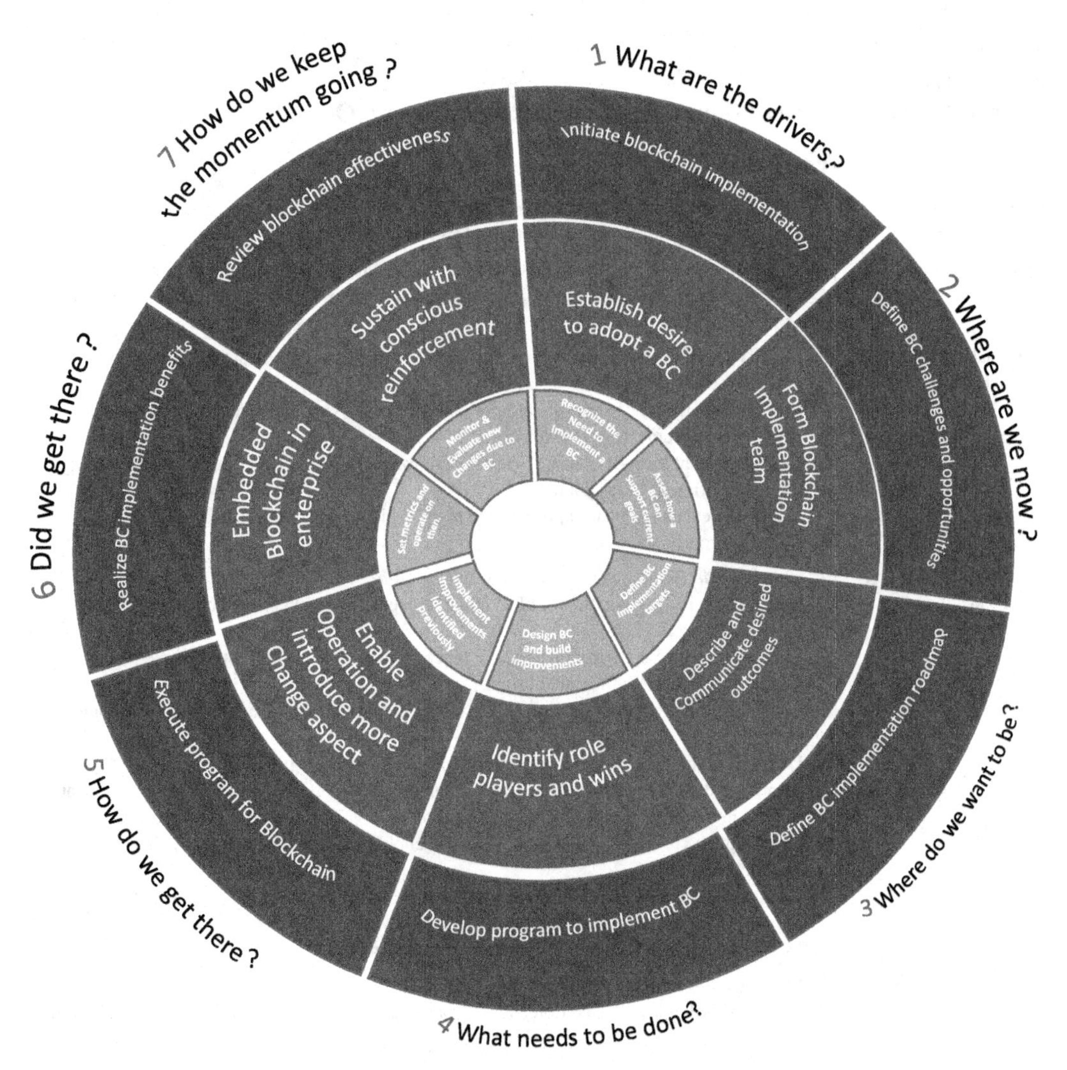

FIGURE 11.4 Blockchain implementation using COBIT 2019 implementation cycle.
Source: [28].

A blockchain roadmap should be developed to achieve the vision of the implementation. Feedback from various team members should be documented and appropriate actions should be taken (Tables 11.5 and 11.6).

Phase 4: What Needs to Get Done?

This phase proposes appropriate methods to implement a blockchain solution. Practical and reasonable solutions to implementing blockchains should be developed in this phase. These should be achieved by defining projects backed up by business use cases and a change plan. Continual monitoring should be done by producing a solid business use case. Change response plans should be developed to empower all the role players. While building the proposed blockchain improvements, key role players should be identified.

TABLE 11.1
Phase 1: Objective, Description, Tasks, Inputs, Resources and Outputs

Description of Phase 1 – What Are the Drivers?	
Phase objective	Obtain an understanding of blockchains' objectives and the background against the current governance approach. Get commitment from all stakeholders.
Phase description	Describe the need to implement a blockchain solution in the organization. To define blockchain background against current governance approaches. Also, obtain commitment and budget from stakeholders.
Continual improvement (CI) tasks	1. Identify current governance approaches and IT pain-points that demand the implementation of a blockchain platform. 2. Identify drivers and compliance requirements for improving enterprise governance of IT (EGIT) by implementing blockchains. 3. Identify current business priorities in the organization within IT that directly impact the implementation of blockchain. 4. Ensure the executive board approves the high-level approach and accepts the risk of not taking appropriate action.
Change enablement (CE) tasks	*Establish the desire to implement a blockchain solution:* 1. Integrate the implementation of blockchain with enterprise-change enablement programs within the organization. 2. Analyze the current environment where blockchain will be implemented. For example, management style, culture and attitudes. 3. Determine other ongoing initiatives that would hinder the implementation of blockchain technology. 4. Understand the impact of implementing blockchain technology. 5. Identify stakeholders for the blockchain implementation project from across the organization, such as in business, development, auditing, etc. 6. Determine the level of commitment from stakeholders and the impact of the change from their contributions. 7. Using the pain-points the I&T governance board should promote and propagate the principles and objectives among all stakeholders. 8. Instill a necessity for change to blockchain environment.
Program management (PM) tasks	*Initiate the implementation:* 1. Set high-level strategic direction and objectives in adherence to the I&T governance board. 2. Define and assign all the stakeholders with the appropriate roles in the implementation project. 3. Develop an outline of a blockchain business case with success factors to check for performance monitoring and report for improvements. Obtain financial commitment from stakeholders.
Input	1. Enterprise policies, strategies, governance, audit reports and business plans. 2. Ongoing/scheduled initiatives within the enterprise may be cause hindrances. 3. I&T governance board performance reports indicating any existing pain-points. 4. Useful industry overviews, case studies and success stories. 5. Enterprise vision and mission statements, market position and servicing strategy.
Output	1. Blockchain business case outline. 2. High-level roles and responsibilities of stakeholders. 3. Financial commitment from stakeholders. 4. Communication among all stakeholders and establishing the need to implement a blockchain.

Source: [28].

TABLE 11.2
Sample RACI Chart for Phase 1

	Responsibilities of Implementation Role Players								
Key Activities	**Board**	**I&T Governance Board**	**CIO**	**Business Executive**	**IT Managers**	**IT Process Owners**	**IT Audit**	**Risk and Compliance**	**Program Steering**
Identify drivers for the need to act	C/I	A	R	R	C	C	C	C	R
Identify business priorities affecting IT	C	A	R	R	C	C	C	C	R
Obtain commitment from stakeholders	C	A/R	R	C	I	I	I	I	R
Maintain the urgency to implement change	I	A	R	R	C	C	C	C	R
Produce outline of blockchain use case	I	A	R	C	C	C	C	C	R

Note: A = accountable; I = informed; R = responsible.
Source: [28].

TABLE 11.3
Phase 2: Objective, Description, Tasks, Inputs, Resources, and Outputs

Description of Phase 2 – Where Are We Now?	
Phase objective	Ensure that the project team understands the enterprise goals and IT function to deliver value from implementing the blockchain. Identify critical processes that may be addressed while implementing blockchains. Obtain an understanding of the enterprise's present and future attitude toward risk and IT risk position's impact on blockchain implementation. Determine the current capability of implementation and the enterprise's capacity for change.
Phase description	Identify and analyze how I&T enables business transformation while implementing a blockchain and still meeting governance-related requirements such as managing risk, security concerns, compliance with legal and regulatory requirements. Understand business and governance drivers' effects on risk assessment, arising from implementing blockchain technology on critical processes while ensuring business/IT goals alignment. The performance level of governance objectives should be established using policies, standards, and procedures.
Continual improvement (CI) tasks	*Assess current state:* 1. Establish the significance of I&T's contribution to the enterprise's goal of implementing the blockchain. 2. Identify governance issues and weaknesses related to the current and future solutions after implementing the blockchain. 3. Identify and select governance and management objectives to implement the blockchain. 4. Assess value enablement, implementation project delivery and operations risk arising from the implementation. 5. Identify objectives to ensure that the implementation risks are avoided or that the risk acceptance position is defined. 6. Define a method for executing the assessment and document the understanding of how current management and governance objectives address practices.
Change enablement (CE) tasks	*Form a blockchain implementation team:* 1. Assemble a team from the business and IT departments with sound knowledge, experience, credibility and authority to implement blockchain. Team members may include blockchain architects, blockchain network administrators, quality analysts, risk and compliance, security managers, CIO and the governance board. 2. Identify a blockchain project manager who can lead the team. 3. Identify potential vested interests to create trust and a proper work environment. 4. Create a shared vision within the team to mandate the change of implementing a blockchain.
Program management (PM) tasks	*Define problems and opportunities:* 1. Review and evaluate the blockchain use case, feasibility and return on investment (ROI). 2. Assign roles and responsibilities to all stakeholders and ensure commitment. 3. Identify challenges while executing these tasks.
Input	1. Outline of the blockchain business case. 2. High-level roles and responsibilities. 3. Program kick-off communication. 4. IT process description, policies and standards. 5. BCPs, impact analysis, regulatory requirements, SLAs and OLAs of BC implementation. 6. Audit reports, risk management policies, performance reports.
Output	1. Agreed alignment goals and impact on I&T. 2. Current performance levels and capability levels. 3. Risk acceptance position and risk profile. 4. Formation of the implementation team. 5. Evaluated blockchain use case. 6. Understanding of issues and challenges due to implementation.

Source: [28].

TABLE 11.4
Sample RACI Chart for Phase 2

	Responsibilities of Implementation Role Players								
Key Activities	**Board**	**I&T Governance Board**	**CIO**	**Business Executive**	**IT Managers**	**IT Process Owners**	**IT Audit**	**Risk and Compliance**	**Program Steering**
Identify IT goals supporting change to blockchain	I	C	R	C	R	C	C	C	A
Identify process critical to IT goals and blockchain		I	R	C	R	C	C	C	A
Assess risk related to BC implementation		I	R	C	R	R	C	R	A
Identify processes to avoid assessed risk		I	R	R	R	C	C	R	A
Assess current processes and their performance		I	R	R	C	C	C	C	A
Assemble team from business and IT		I	R	R	C	C	C	C	A
Review and evaluate business case	I	A	R	R	C	C	C	C	R

Source: [28].

TABLE 11.5
Phase 3: Objective, Description, Tasks, Inputs, Resources and Outputs

Description of Phase 3 – Where Do We Want to Be?	
Phase objective	Determine the capability for processes within the objectives of the blockchain project. Determine the gaps between "as-is" and "to-be" and translate these into improvement opportunities. Use this information to create a detailed blockchain business case and high-level program plan for implementing blockchain.
Phase description	Based on the assessment of current-state process capabilities and aligning goals to enterprise goals, appropriate target capabilities should be determined for implementing the blockchain. The gaps between the current state and future state implementation should be evaluated and improvements should be identified.
Continual improvement (CI) tasks	*Define targets for improvement:* 1. Based on enterprise performance and conformance, decide target capability levels within processes for implementation. 2. Benchmark internally to identify best practices for adoption. 3. Do sanity checks on target levels at what is desirable/achievable to create a positive impact. *Analyze gaps:* 1. Compare current capability level to target capability level. 2. Look for patterns to identify root causes for improvement that can be addressed through blockchain implementation. *Identify potential improvements:* 1. Collate gaps between current and proposed future states through implementation. 2. Identify risks that may arise through the implementation of blockchain and ensure that it is accepted to a certain level.
Change enablement (CE) tasks	*Describe and communicate outcomes:* 1. Secure the willingness from stakeholders to participate. 2. Develop a communication strategy to optimize the awareness of change. 3. Emphasize the vision and benefits of implementing blockchains. 4. Link the objectives of the initiative in communications and how it will be realized in the implementation.
Program management (PM) tasks	1. Set program direction, and the scope of the direction to implement the blockchain. 2. Identify the risks that might arise due to the implementation. Adjust the scope to the optimal level. 3. Consider the implications that may arise due to this change enablement 4. Secure budgets and define roles and responsibilities.
Input	1. Agreed goals and impact on alignment goals. 2. Risk acceptance and risk profile. 3. Stakeholder analysis. 4. Internal and external capability benchmarks. 5. The core team is assembled and assigned roles and responsibilities. 6. Challenges and success factors.
Output	1. Description of improvement opportunities. 2. High-level program plan. 3. Metrics to track the program and performance. 4. Detailed business case.

Source: [28].

TABLE 11.6
Sample RACI Chart for Phase 3

	Responsibilities of Implementation Role Players								
Key Activities	Board	I&T Governance Board	CIO	Business Executive	IT Managers	IT Process Owners	IT Audit	Risk and Compliance	Program Steering
Agree on blockchain implementation for improvement	I	A	R	C	R	R	C	C	R
Analyze shortcomings		I	R	C	R	R	C	C	A
Identify potential improvements		I	R	C	R	R	C	C	A
Communicate the vision charge		A	R	R	C	I	I	I	R
Set direction and prepare detailed blockchain business care	I	A	R	C	C	C	I	I	R

Source: [28].

These response plans may span over various factors such as organizational design change, team structures, changes like process flows and people management. Identified role players in blockchain implementation should be identified for quick wins. A clear blockchain project plan should be developed (Tables 11.7 and 11.8).

TABLE 11.7
Phase 4: Objective, Description, Tasks, Inputs, Resources and Outputs

Description of Phase 4 – What Needs to Get Done?	
Phase objective	Translate improvement opportunities due to the blockchain implementation into several quick wins to boost stakeholder morale.
Phase description	When all the potential improvements are identified due to the implementation of blockchains, these potential improvements should be further organized and prioritized into subsequent justifiable projects. The projects with more benefits and easier implementations should be selected first. Each project must have a project plan describing the contribution to program objectives. These projects should be included in the updated blockchain business case. Unofficial recommendations should be documented for reference. An opportunity grid with elements such as project definitions, resource plans, IT budget and prioritized improvements are turned into documented projects. The impact of executing these projects is determined and a change plan is prepared to describe the program activities that are sustainably fused into the enterprise. Metrics should be established to gauge performances and deliver the business benefits. The blockchain project schedule should be documented on a Gantt chart.
Continual improvement (CI) tasks	*Design and build improvements due to blockchain implementation:* 1. Consider the benefits and difficulty level of implementation. 2. Prepare an opportunity grid listing priority actions based on benefit and ease of implementation. 3. Decompose improvements that are not easy to implement into smaller projects. 4. Prioritize and select the improvements. 5. Analyze the selected projects based on approach, deliverables, resources, etc., and refine improvement requirements.
Change enablement (CE) tasks	*Empower role players and identify quick wins:* 1. Obtain commitment from teams and take feedback through workshops and review processes. 2. Prepare change response plans to track and engage throughout the blockchain implementation. 3. Single out the wins, indicating the positive impact of blockchain implementation on the environment. 4. Identify the strengths of Phase 2 of blockchain implementation and avoid revamping the identified strengths with a brand-new implementation. Rather, align the strengths with current improvements.
Program management (PM) tasks	*Develop blockchain implementation program plan:* 1. Organize the potential programs in a sequence. 2. Employ portfolio management techniques to ensure the program conforms to strategic goals. 3. Identify the impact of blockchain implementation on IT and business and see how the momentum can be maintained. 4. Create a change plan containing migration, tenting, processes, etc., included in the blockchain implementation program. 5. Identify the performance metrics against the original success factors. 6. Assist in the allocation and prioritization of processes and resources to execute and achieve the program plan. 7. Establish project plans and reporting procedures to keep track of the progress of the blockchain implementation project.
Input	1. Target maturity rating for selected blockchain processes. 2. Description of the improvement processes. 3. Risk response document. 4. CE plan and objectives. 5. Communication of the vision through the 4 Ps: picture, purpose, plan and part of blockchain implementation. 6. Detailed blockchain business case. 7. Opportunity worksheets, good practices, external assessment and technical assessment. 8. Strengths from the previous phase.

TABLE 11.7 (Continued)
Phase 4: Objective, Description, Tasks, Inputs, Resources and Outputs

Description of Phase 4 – What Needs to Get Done?	
Output	1. Implementation of improvements' project definitions. 2. Change response plans. 3. Identified quick wins. 4. Documented unofficial projects. 5. Project plan with allocated resources, priorities and blockchain deliverables. 6. Success metrics.

Source: [28].

Phase 5: How Do We Get There?

This phase implements the proposed blockchain solution. This should be done through everyday activities and through the establishment of measures to ensure that business goals are aligned with blockchain implementation processes and that their performances are measured.

The blockchain is implemented in the core continual improvement cycle, with the change response plans also being implemented. During the implementation of the blockchain solution, coaching and mentoring of the team is important to realize the commitment to the vision.

The initially set blockchain requirements and objectives should be revised and analyzed if the targets have been met. Success should be maintained through awareness, engagement, communication and commitment of the management to the vision of the blockchain implementation (Tables 11.9 and 11.10).

Phase 6: Did We Get There?

A sustainable transition of the blockchain improvements for governance and management should be the focus of this phase. Phase 3 should be reassessed to verify that all the goals have been met. Performance metrics should be used to monitor the achieved targets. Integrate the blockchain implementation by normalizing the new activities into the regular business processes, procedures and enterprise standards. A communication channel should be maintained to sustain the awareness of the new improvements (Tables 11.11 and 11.12).

Phase 7: How Do We Keep the Momentum Going?

The last phase of the implementation lifecycle is to help sustain the implemented blockchain solution. It should be done through conscious reinforcement, ongoing communication channels and commitment from all stakeholders. Corrective actions and plans are implemented when necessary, and lessons learned are documented to share the knowledge for future reference (Tables 11.13 and 11.14).

SUMMARY AND CONCLUSIONS

Blockchain technology is undoubtedly a revolutionary concept. It has the potential to change the way many industries conduct their day-to-day business activities. With changing times, it is wise to test and try new technologies to stay updated with IT industry standards.

Several organizational use cases were discussed in this chapter, including how blockchain can be used as a tool in disaster recovery planning to help minimize collateral damage. Specifically, in

TABLE 11.8
Sample RACI Chart for Phase 4

	Responsibilities of Implementation Role Players								
Key Activities	Board	I&T Governance Board	CIO	Business Executive	IT Managers	IT Process Owners	IT Audit	Risk and Compliance	Program Steering
Priorities and select improvements		A	R	C	C	R	C	C	R
Define and justify blockchain projects		I	R	C	R	R	C	C	A
Design change to blockchain response plans		I	R	R	C	C	C	C	A
Identify achievements and build strengths		I	C	C/I	R	R	C/I	C/I	A
Develop blockchain project and program plans		A	C	C	R	C	I	I	R

Source: [28].

TABLE 11.9
Phase 5: Objective, Description, Tasks, Inputs, Resources and Outputs

Description of Phase 5 – How Do We Get There?	
Phase objective	Implement the blockchain improvement projects. Monitor, measure and report on project progress.
Phase description	The approved blockchain project can be built/acquired and integrated into the enterprise. The projects are infused into the regular developmental lifecycle, through project management governance. The project should align with project definitions and change plan improvements. Despite being the longest phase, this phase should be manageable, and its benefits should be delivered within a specific time frame. The performance of the blockchain implementation project should be monitored and its progress reported to the relevant stakeholders.
Continual improvement (CI) tasks	*Blockchain implementation of improvements:* 1. Develop the full scope of activities required for the blockchain implementation. For example: skills, procedures, organizational structure, etc. 2. Use good or best practices to adapt and adopt the guidance to best fit the enterprise's policies and procedures. 3. Test the practicality and sustainability of the blockchain solution in real time.
Change enablement (CE) tasks	*Enable operation and use of blockchain:* 1. Leverage the momentum gained by identifying quick wins and introduce complex change aspects later. 2. Communicate quick wins and recognize/reward the team. 3. Implement change response plans. 4. Ensure the role players have the right skillset and resources. Obtain buy-in and commitment from all stakeholders. 5. Communicate roles and responsibilities. 6. Ensure all change requirements have been met. 7. Monitor the change enablement process and take corrective measures when necessary.
Program management (PM) tasks	*Execute the blockchain project:* 1. Execution processes should depend on an integrated plan of the project. 2. Direct and monitor the contribution of the project into various programs to ensure delivery of expected outcomes. 3. Provide timely and regular reports to stakeholders. 4. Document risks and issues and brainstorm remediation actions. 5. Approve the initiation of each major project and communicate to stakeholders. 6. Approve major changes to program and project plans.
Input	1. Improvement project definitions. 2. Definitions of change response plans. 3. Quick wins documentation. 4. Records of unapproved projects. 5. Program plans with allocated resources. 6. Success metrics. 7. Project definitions, Gantt charts, change response plans, change strategies. 8. Integrated program and project plans.
Output	1. Implemented improvements. 2. Implemented change response plans. 3. Realized quick wins. 4. Success communication. 5. Benefits tracked to realization

Source: [28].

TABLE 11.10
Sample RACI Chart for Phase 5

	Responsibilities of Implementation Role Players								
Key Activities	**Board**	**I&T Governance Board**	**CIO**	**Business Executive**	**IT Managers**	**IT Process Owners**	**IT Audit**	**Risk and Compliance**	**Program Steering**
Develop; if necessary, acquire solutions		A	C	C	R	R	C	C	R
Adapt and adopt good practices		I	R	C	R	R	C	C	A
Test and roll out solutions		I	R	C	R	R	C	C	A
Capitalize on quick wins		I	C	C/I	R	R	C/I	C/I	A
Implement change response plans	I	I	R	C	R	R	C	C	A
Direct and monitor projects in the program	I	A	C	C	R	C	C	C	R

Source: [28].

TABLE 11.11
Phase 6: Objective, Description, Tasks, Inputs, Resources and Outputs

Description of Phase 6 – Did We Get There?	
Phase objective	To integrate the blockchain performance metrics and output benefits of blockchain implementation into a performance management system.
Phase description	It is important to monitor the improvements in the project (blockchain implementation) through IT and process goals using tools such as the Balanced Score Card (BSC). A benefits register should also be maintained to demonstrate that the intended goals have been met after the blockchain implementation. For every metric, targets need to be set and constantly compared against real-time checks during implementation. Both positive and negative results of the performance management system (PMS) must be recorded for stakeholder transparency to build trust and enable corrective measures when necessary. Projects should be monitored for any deviations from milestones and be called out when off track.
Continual improvement (CI) tasks	*Operate and measure:* 1. Targets for each performance metric of blockchain implementation should be set and timed. These targets should be monitored to determine the successful implementation of the blockchain. 2. Record metrics data. 3. Pit targets against actual measures and compare for deviations. 4. Convey positive and negative results of the performance management to stakeholders and record corrective measures.
Change enablement (CE) tasks	*Embed new approaches to blockchain implementation:* 1. Ensure that the blockchain is a working part of the organization's working culture for better results to be populated. 2. Transition from a project to a day-to-day aspect of business where the organization incorporates new changes such as revised job descriptions, performance criteria, KPIs and operating procedures. All these should be implemented through change enablement plans. 3. Monitor the new roles that have been taken up. 4. Monitor the changes and analyze the change-response plans that have effectively met the goals and objectives, such as, through feedback surveys, etc. 5. Continue communication between various departments for awareness and celebration of quick wins. 6. Document the change enablement lessons learned through the blockchain implementation for future reference.
Program management (PM) tasks	*Realize post-implementation benefits of blockchain:* 1. Compare the performance of the blockchain implementation program against business objectives. 2. Observe cost against budget and benefits realized through implementation. 3. Document lessons learned throughout the implementation for future reference.
Input	1. Implemented suggested improvements. 2. Change response plans. 3. Identified quick wins 4. Blockchain project change logs. 5. Blockchain business case benefits. 6. Change response logs.
Output	1. Blockchain program scorecards. 2. Change enablement measures due to blockchain implementation.

Source: [28].

TABLE 11.12
Sample RACI Chart for Phase 6

	Responsibilities of Implementation Role Players								
Key Activities	**Board**	**I&T Governance Board**	**CIO**	**Business Executive**	**IT Managers**	**IT Process Owners**	**IT Audit**	**Risk and Compliance**	**Program Steering**
Operate solutions and obtain performance feedback		I	A	R	R	R	I	I	I
Monitor blockchain performance against metrics		I	A	C	R	R	C	C	I
Communicate positive and negative results		I	A	C	R	C	I	I	I
Monitor roles and responsibilities	I	A	R	C	C	C	C	C	I
Monitor blockchain program results	I	A	C	C	C	C	C	C	R

Source: [28].

TABLE 11.13
Phase 7: Objective, Description, Tasks, Inputs, Resources and Outputs

Description of Phase 7 – How Do We Keep the Momentum Going?	
Phase objective	Assess the results of the blockchain implementation program. Document and share lessons learned. Based on lessons learned, improve the organizational structure, processes, roles and responsibilities so that the enterprise can strengthen its IT goals and objectives. This should enable the organization to resume business optimally post blockchain implementation. Keep the momentum going with iterations in the implementation cycle through continuous monitoring of performance and the results reported at regular intervals.
Phase description	This phase of the implementation cycle determines if the expected blockchain deliverables have been met. This can be done by conducting feedback sessions, workshops or surveys, requesting responses from the blockchain implementation team and stakeholders. The collective feedback should be compared against the original success criteria. This can be used for new initiatives and in developing new improvement projects. Improvements identified are used in the next step of iterations. The enterprise should identify the lessons learned, build on the success and be reinvested in IT governance in the future. Roles, responsibilities, policies and procedures should be developed and optimized to enable EGIT to operate effectively and be integrated into normal day-to-day operations.
Continual improvement (CI) tasks	*Monitor and evaluate:* 1. Identify new blockchain governance objectives from previous experiences, current business objectives and other trigger events. 2. Obtain feedback from all stakeholders. 3. Compare the original objectives of the blockchain implementation to actual results. Embed CI tasks. 4. Conduct a project review with the blockchain project team and stakeholders to record lessons learned. 5. Identify any high-impact and low-cost solutions to strengthen EGIT. 6. Report the identified solutions to stakeholders so they can be used as input for the next iteration of the blockchain implementation cycle.
Change enablement (CE) tasks	*Sustain the blockchain implementation:* 1. Confirm conformance of blockchain implementation objectives and requirements. 2. Monitor the effectiveness of blockchain implementation and the resultant change enablement activities. 3. Implement corrective actions where required. 4. Provide feedback on blockchain's performance and celebrate wins. 5. Work on the lessons learned for the future.
Program management (PM) tasks	*Review blockchain effectiveness:* 1. At the end of the blockchains' implementation make sure there is a review process to conclude approval. 2. Review the effectiveness of the blockchain implementation.
Input	1. Updated blockchain project and scorecards. 2. Changes to blockchain and its effectiveness measures. 3. Scorecard reports. 4. Review reports of post-blockchain implementation. 5. Blockchain performance reports. 6. Business and IT strategy. 7. New regulatory requirements.
Output	1. Recommendation for the next blockchain project after a normalization period. 2. Stakeholder survey to record satisfaction. 3. Record of lessons learned. 4. Communication plan for ongoing activities for the stakeholders. 5. Performance reward schemes for the blockchain implementation team.

Source: [28].

TABLE 11.14
Sample RACI Chart for Phase 7

Key Activities	Responsibilities of Implementation Role Players								
	Board	I&T Governance Board	CIO	Business Executive	IT Managers	IT Process Owners	IT Audit	Risk and Compliance	Program Steering
Identify new blockchain governance objectives	C	A	R	R	C	C	C	C	I
Identify lessons learned		I	A	C	R	R	C	C	I
Sustain and reinforce changes		A	R	R	R	R	C	C	I
Confirm blockchain conformance to objectives and requirements	I	A	R	C	R	R	R	I	R
Close program after effectiveness review	I	A	C	C	C	C	C	C	R

Source: [28].

a post-incident environment, blockchain can be used to verify the integrity and ensure the availability of resources in an enterprise. In the case of contract management, blockchain can be used to automate a procurement process while eliminating third-party verification. Concerning product monetization, the availability of cashless payment systems with much lower transaction fees is what blockchain can bring to the table. Finally, the supply chain sector is also showing great strides in the adaptation of blockchain technology into its processes.

The implementation of blockchain technology using an established framework such as COBIT 2019 can provide an effective and efficient approach for optimizing the benefits of a blockchain organizational use case while also optimizing project risk and resource utilization.

CORE CONCEPTS

- There are numerous benefits that blockchain technology can provide to an enterprise for disaster recovery. One of blockchain's greatest attributes is transparency, which in turn can further foster responsibility and accountability within organizations. A blockchain system is far more secure than other record-keeping systems since it is a cryptographically secure, distributed ledger system aimed at avoiding single points of failure.
- Contract management in organizations revolves around negotiating procurement terms and conditions. Procurement fraud occurs in several phases and blockchain can potentially eliminate these types of fraud through self-executing smart contracts. Blockchains can also eliminate third-party transaction fees. The immutable nature of blockchains provides a tamper-proof audit trail for any revisions made in the contract's terms and conditions.
- To gain consumer trust and achieve value proposition, manufacturers and merchandisers seem to welcome a decentralized, distributed ledger that can be used by both wholesale and retail entities to gain optimal efficiency, provenance, transparency and ease of doing business. The areas where blockchain can have a significant impact on supply chains include minimizing counterfeits, reduced the cost of doing business, and more effective compliance and quality management.
- The two main objectives of asset management are the appreciation of client assets over the long haul, and the effective mitigation of clients' portfolio risk. Blockchain technology can offer many solutions that can accommodate rapid and radical changes to the asset management industry by enabling open collaborations which make adding new business partners easier. By adopting blockchain for asset management, increased consistency is ensured by the technology since all data blocks must be verified by all the processing nodes based on consensus. The distributed structure of blockchain technology will also offer an increase in operational efficiency for B2B transactions within the asset management industry.
- Third-party vendors have access to customer records and subletting this information to other parties will pose a threat of information leak. Blockchain technology eliminates the need for an intermediary to authorize/authenticate a certain subject cryptographically without revealing any identifying information, thus bridging a security gap. The encryption attributes of blockchain ensure that data are not accessible to unauthorized parties. An immutable blockchain is secure and tamperproof, reducing or eliminating opportunities for crimes like money laundering. Anti-Money Laundering (AML) monitoring can be automated with smart contracts. Since the entire history of transactions is updated in real time, audits can be performed more effectively.
- Implementing a blockchain using the COBIT 2019 framework provides value through efficient governance and management. Several challenges should be considered before implementing blockchains. These drawbacks include cost, acceptance within the organization, as well as a lack of usability, and interoperability. The COBIT 2019 approach uses the continual improvement lifecycle. There are three components within the cycle, namely, continual improvement,

change enablement and program management. Successful blockchain implementation can be more effectively accomplished by following the COBIT implementation phases.

ACTIVITY FOR BETTER UNDERSTANDING

Readers who are interested in developing a blockchain use case for their organization are encouraged to acquire a copy of ISACA's *COBIT 2019 Framework: Introduction and Methodology* as a general IT governance and management study guide toward creating a blockchain business case.

Pay particular attention to Chapter 9 (*Getting Started with COBIT: Making the case*) to familiarize yourself with the various considerations that must be part of a blockchain business case. Consider the following questions as you further evaluate the implementation of a blockchain system in your organization:

- In what ways can a blockchain implementation provide more 'value' to your organization?
- What are some likely business challenges and 'pain points' that a blockchain implementation can alleviate in your organization?
- What are some specific blockchain implementation challenges that you are likely to encounter?
- What is the 'current state' related to a blockchain implementation? What about the 'future state'? What is the vision for this project?
- How would the project risk elements be managed?
- How will the blockchain implementation project stand from a cost/benefit perspective?
- Who is going to 'champion' (should be an upper management member and a major supporter of the initiative) the blockchain implementation project in your organization?

REFERENCES

[1] Mcisaac, J., Brulle, J., Burg, J., Tarnacki, G., Sullivan, C., & Wassel, R. (2019). Blockchain Technology for Disaster and Refugee Relief Operations. *Prehospital and Disaster Medicine, 34*(S1). doi:10.1017/s1049023x1900222x

[2] G., & P. (2019). Use case of Blockchain in Disaster Management: A Conceptual View. Retrieved January 29, 2021, from http://www.aims-international.org/aims17/17ACD/PDF/A374-Final.pdf.

[3] G., & I. (2020, April 14). Improving Disaster Relief Efforts with Blockchain Technology. Retrieved January 31, 2021, from https://www.hyperledger.org/blog/2020/04/14/improving-disaster-relief-efforts-with-blockchain-technology

[4] Gupta, V., Kumar, D., & Kumar, P. (2016). Exploring Disaster Recovery Parameters in an Enterprise Application. Retrieved February 1, 2021, from https://ieeexplore.ieee.org/stamp/redirect.jsp?arnumber=7542345

[5] EC-Council. (2020, June 16). The Importance of a Disaster Recovery Plan for Business Continuity. Retrieved January 25, 2021, from https://blog.eccouncil.org/the-importance-of-a-disaster-recovery-plan-for-business-continuity/

[6] Mks&h. (2021, February 11). 5 Elements of a Disaster Recovery Plan – Is Your Business Prepared? Retrieved January 27, 2021, from https://mksh.com/5-elements-of-a-disaster-recovery-plan-is-your-business-prepared/

[7] Flesch, M. (2019, April 30). Developing a Disaster Recovery Plan – 5 Essential Elements. Retrieved February 6, 2021, from https://www.gflesch.com/elevity-it-blog/essential-elements-for-developing-a-disaster-recovery-plan

[8] Schiff, J. L. (2016, July 5). 8 Ingredients of an Effective Disaster Recovery Plan. Retrieved February 8, 2021, from https://www.csoonline.com/article/3091716/8-ingredients-of-an-effective-disaster-recovery-plan.html?page=2

[9] Posey, B. (2019, June 5). Can You Use Blockchain for Disaster Recovery Purposes? Retrieved February 24, 2021, from https://searchdisasterrecovery.techtarget.com/answer/Can-you-use-blockchain-for-disaster-recovery-purposes

[10] CB Insights. (2021, February 11). 12 Tech Trends to Watch Closely in 2021. Retrieved from https://www.cbinsights.com/research/industries-disrupted blockchain/#: ~:text=CLOUD COMPUTING text= Blockchain technology can help facilitate, security, and computational power. Text=The product builds on the, and blockchain-based data sharing
[11] ACFE. (2010). Report to the Nations on Occupational Fraud and Abuse. pp. 1–84. Retrieved January 30, 2021, from https://www.acfe.com/uploadedFiles/ACFE_Website/Content/documents/rttn-2010.pdf.
[12] Martyn, P. (2018, June 14). Blockchain Contract Management: A Perfect Application. Retrieved January 26, 2021, from https://www.forbes.com/sites/paulmartyn/2018/06/14/blockchain-contract-management-a-perfect-application/?sh=3ffe33a349a8
[13] Nikoways. (n.d.). Contract Management. Retrieved from https://www.nikoways.com/cms/
[14] How Contract Management Solutions and Blockchain Work Together. (2020, February 12). Retrieved February 26, 2021, from https://101blockchains.com/contract-management-solutions-and-blockchain/
[15] Team, X. (2021, January 20). Top 5 Areas for Blockchain in Distribution. Retrieved January 30, 2021, from https://xcelpros.com/top-5-areas-for-blockchain-in-distribution/
[16] Sharma, T. K. (2020, November 20). The Largest Music Festivals in Thailand Is Adopting Blockchain for Cashless Payment. Retrieved February 3, 2021, from https://www.blockchain-council.org/blockchain/the-largest-music-festivals-in-thailand-is-adopting-blockchain-for-cashless-payment/
[17] Si, C., Rui, S., Zhuangyu, R., Jiaqi, Y., Yani, S., & Jinyu, Z. (2017, November 23). A Blockchain-Based Supply Chain Quality Management Framework. Retrieved February 3, 2021, from https://ieeexplore.ieee.org/document/8119146
[18] Blockchain Technology for Supply Chains: A Must or a Maybe? (2020, October 20). Retrieved February 2, 2021, from https://www.mckinsey.com/business-functions/operations/our-insights/blockchain-technology-for-supply-chainsa-must-or-a-maybe#
[19] Wolfson, R. (2020, September 9). Walmart Canada's Blockchain Freight Supply Chain Proving Its Value. Retrieved September 25, 2020, from https://cointelegraph.com/news/walmart-canada-s-blockchain-freight-supply-chain-proving-its-value
[20] A-B Trust (2020, February 18). Overview, Purpose, How It Works, Advantages. Retrieved February12, 2021, from https://corporatefinanceinstitute.com/resources/knowledge/finance/asset-management/a
[21] Maayan, G. D. (2019, October 10). How Is Blockchain Changing the Face of Asset Management? Retrieved February 4, 2021, from https://www.dataversity.net/how-is-blockchain-changing-the-face-of-asset-management/#
[22] Martin, R. (2019, February 8). How Blockchain Will Transform the Asset Management Industry. Retrieved October 8, 2020, from https://igniteoutsourcing.com/blockchain/blockchain-asset-management/
[23] Rijmenam, D. M. (2019, August 14). How Blockchain Will Give Consumers Ownership of Their Data. Retrieved February 21, 2021, from https://markvanrijmenam.medium.com/how-blockchain-will-give-consumers-ownership-of-their-data-3e90020107e6.
[24] Drinkwater, D. (2018, February 6). 6 Use Cases for Blockchain in Security. Retrieved February 21, 2021, from https://www.csoonline.com/article/3252213/6-use-cases-for-blockchain-in-security.html
[25] Platform, K. (2020, July 1). How Blockchain Will Impact the Legal and Compliance? Retrieved February 19, 2021, from https://medium.com/@kratosplatform/how-blockchain-will-impact-the-legal-compliance-7122818cc9f6
[26] Bryan, J. (2019, September 6). What Assurance Leaders Need to Know about Blockchain. Retrieved February 21, 2021, from https://www.gartner.com/smarterwithgartner/what-assurance-leaders-need-to-know-about-blockchain/#:~:text=and new risks
[27] Sinhasane, S. (2020, November 25). Top 10 EHR Implementation Challenges and How to Overcome Them. Retrieved February 21, 2021, from https://mobisoftinfotech.com/resources/blog/top-10-ehr-implementation-challenges-and-how-to-overcome-them/
[28] COBIT 2019 Implementation Guide: Implementing and Optimizing an Information and Technology Governance Solution. (2018). Schaumberg, IL: ISACA.
[29] Singh, N. (2019, August 8). Blockchain Best Practices – Enterprise Blockchains Principles. Retrieved March 1, 2021, from https://101blockchains.com/blockchain-best-practices/#2
[30] Blockchain Technology for Digital Contracting. (n.d.). Retrieved January 28, 2021, from https://www.accenture.com/se-en/case-studies/about/blockchain-contracts-harnessing-new-technology

[31] PricewaterhouseCoopers. (n.d.). Blockchain: A New Tool to Cut Costs. Retrieved March 15, 2021, from https://www.pwc.com/m1/en/media-centre/articles/blockchain-new-tool-to-cut-costs.html

[32] Palmer, D. (2018, January 17). Cyber-Attacks Are a Top Three Risk to Society, Alongside Natural Disaster, and Extreme Weather. Retrieved March 23, 2021, from https://www.zdnet.com/article/cyber-attacks-are-a-top-three-risk-to-society-alongside-natural-disaster-and-extreme-weather/

12 Blockchain Risk, Governance Compliance, Assessment and Mitigation

Bikramjit Pandher, Manoj Kumar Nagavamshi, Poojaben Prajapati and Vijay Kundru

ABSTRACT

This chapter discusses a total of seventeen blockchain-related risks, four major regulations/standards and ten COSO-based blockchain good practices. The risks, regulations and good practices presented in this chapter should enable assurance professionals to conduct more relevant and effective blockchain-related risk assessments.

Topics covered in this chapter include:

- *High-level risk management considerations, such as:*
  - *Ledger transparency risks*
  - *Security risks*
  - *Operational risks*
  - *Application development risks*
  - *Cryptocurrency and payment risks*
  - *Regulatory compliance risks*
- *An overview of some major regulatory compliance requirements, including,*
  - *PCI DSS*
  - *HIPAA*
  - *GDPR*
  - *PIPEDA*
- *A discussion of blockchain good practices based on the COSO framework*

With the wide use of blockchain technology, various types of attacks have been reported in recent years. *Slowmist Hacked* reports that, in 2020, 122 blockchain-related attacks resulted in a total economic loss of more than 3.7 billion USD. Most of the attacks targeted three main applications of blockchain – cryptocurrency exchanges, blockchain wallets and the Ethereum platform-based decentralized applications [10]. Although blockchain is still a young technology, the rapid development of this technology has been accompanied by some security risks. Blockchain-related security risks can originate from any internal factor or external entity [21].

BLOCKCHAIN TECHNOLOGY RISKS

Blockchains store data by distributing it across several nodes and the data stored on the ledger may be accessed by node operators. There exist three major types of potential liability risks linked to blockchain technologies use, namely ledger transparency, security and operational risks. Table 12.1 shows the main risks associated with blockchains and their sub-components.

DOI: 10.1201/9781003211723-12

TABLE 12.1
Major Blockchain Risks

Main Risk Categories	Risk Sub-components
Ledger transparency risk	a) Violation of data privacy b) Insider trading and market abuse c) Identity theft
Security risks	a) Faulty data input caused by employee error or fraud b) Hacker attacks c) Vendor/supply chain risks
Operational risks	a) Faulty coding b) Key person risk c) Disaster recovery/business continuity risks d) Consensus risks e) Smart contract risks
Application development risks	a) Hard to integrate protocols b) Lack of standardization c) Blockchain development talent scarcity
Cryptocurrency and payment risks	a) Cryptocurrency valuations and volatility b) Lack of an intermediary to provide customer support in case of lost money transfers
Regulatory compliance risks	a) New laws and/or current updated or expanded regulation may cause compliance risks.

Source: [12].

LEDGER TRANSPARENCY RISKS

One of the most important and useful attributes of blockchain is the extent of transparency that the technology can provide through its various encryption and control mechanisms. Be that as it may, this ledger transparency can also translate to a number of risks as follows:

Violation of data privacy: There is an inherent conflict between the transparency features of distributed ledgers and data privacy. The disclosure of private data on a blockchain will most likely breach jurisdictional data privacy laws. Penalties for breaking data privacy laws are more severe in some jurisdictions than others. As such, entities using blockchain will need to carefully evaluate their data protection responsibilities in order to remain compliant with various applicable privacy laws.

Insider trading and market abuse: A variety of financial crimes – including insider trading, front running, tipping and market manipulation – can be encouraged if blockchains are used to store confidential, market-related information. The European Securities and Markets Authority (ESMA) is concerned that blockchain's shared and public features could encourage manipulation of financial markets and other unfair practices.

Identity theft: Although transparency is useful when ensuring data integrity, the threat of identity theft is also introduced with ledger transparency. Specifically, if only the private key is needed to redirect assets and no central ledger authority can block access upon notification of failure, the private key itself becomes the object of illegal activities [23].

BLOCKCHAIN SECURITY RISKS

Maintaining system confidentiality, integrity and availability (the CIA triad) is a significant concern for information technologies. With the high-profile nature of blockchain technology, precautionary steps must be taken to ensure a safer platform for participating companies. blockchain security risks include:

Faulty data/employee security risks: Although decentralized in nature, blockchains still require some human interactions to produce accurate outcomes. For instance, a business using blockchain systems uses servers or automated systems which are operated by employees. Such systems are password-protected; thus, user credentials can be stolen or compromised. Other employee-related risks include negligence, fraud or security issues caused by blockchain security operators not being able to perform their duties, due to circumstances such as absenteeism. Additionally, some data integrity issues, such as fraudulent or manipulated input data may not get flagged during the block creation consensus procedure and thus be stored as such, leading to ensuing data integrity problems or attacks.

Hacker risks: Malicious attackers are capable of launching a number of blockchain-related attacks, involving data at rest, in transmission or end-point wallet attacks. Chapters 13 and 14 provide a detailed discussion related to major blockchain and smart contracts attacks. Chapter 3 discusses crypto-wallet features, advantages and vulnerabilities.

Vendor and supply chain security risks: Third-party entities often provide vendor solutions such as payment processing, smart contracts, wallets, etc. Such third-party-created tools and platforms can have any number of vulnerabilities like untested or weak code, inadequate security or software [12, 19].

Operational Risks

Operational risks refer to the potential losses resulting from faulty or inadequate internal processes, people, systems controls or even major external events. It stands to reason that a major operational failure may also result in a significant security breach and/or significant reputational loss for the enterprise. Operational risks associated with blockchains include:

Faulty coding: Once code errors get implemented on the blockchain platform, the error can propagate throughout the network nodes. Perfect coding is extremely rare; as such, errors or bugs will always negatively impact an application's performance or security. As discussed in Chapter 14, faulty coding should be considered a major operational (also security) risk for smart contract development and implementation.

Key person risk: In any business organization, there are key people or experts who understand the structure of software codes. As mere mortals, such key persons could become ill, tired, mentally unwell or corrupted. Irrespective of the reason, if the trust placed on the key people is broken, the ledger's security and reliability will be at risk [23].

Disaster recovery/business continuity risks: Blockchain technologies are flexible due to the built-in redundancy resulting from the distributed nature of the technology. However, some of the business processes built around a blockchain may have vulnerabilities that may lead to cyberattacks. To reduce such risks, enterprises need to include appropriate blockchain-related incident response considerations into their disaster recovery and business continuity plans.

Smart contracts risks: Smart contracts implemented on a blockchain network will apply rules and procedures in a consistent manner to participants across the network. Smart contracts have the potential to be used in encoding complex business, legal and financial arrangements on the blockchain network. When mapping these arrangements from a physical to a digital framework, several security risks could surface. Thus, smart contracts must be capable of exception handling. The consequences of exception handling when implemented in programmatic outputs on the blockchain framework must be tested across the world for all other smart contracts within the network. Chapter 14 discusses smart contract vulnerabilities and attacks in detail [6].

TABLE 12.2
Consensus Risks in a Distributed Infrastructure

Risks	Description
Crash failure	The risk of a process or system stopping midway and in a manner that is irreversible in direction.
Omission failure	The risk of omission happens for various reasons, such as buffer overflow, transmitter malfunction, collisions at the MAC layer, receiver out of range (impacts and transmission defects).
Security failure	The risk of a security failure following a security attack. Consensus information may become corrupted as a result of security failures.
Software failure	The risk of software failure is introduced as a result of software design and modeling defects. Also, other risks such as crash or omission risks can arise from such failures.
Byzantine failure	The risk of failure in systems requiring consensus that may translate into different symptoms or issues for various system participants. Byzantine failures may cause detection systems to malfunction, thereby reducing fault tolerance.
Temporal failure	Temporal risks happen due to delays in meeting cutoff times. The right results may be generated, but the delay makes the results outdated and thus useless.
Environmental perturbations	The risk arising from an inability to respond to environmental changes. As a result, the right outcome may lose its relevancy. Examples of environment parameters include time of day, network topology, user demand, etc. Distributed systems are predicted to adapt to the environment.

Source: [4].

Blockchain consensus risks: The *consensus* mechanism acts as a backbone for distributed systems. In particular, the consensus is applicable when multiple nodes or processes agree upon data items or states. The formation of blocks occurs when thousands of computers scattered worldwide form an agreement [19]. It is also important to note that distributed consensus protocols – like proof-of-work – provide a probabilistic finality. A probabilistic finality occurs when a transaction's finality increases as more blocks are added to the blockchain after the transaction [3].

Consensus applies to situations in which nodes must maintain a common state of a data item. In *permissionless* blockchain, nodes are mostly anonymous. Sometimes, a valid transaction mismatches with an invalid transaction and results in a *fork*. A blockchain fork is caused by introducing changes to the software protocol – like adding a new tampered transaction blocks. Table 12.2 depicts the risks related to consensus in a distributed infrastructure [4].

BLOCKCHAIN APPLICATION DEVELOPMENT RISKS

This category of risk stems from the fact that blockchain technology is still in its infancy, and as such, it remains a 'work in process'. As such, blockchain application development risks merit special attention. Chapters 8, 9 and 10 discuss blockchain application design, development and testing considerations in more detail. Blockchain development risks include:

Blockchain protocols are hard to integrate: According to Deloitte, implementing blockchain protocols into a project presents some significant challenges. For instance, the integration of Hyperledger Fabric and Ethereum can be problematic, as an integration layer should be present between the Hyperledger Fabric Protocol and the Ethereum Protocol in order to facilitate the sharing of information. However, implementing an integration layer between two different enterprise systems may be a challenging task [6].

Lack of standardization: When a technology has a wide variety of frameworks, it results in a lack of standardization. Currently, all blockchain platforms – including cryptocurrency platforms – are suffering from a lack of standardization [12].

Blockchain talent scarcity: According to a 2019 new job market report from *Janco Associates*, numerous blockchain positions remain unfilled. Salaries for blockchain developers range from $119,000 to $176,000. According to *Januaitis*, 'In 2019, 96,000 new IT jobs will be created, and blockchain will represent 5% to 8% of those jobs'. LinkedIn also identified blockchain development as the no. 1 IT emerging position. According to Arun Ghosh, KPMG's U.S. blockchain leader, 'One of the reasons for the sudden increase in blockchain related postings is because many enterprise projects are moving from proof of concept in 2017 to pilots in 2018 to production systems this year' [16].

CRYPTOCURRENCY AND PAYMENT CONSIDERATIONS

Inflated valuation of cryptocurrencies: Currently cryptocurrency valuation is marked by wild price fluctuations, indicative of the current investor psychology, as more entities and investors are embracing cryptocurrencies as an alternative asset class. This current instability and volatility of cryptocurrency pricing is among the most significant risks to new cryptocurrency projects [12].

Lack of an intermediary to provide network-level customer support: As discussed in previous chapters, in the case of most public blockchain networks, no network-level customer support is available to address issues concerning 'lost in transfer' payments (payments made or received by error).

BLOCKCHAIN REGULATORY COMPLIANCE RISKS

Regulatory requirements affecting blockchain use cases are going to evolve over the coming decades. As such, audit professionals need to remain apprised of future regulatory mandates, changes and/or updates. The uncertain IT regulatory landscape presents its own blockchain use case challenges and risks. The next section discusses four major regulatory mandates that may be applicable to a blockchain use case.

REGULATORY COMPLIANCE CONSIDERATIONS

Regulatory requirements refer to an organization's mandated adherence to laws, regulations, guidelines and specifications relevant to its business processes. Violations of regulatory compliance often result in legal sanctions, which may include substantial fines. This chapter discusses four important regulatory requirements that include Payment Card Industry Data Security Standards (PCI DSS), Health Insurance Portability and Accountability Act (HIPAA), General Data Protection Regulation (GDPR), and Personal Information Protection and Electronic Documents Act (PIPEDA) [18].

PAYMENT CARD INDUSTRY DATA SECURITY STANDARDS

Any business which involves cash payments for trading goods and services needs to have a secure chain of trust. Each customer in the system trusts that user data are used and stored securely. Consequently, to maintain customer trust and security control to prevent security breaches and credit card fraud, most payment card network operators collectively formed the Payment Card Industry Security Standard Council (PCI DSS) in 2004. Subsequently, the PCI DSS act was released on December 15, 2004 [15].

Under this act, a standardized set of practices and security measures are followed to secure credit card holders' data. Moreover, the user data are overseen by the service provider and merchant. For instance, whenever a purchase between a customer and merchant occurs, the retailer is required to store the payment card information and transactional data and send them to the card issuer over a network. Meanwhile, keeping the data within the merchant's infrastructure and authenticating it by transmitting it over the network to the card issuer increases the likelihood of a network security breach. Such breaches can severely impact card users, merchants and card issuers.

To overcome such issues, PCI DSS implements 12 main security requirements, which are classified into six groups [15]:

1. Build and maintain a secure network
   - Protecting cardholder data can be achieved by installing and maintaining a firewall configuration.
   - Update vendor-provided default passwords as an attacker can easily access them.
2. Protect cardholder data
   - Cardholder data stored within the merchant's infrastructure needs to be protected.
   - Cardholder data transmitted over public networks to the card issuer needs to be encrypted.
3. Maintain a vulnerability management program
   - Use anti-virus software with updated definitions.
   - Develop and maintain a secure environment around the system and application.
4. Implement strong access control measures
   - Only the business need should have access to the cardholder data.
   - Every individual who gets access to a computer must get a unique ID.
   - Physical access to cardholder data servers must be restricted.
5. Regularly monitor and test networks
   - All access to cardholder data needs to be tracked and monitored all the time.
   - All security procedures should regularly be tested in systems and processes.
6. Maintain an information security policy
   - Policies must be developed for employees and contractors addressing information security.

Blockchain Considerations Related to Payment Card Processing

In any payment card transaction, the most crucial step involves the identification of the cardholder. The identification and authorization of the cardholder entail a multi-step procedure that involves multiple third-party entities. Therefore, a default digital identity verification technology is used in private blockchain and is known as *public-key cryptography*. Digital signatures are used in public-key cryptography to validate a person, indicating whether an individual has the right key to sign the digital assets and to prove authenticity.

In the process of validation, two essential keys are involved, *public* and *private* keys. The cardholder uses the private key – which is kept secret (such as a credit card pin). The public key's role is to verify a transaction related to the private key mathematically, proving the signature using the private key.

Steps for identification of an owner using a public key.

- The card owner, while purchasing a product, submits a card for payment and signs with the help of a private key (PIN), which generates a digital token.
- For the verification process, the public key and digital tokens are sent to the merchant's bank.
- Both public and private keys help the merchant's bank to verify digital tokens [11].

Chapter 2 discusses various cryptographic concepts in more detail.

Health Insurance Portability and Accountability Act

To ensure the confidentiality of medical records and to increase the quality of health care services, the American Congress released an act in 1996 named the *Health Insurance Portability and*

Accountability Act (HIPAA) [23]. The act for medical service made the protection of patient information of the utmost importance. In general, securing information in the medical industry is challenging due to the constant data transmissions, making it vulnerable to attacks. Therefore, blockchain technology can help protect patient information.

Significant provisions in Health Insurance Portability and Accountability Act include:

1. *Portability*: Available and renewable health coverage is provided under portability provisions, and within the defined guidelines, pre-existing clauses are removed for an individual who switches employer-sponsored health plans.
2. *Medicaid Integrity Program/fraud and abuse*: The Medicaid Integrity Program (MIP) ensures a source of funding for integrity programs at the Centers for Medicare & Medicaid Services (CMS) and extends its authority to employ anti-fraud contractors.
3. *Administrative simplification*: Within the administrative simplification standard, transaction and code sets, identifiers, security and privacy rules are implemented across the healthcare industry [1].

Blockchain Considerations Related to HIPAA

The traditional method of owning information still applies when information is stored on a single server or entity. This is in contrast with the typical attributes of blockchain technology, which often promotes data availability to the public. Moreover, blockchains may be controlled by multiple entities, which serves as a way to verify the unaltered state of stored information. Also, the addition of a transactional block of data to the chain is performed using encryption and verification with a cryptographic algorithm provided by other computers on the network.

The block is considered valid if a general agreement is reached among most nodes on the network. The following agreement results in the addition of a new block to the chain, which is then updated throughout the network. Since blockchains are typically set up as decentralized networks, they serve as a security mechanism to ensure that the blockchain's data are not manipulated or compromised. The transactions on the blockchain are secured, trusted, auditable and immutable. Tampering with data would become much simpler if the blockchain were to work under a single entity system when considering a 51 percent attack scenario.

Blockchain is not just a reliable tool because of the decentralization but can also offer access controls to ensure that the information which is contained in its blocks is not compromised. Blockchains do not solely focus on single decentralization techniques but also incorporate cryptographic algorithms, mathematics and economic models, in collaboration with peer-to-peer networks which collectively assist in securing the data [5]. Be that as it may, it is extremely important to emphasize the need for proper and highly effective access controls, as well as encryption and compliance mandates when considering a health information-related blockchain use case.

General Data Protection Regulation

In May 2018, the *General Data Protection Regulation Act* came into existence in the European Union (EU) based on the *1995 Data Protection Directive*. The GDPR achieves two goals with a single act; namely, the free movement of personal data between the EU's member states, and second, the development of a framework of fundamental rights' protection which is based on the right to data protection in Article 8 of the Charter of Fundamental Rights [9].

GDPR sets out seven main principles as follows [2]:

Lawfulness, fairness, and transparency: When data are collected by an organization, the organization must state why these data are being collected and where they are going to be processed.

For the processing of personal data to be lawful, it needs to specify ground rules called 'lawful basis' for processing. Personal data processing must always be fair as well as lawful. Transparency is linked to the fundamentality of fairness. Data processing should be clear, open and honest with the people whose data are going to be processed.

Purpose limitation: The usage of personal data should be limited and processed for a specific purpose. In other words, individuals should know how and for what purposed their data are going to be processed.

Data minimization: For a specific purpose, only the minimum amount of data needed to fulfill a specified task should be used. According to GDPR, individuals have the *right to update* incomplete data that are insufficient for a purpose. An individual can also have data deleted whenever they want it to be deleted based on the *right to be forgotten*.

Accuracy: Individuals have a right to request that incomplete data should be cleared or rectified within 30 days.

To uphold accuracy, the following should be considered:

- To ensure the accuracy of data, reasonable steps should be considered.
- The source and status of the personal data should be legitimate.
- Consider if it is necessary to regularly update the information.
- For the accuracy of data, carefully consider any challenges.

Storage limitation: The personal data which are collected fairly and lawfully cannot be stored for an indeterminate period if it is not processed within a specified timeframe. Storing the unused data may create a risk.

Integrity and confidentiality: Poor information security leaves the organizational system and service at risk. The security measures which provide integrity and confidentiality should seek to ensure the following:

- Data should be accessed, changed, deleted by the authorized individual only.
- Data should hold accuracy and completeness concerning the processing of data.
- Data should remain accessible and usable (i.e., if anything happens to the personal data, such as alteration or deletion, the organization should recover and restore the original data).

Accountability: The controller must be responsible if anything happens to the personal data of an individual. Accountability is an opportunity for an organization to show respect toward individuals' personal data.

Blockchain considerations in GDPR: Due to the implementation of encryption technology in blockchain, data breaches in healthcare systems can be prevented, or at least significantly reduced. Smart contracts may be implemented as record management systems for electronic medical records (EMR) in multi-institutional settings. The result of this implementation proved that blockchain could ensure high availability and full control over personal data by providing solutions whereby users keep pointers to the origin of the data. Accessing and further processing of personal data can be managed with the distributed ledger technology via third parties. The private key provided to the data subject can implement access control on users' data to third parties based on categories. A multi-layered system offers more control over personal data. Data exchange between a user and a service provider or a purchaser is stored on the smart contracts, with the access layer serving as the connection between the blockchain and offline storage. The following framework would lead users to control and own personal data while service providers are guests with delegated permissions. The user only changes the set of permissions, and by implementing changes access to corresponding data is given. In the final step, the hash value of the data is stored in the hash storage layer generated only when the personal data are verified by certain trusted authorities – such as the government – who could verify an individual's data. In the on-going process, the hash of the verified data is stored on the blockchain [9].

The Personal Information Protection and Electronic Documents Act

To secure individual information within a private sector organization, a federal law – PIPEDA – was passed in Canada in 2001 and fully implemented on January 1st, 2004. To protect information under PIPEDA, a business must follow ten fair principles as stated below [8, 20]:

Accountability: An organization is responsible for securing an individual's personal information and shall appoint an individual who is accountable for an organization's compliance. The Chief Privacy Officer will deal with issues and concerns regarding the policies and procedures surrounding these data.

Identify purposes: The purpose for which the information is collected should be defined at the time of collection. A purpose statement should be created by the organization.

Consent: The knowledge or consent of an individual at the time of the collection of information must be taken. The consented use will avoid issues that may arise in the future.

Limiting collection: Information collected for a specific purpose should be used only for the defined purpose. Misleading or deceiving the individual is not acceptable.

Limiting use, disclosure, and retention of personal information: An organization should only use the information for the specific purpose for which collection was done in the first place, and information must be discarded once the particular use is over.

Accuracy: The information of the individual should be accurate, complete and up to date.

Safeguard: The organization should be responsible for securing individual information from loss or theft and unauthorized access. An agreement of confidentiality must be signed between the third party having access to the information.

Openness: The privacy policy of an organization should be readily available to anyone.

Individual access: The individual has a right to know how the data are accessed, used, and disclosed.

Challenging compliance: Individuals should be able to resolve any challenges about compliance with the Chief Privacy Officer of the organization .

Blockchain considerations in PIPEDA: Visibility is one of the key features of a public blockchain that allows everyone to see the information on the blockchain. As the PIPEDA act focuses on privacy rights, the fundamental aspect of the public blockchain (i.e., visibility) makes it challenging to meet data subjects' rights related to use, disclosure, collection and consent.

The answer to the above problem is to use the cryptographic techniques (discussed in Chapter 2) to safeguard users' privacy rights by anonymizing individuals' identities on the blockchain, i.e., storing all personally identifiable information off the chain to secure private information on public blockchains.

The following approach provides privacy by restricting access to the data except for the *trusted third parties (TTP)*, which are required. The personal information stored on the chain requires a hash of that transaction's details on the chain by trusted third parties. During the hashing process, the observer would be unable to collect personal information from the transaction itself. However, the above process requires both counterparties to verify the hash of data on-chain and match it with records off-chain. By the following method, a public blockchain would be able to keep the transactional details private [8].

THE COSO FRAMEWORK

In 1985, the *Committee of Sponsoring Organizations of the Treadway Commission* (COSO) was formed to sponsor the National Commission on Fraudulent Financial Reporting, jointly sponsored by the *American Accounting Association (AAA)*, the *American Institute of Certified Public Accountants*

(AICPA), Financial Executives International (FEI), the *Institute of Internal Auditors (IIA),* and the *National Association of Accountants,* now the *Institute of Management Accountants (IMA).*

The key objective of the COSO framework is to efficiently and effectively develop and manage internal control mechanisms that can improve the probability of achieving the goals of the organization while responding to changes in the market and operating environments. The goal in updating the framework was to increase its relevance in the increasingly complex and global business environment so that organizations worldwide can better design, implement and assess internal controls. The COSO framework provides organizations significant benefits, for example, increased confidence that controls mitigate risks to acceptable levels and reliable information supporting sound decision-making.

COSO's key objective is to provide thought leadership on three interrelated topics: enterprise risk management (ERM), internal control and fraud deterrence. Internal control is a process carried out by the board of directors, management and staff of an organization, planned to provide a minimum amount of assurance to achieve the objectives relating to operations, reporting and compliance.

The objectives of internal controls per COSO include operational objectives to ensure the effectiveness and efficiency of the entity's operations, in addition to financial reporting and compliance objectives.

The COSO framework is comprised of five components as follows:

Control environment: A set of standards, processes and structures is described under the control environment, which provides a basis for implementing internal controls across the organization. A right control environment provides reliable financial reporting to internal and external stakeholders and helps ensure ethical behavior.

Risk assessment: In the process of achieving the entity's objectives, an organization will face numerous risks internally and externally. Risk assessment requires a dynamic and iterative process for identifying and assessing risks in an organizational environment. Risk assessment requires management to consider the impact of possible changes in the internal and external environment and take necessary action to manage the potential impact.

Control activities: Control activities are described in policies, procedures and standards which help management at various levels in mitigating the identified risks to achieve the organizational objectives. Control activities can be preventive or detective in nature, depending on the circumstances.

Information and communication: Proper *i*nformation flow is necessary within an entity when meeting its responsibilities and carrying out its objectives. Management utilizes significant information from both internal and external sources to support the functioning of other components in the internal environment while communication is required for sharing the necessary information.

Monitoring activity: Monitoring activities are the separate and on-going evaluations to verify the five components of internal control, including the controls that affect the principles within each component. On-going evaluations are built at different levels of an entity on business processes that provide timely information. Separate evaluations are conducted periodically depending on the assessment of risk [17].

Table 12.3 lists the five components of COSO's internal controls, subdivided into seventeen principles [17].

A PROPOSED LIST OF BLOCKCHAIN GOOD PRACTICES BASED ON COSO

The proposed list of blockchain-related good practices presented in Table 12.4 was put together based on COSO's principles, and COSO's recent guidance document entitled *Blockchain and Internal Control.*

TABLE 12.3
The Five Components and Seventeen Principles of Internal Controls

Components	Principles
Control environment	1. Demonstrate commitment to integrity and ethical values. 2. Exercise oversight responsibility. 3. Establish structure, authority and responsibility. 4. Demonstrate commitment to competence. 5. Enforces accountability
Risk assessment	6. Specify suitable objectives. 7. Identify and analyze risk. 8. Assess fraud risk. 9. Identify and analyze significant changes.
Control activities	10. Select and develop control activities. 11. Select and develop general controls over technology. 12. Deploy control activities through policies and procedures.
Information and communication	13. Use relevant information. 14. Communicate internally. 15. Communicate externally
Monitoring activities	16. Conduct ongoing and/or separate evaluations. 17. Evaluate and communicate deficiencies.

Source: [17].

TABLE 12.4
Recommended Blockchain Good Practices

COSO Component	Related Principles	Implications of Blockchain	Recommended Blockchain-Related Good Practices
Informative Reference	*COSO Internal Control-Integrated Framework Principles [7]*	*Blockchain and Internal Control, 2020, p.8 [7]*	*Blockchain and Internal Control, 2020 [7]* *COBIT 2019: Introduction and Methodology [13]*
Control environment	1. The organization demonstrates a commitment to integrity and ethical values. 2. The board of directors demonstrates independence from management and exercises oversight of the development and performance of internal controls. 3. Management establishes – with board oversight – structures, reporting lines and appropriate authorities and responsibilities in the pursuit of objectives.	Blockchain may be a tool to help facilitate an effective control environment (e.g., by recording transactions with minimal human intervention). However, many of the principles within this component deal primarily with human behavior, such as management promoting integrity and ethics, which – even with other technologies – blockchain is not able to assess. The greater	Inherent blockchain technology attributes such as its immutability and provenance should be properly leveraged – whenever appropriate – to help further foster the organization's commitment to integrity and ethical values. *[Relates to operational risks – see Table 12.1]* --- The organization should strive to hire and retain highly competent blockchain professionals, including blockchain solutions architects, blockchain application developers, as well as blockchain security and assurance specialists. The hiring

(continued)

TABLE 12.4 (Continued)
Recommended Blockchain Good Practices

COSO Component	Related Principles	Implications of Blockchain	Recommended Blockchain-Related Good Practices
Informative Reference	*COSO Internal Control-Integrated Framework Principles [7]*	*Blockchain and Internal Control, 2020, p.8 [7]*	*Blockchain and Internal Control, 2020 [7]* *COBIT 2019: Introduction and Methodology [13]*
	4. The organization demonstrates a commitment to attract, develop and retain competent individuals in alignment with objectives. 5. The organization holds individuals accountable for their internal control responsibilities in the pursuit of objectives.	an entity with other entities or persons participating in a blockchain and how to manage the control environment as a result.	challenge relates to the intertwining of process should include processes to help ensure that those hired not only possess the required blockchain-related skillsets but also share the organization's ethical values and future vision. *[Relates to application development risks – see Table 12.1]* -- As with all other organizational positions, each blockchain position must include a precise list of responsibilities and authorities which in turn should be used as metrics for blockchain professionals' annual performance reviews. Each blockchain position must also clearly indicate in what way(s) blockchain specialists are to be held accountable while achieving the intended blockchain-related organizational objectives. *[Relates to operational risks – see Table 12.1]*
Risk assessment	6. The organization specifies objectives with sufficient clarity to enable the identification and assessment of risks relating to objectives. 7. The organization identifies risks to the achievement of its objectives across the entity and analyzes risks as a basis for determining how the risks should be managed. 8. The organization considers the potential for fraud in assessing risks to the achievement of objectives.	Blockchain creates new risks and simultaneously helps to mitigate extant risks by promoting accountability, maintaining record integrity and providing an irrefutable record (i.e., a person or an organization cannot deny or contest their role in authorizing/sending a message or record).	The organization should develop a comprehensive blockchain use case for each intended blockchain application. The use case should include the project's intended objectives, its architectural considerations, as well as an initial use case risk assessment. *[Relates to operational risks – see Table 12.1]* -- Following its initial risk assessment, the blockchain program should undergo frequent (annual) risk assessments to determine its potential risk vulnerabilities.

TABLE 12.4 (Continued)
Recommended Blockchain Good Practices

COSO Component	Related Principles	Implications of Blockchain	Recommended Blockchain-Related Good Practices
Informative Reference	*COSO Internal Control-Integrated Framework Principles [7]*	*Blockchain and Internal Control, 2020, p.8 [7]*	*Blockchain and Internal Control, 2020 [7]* *COBIT 2019: Introduction and Methodology [13]*
	9. The organization identifies and assesses changes that could significantly affect the system of internal control.		*[Relates to operational risks – see Table 12.1]* --- The on-going blockchain risk assessment must also consider blockchain-related fraud risks. Fraud risks refer to the intentional misuse of the technology to derive an economic benefit. Fraud risks apply to internal and external attacks. *[Relates to security risks – see Table 12.1]* As part of the risk management processes, the organization must also implement an effective change control process to help ensure that all blockchain-related changes are properly approved by the IT steering committee and that all changes are properly documented. *[Relates to operational and security risks – see Table 12.1]*
Control activities	10. The organization selects and develops control activities that contribute to the mitigation of risks as well as to the achievement of objectives to acceptable levels. 11. The organization selects and develops general control activities over technology to support the achievement of objectives. 12. The organization deploys control activities through policies that establish what is expected and procedures that put policies into action.	Blockchain can act as a tool to help facilitate control activities. Blockchain and smart contracts can be powerful means of effectively and efficiently conducting global business (e.g., by minimizing human error and opportunities for fraud). However, the collaborative aspects of blockchain can introduce additional complexity, particularly when the technology is decentralized, and there is no single party accountable for the systems that fall under internal control over financial reporting (ICFR).	Blockchain platforms must possess appropriate internal control mechanisms to help ensure that each organizational blockchain provides 'value' by achieving its intended objective(s) effectively and efficiently. According to COBIT 2019, 'value' can be defined as providing maximum intended benefit to the organization while optimizing risks and resource utilization (see Chapter 11 for a COBIT discussion. *[Relates to operational risks – see Table 12.1]* ---

(*continued*)

TABLE 12.4 (Continued)
Recommended Blockchain Good Practices

COSO Component	Related Principles	Implications of Blockchain	Recommended Blockchain-Related Good Practices
Informative Reference	*COSO Internal Control-Integrated Framework Principles [7]*	*Blockchain and Internal Control, 2020, p.8 [7]*	*Blockchain and Internal Control, 2020 [7]* *COBIT 2019: Introduction and Methodology [13]*
			The organization must ensure that its general IT controls (e.g., firewalls, intrusion detection and/or prevention technologies, identity management tools, etc.) are properly implemented and functioning to be able to implement appropriate blockchain-specific security controls. *[Relates to security risks– see Table 12.1]* Appropriate blockchain-related policies and procedures must be put in place and reviewed regularly to help ensure blockchain is adequately secure and properly functioning through the relevant and effective implementation of relevant controls. *[Relates to security risks – see Table 12.1]*
Information and communications	13. The organization obtains or generates and uses relevant, quality information to support the functioning of internal controls. 14. The organization internally communicates information, including objectives and responsibilities for internal control, necessary to support the functioning of internal controls. 15. The organization communicates with external parties regarding matters affecting the functioning of internal controls.	The inherent attributes of blockchain promote enhanced visibility of transactions and availability of data and can create new avenues for management to communicate financial information to key stakeholders in a faster and more effective manner. One aspect for management to consider in applying blockchain is the availability of information to support the financial books and records, and related auditability of information transacted on a blockchain.	The organization should strive to use its blockchain technology to further improve the flow of data and information – whenever appropriate to do so – with both internal and external constituencies. *[Relates to operational and regulatory compliance risks – see Table 12.1]* -- Specifically, objectives and assigned responsibilities to help ensure the proper functioning of blockchain-related controls must be effectively communicated in both a top-down and bottom-up management approach, to employees as well as internal and external audit functions, etc. *[Relates to operational and regulatory compliance risks – see Table 12.1]*

TABLE 12.4 (Continued)
Recommended Blockchain Good Practices

COSO Component	Related Principles	Implications of Blockchain	Recommended Blockchain-Related Good Practices
Informative Reference	*COSO Internal Control-Integrated Framework Principles [7]*	*Blockchain and Internal Control, 2020, p.8 [7]*	*Blockchain and Internal Control, 2020 [7]* *COBIT 2019: Introduction and Methodology [13]*
Monitoring activities	16. The organization selects, develops, and performs ongoing and/or separate evaluations to ascertain whether the components of internal control are present and functioning. 17. The organization evaluates and communicates internal control deficiencies promptly to those parties responsible for taking corrective action – including senior management and the board of directors, as appropriate.	The promise of blockchain to facilitate monitoring more often, on more topics, in more detail, may change practice considerably. The use of smart contracts and standardized business rules, in conjunction with Internet of Things (IoT) devices, may alter how monitoring is performed.	The organization should strive to optimize and automate blockchain-related monitoring activities through the use of blockchain components – such as smart contracts, blockchain's various inherent cryptographic tools, and blockchain's ability to connect with IoT interface devices. *[Relates to operational and security risks – see Table 12.1]* --- The internal and/or external audit functions should aim to develop an effective audit approach to aid in detecting and reporting various blockchain-related deficiencies and related recommendations. This will help expediently address those deficiencies. *[Relates to operational, security and regulatory compliance risks – see Table 12.1]*

CONCLUSIONS AND RECOMMENDATIONS

In this chapter, a discussion of various blockchain risks has been presented. The reader should note that the risk discussions in this chapter are by no means comprehensive in nature, as we are dealing with a rapidly developing technology that is currently being experimented with through numerous use cases. Adopting new technologies always comes with a fear of unknown threats. Blockchain technology creates an immutable audit trail of transactions and improves efficiencies for entities. Besides the benefits, there are numerous risks caused by blockchain technology. To respond to such risks, firms should consider establishing a robust risk management strategy, governance and controls framework. Some of the major risks include the following:

- *Ledger transparency risk*: One of the most important and useful attributes of blockchain is the extent of transparency that the technology can provide through its various encryption and control mechanisms. Be that as it may, this ledger transparency can also translate to a number of risks.
- *Security risks*: Distributed ledger technologies, although known for superior security, are not 100 percent fool-proof. Blockchain needs precautionary steps to function safely with fewer risks.

- *Operational risks*: Operational risks refer to the potential loss resulting from faulty or inadequate internal processes, people or systems controls or even major external events.
- *Application development risks*: When implementing blockchain into new technologies (like IoT and AI) unknown development risks also crop up.
- *Cryptocurrency and payment risks*: The current instability and volatility of all cryptocurrencies as a store of value is a significant risk of new cryptocurrency projects.
- *Regulatory compliance risks*: As a developing technology still in its infancy, regulatory requirements affecting blockchain use cases are going to evolve over the coming decades. As such, audit professionals need to remain apprised of future regulatory mandates, changes and/or updates.

This chapter also focuses on four different regulatory requirements – PCI-DSS, HIPAA, GDPR, PIPEDA – and how blockchains may help enforce or stay compliant with these regulations. Within blockchain implementations, each regulatory requirement is processed differently. For example, PCI-DSS makes use of asymmetric cryptography keys, GDPR requires smart contracts implementation, whereas PIPEDA and HIPAA use trusted third parties (TTP) and peer-to-peer networks, respectively.

In addition to the above, the chapter also presented several COSO-mapped practices for efficiently managing and developing various internal control mechanisms for an organization in order to achieve its intended objectives and improve performance.

EXERCISE FOR BETTER UNDERSTANDING

Download and study Deloitte's *Blockchain and Internal Control* document from the Internet (use the Google search engine to locate this document).

1. If your organization has adopted the COSO governance model and is interested in adopting blockchain technology, what additional key points can you take away from studying this document?
2. In what way(s) does the review of the *Blockchain and Internal Control* document help you better understand the proposed COSO risk framework discussed in this chapter?
3. If contemplating a blockchain use case or a blockchain audit program, in what ways do the 17 risks discussed in this chapter influence or shape your blockchain risk assessments?

REFERENCES

[1] Advise Tech. (2021). *3 Major Provisions*. Advise Tech. Retrieved February 24, 2021, from https://advisetech.com/3-major-provisions/

[2] Arampatzis, A. (2020). *What Are the 7 Principles of GDPR?* (ITEGRITI) Retrieved February 24, 2021, from https://itegriti.com/2020/blog/what-are-seven-principles-gdpr/?gclid=Cj0KCQiA0-6ABhDMARIsAFVdQv8OhPGa87njb1h2upGOiqgRjLjPhXTffG9WRt3O0tW2Ia4Rve8Q-EAaAi7LEALw_wcB

[3] Caron, F. (2017). *Blockchain: Identifying Risk on the Road to Distributed Ledgers*. Retrieved from https://www.isaca.org/resources/isaca-journal/issues/2017/volume-5/blockchain-identifying-risk-on-the-road-to-distributed-ledgers

[4] Chaudhry, N., & Yousaf, M. M. (2018). Consensus Algorithms in Blockchain: Comparative Analysis, Challenges and Opportunities. *2018 12th International Conference on Open Source Systems and Technologies (ICOSST), Lahore, Pakistan*, 54–63. doi:10.1109/ICOSST.2018.8632190

[5] DeLeon, C., Choi, Y., & Ryoo, J. (2018). Blockchain and the Protection of Patient Information: Using Blockchain to Protect the Information of Patients in Line with HIPAA (Work-in-Progress). *2018 International Conference on Software Security and Assurance (ICSSA), Seoul, Korea (South)*, 34–37. doi:10.1109/ICSSA45270.2018.00017

[6] Deloitte. (2017, January). *Risk functions Need to Play an Active Role in Shaping Blockchain Strategy.* Retrieved March 3, 2021, from https://www2.deloitte.com/content/dam/Deloitte/us/Documents/financial-services/us-fsi-blockchain-risk-management.pdf

[7] Deloitte. (2020, July). Blockchain and Internal Control: The COSO Perspective. Retrieved March 2021, from https://www.aicpa.org/content/dam/aicpa/interestareas/informationtechnology/downloadabledocuments/blockchain-and-internal-control-the-coso-perspective.pdf

[8] Edwards, D. (2005, August). *Personal Information Protection and Electronic Documents Act.* Ministry of Agriculture, Food and Rural Affairs. Retrieved February 2, 2021, from http://www.omafra.gov.on.ca/english/nfporgs/05-049.htm

[9] Finck, M. (2019, July 24). *Blockchain and the General Data Protection Regulation.* European Parliament Research Service. Retrieved February 2, 2021, from https://www.europarl.europa.eu/thinktank/en/document.html?reference=EPRS_STU%282019%29634445

[10] Foremski, T. (2021, January 19). *Billions Were Stolen in Blockchain Hacks Last Year.* Retrieved from https://www.zdnet.com/article/billions-were-stolen-in-blockchain-hacks-in-2020/

[11] Godfrey-Welch, D., Lagrois, R., Law, J., Anderwald, R., & Engels, D. (2018). Blockchain in Payment Card Systems. *SMU Data Science Review, 1*: No. 1, Article 3. Retrieved from https://scholar.smu.edu/datasciencereview/vol1/iss1/3

[12] Iredale, G. (2021, February 21). *Blockchain Risks Every CIO Should Know.* 101 Blockchains. Retrieved March 3, 2021, from https://101blockchains.com/blockchain-risks/#

[13] Lanter, D. (2019). COBIT 2019 Framework: Introduction and Methodology. Retrieved March, 2021, from https://community.mis.temple.edu/mis5203sec001sp2019/files/2019/01/COBIT-2019-Framework-Introduction-and-Methodology_res_eng_1118.pdf

[14] Li, J., Lee, J. S., & Chang, C. C. (2008). Preserving PHI in Compliance with HIPAA Privacy/Security Regulations Using Cryptographic Techniques. *2008 International Conference on Intelligent Information Hiding and Multimedia Signal Processing.* doi:10.1109/iih-msp.2008.38

[15] Liu, J., Xiao, Y., Chen, H., Ozdemir, S., Dodle, S., & Singh, V. (2010). A Survey of Payment Card Industry Data Security Standard. *IEEE Communications Surveys & Tutorials, 12*, 287–303. doi:10.1109/SURV.2010.031810.00083

[16] Mearian, L. (2019, April 8). Blockchain Jobs Remain Unfilled, While Skilled Workers Are Being Poached. Retrieved April 1, 2021, from https://www.computerworld.com/article/3387441/blockchain-jobs-remain-unfilled-while-skilled-workers-are-being-poached.html

[17] Schandl, A., & Foster, P. L. (2019). *Internal Control – Integrated Framework.* Retrieved February 20, 2021, from https://www.coso.org/Documents/COSO-CROWE-COSO-Internal-Control-Integrated-Framework.pdf

[18] Seaman, J. (2020). *An Integrated Data Security Standard Guide.* Retrieved from https://doi.org/10.1007/978-1-4842-5808-8

[19] Singh, N. (2019). *Blockchain Risks Every CIO Should Know.* Retrieved from https://101blockchains.com/blockchain-risks/

[20] Walters, N. (2019, April 20). Privacy Law Issues in Blockchains: An Analysis of PIPEDA, the GDPR, and Proposals for Compliance. *17 Canadian Journal of Law and Technology*, 276. Retrieved from https://ssrn.com/abstract=3481701

[21] Wang, H., Wang, Y., Cao, Z., Li, Z., & Xiong, G. (2019). An Overview of Blockchain Security Analysis. Yun X. et al. (eds), *Cyber Security. CNCERT 2018. Communications in Computer and Information Science*, 970. Retrieved from https://doi.org/10.1007/978-981-13-6621-5_5

[22] Wang, Q., Huang, J., Wang, S., Chen, Y., Zhang, P., & He, L. (2019). A Comparative Study of Blockchain Consensus Algorithms. *2nd International Symposium on Big Data and Applied Statistics*, 1437.

[23] Zetzsche, D., Buckley, R., & Arner, D. (2017). The Distributed Liability of Distributed Ledgers: Legal Risks of Blockchain. *SSRN Electronic Journal.* doi:10.2139/ssrn.3018214

13 Blockchain User, Network and System-Level Attacks and Mitigation

Nishtha Baria, Dharmil Parmar and Vidhi Panchal

ABSTRACT

This chapter is the first of two chapters providing a security and assurance technical discussion of currently known blockchain technology attacks. Whereas, in the next chapter, the focus is placed solely on smart contract vulnerabilities and attacks, this chapter covers blockchain vulnerabilities and ensuing attacks at the user, system and network level.

The user-level attacks include stolen private keys and the use of blockchain-specific malware to initiate a security breach at the user-level. Readers are reminded that while all blockchain nodes are users, not all users are nodes. Node-level attacks stem primarily due to vulnerabilities associated with shared vulnerabilities and membership service provider (MSP) vulnerabilities.

Furthermore, a total of eight node/network-level attacks and four system-level attacks are also presented and discussed. The chapter also provides readers with some suggested mitigation techniques for the various attacks discussed at each level (user/node/network/system).

The chapter concludes with a discussion of some security best practices for each discussed attack, as well as a quick discussion of Ethereum, Hyperledger and Corda inherent security measures.

INTRODUCTION

Blockchain technology has been expanding and changing since its beginnings, circa 2009. By generating additional value, blockchain has the potential to disrupt sectors and reshape business structures in the same way that the Internet did in the 1990s. The positive result of blockchain is due to a large extent to its distributed and decentralized ledger feature; however, every coin has two sides, and that is also the case with blockchain technology as well. Attackers have used various means to conduct a variety of assaults due to the decentralized nature of blockchain's operating environment. As blockchain technology is evolving so are the attacks related to this technology. In short, the growing popularity of blockchain raises new security and privacy concerns.

Blockchain attacks are divided into four levels. User-level attacks include attacks initiated by or due to users' mistakes. Network-level attacks are attacks that are due to errors in the network's properties or improper controls over the network. Node-level attacks are attacks that are caused due to corrupt tasks on nodes. Finally, system-level attacks are the result of common system coding errors. Attacks and mitigation go hand in hand and so, the necessary mitigation techniques are also discussed in this chapter.

DOI: 10.1201/9781003211723-13

BLOCKCHAIN USER, NETWORK AND SYSTEM-LEVEL ATTACKS AND MITIGATION

USER-LEVEL ATTACKS

User-level attacks are attacks that are caused due to interactions between users. Often, these user-level attacks occur due to the carelessness of the users, hence resulting in compromise of the private key. User-level attacks may also be a result of the presence of malware. In addition, user-oriented attacks may occur when software is not regularly updated, resulting in unwanted bugs. Placing adequate focus on user-level security may help prevent some of these user-level attacks [16].

STOLEN PRIVATE KEY

As discussed in Chapter 2, two keys are needed to encrypt data. One is the public key (available to everyone), and the other is the private key (restricted only to a specific user). Public-key cryptography uses a pair of keys for the encryption process, beginning with the encryption of the data with the help of the public key, after which the encrypted data can only be decrypted with the help of the private key. Private-key cryptography works the other way around, by encrypting the information using the private key and decrypting the data using the public key. In the event of a key's compromise or theft, there is an increased likelihood that the user's blockchain account will also be compromised, thereby facilitating unauthorized transactions from the compromised account [16].

MITIGATION

Hardware wallets are hardware devices that maintain keys and the various wallet public addresses related to blockchains. Technically, hardware wallets are USB-type drives used in storing transaction details which can be operated without the use of batteries [33]. Hardware wallets are one of the more secure and appropriate methods of storing a private key. Some of the hardware wallet brands include *Ledger*, *Ledger Nano S*, *Ledger Nano X*, *Trezor*, etc. [21].

Software wallets are computer programs that can be installed on an operating system. Software wallets are sometimes referred to as *desktop wallets* and offer improved privacy, ease of use, absence of third-party involvement and secrecy [33]. Software wallets can also be downloaded to a mobile device, a desktop or a laptop computer. The keys maintained in the software wallets are connected to the Internet and therefore are susceptible to theft if the wallet gets hacked. Desktop wallets are typically used for holding smaller amounts of cryptocurrency needed for frequent trading or other transactional purposes, due to their higher user-friendliness. Software wallet brands include *Coinbase*, *Gemini* and *Blockfi* [20].

Paper wallets are wallets that are typically printed on paper in the form of QR codes. Paper wallets are prone to many dangers due to several inherent flaws; some notable ones being the inability to transfer partial funds and reusability. Typically, paper wallets are mostly used for cold storage purposes [33]. Some of the paper wallet brands include *Bitcoin Paper wallet*, *Wallet Generator* and *Mycelium*, etc. [4].

Cloud wallets are available online and can be accessed over the Internet with the private keys being stored on a cloud. Cloud wallets are of two types – *hosted* and *non-hosted*. Non-hosted wallets keep the funds and assets under the program's control. Cloud wallets, although vulnerable to DDoS attacks, are useful for handling smaller balances and transactions. Some cloud wallet brands include *Coinbase*, *Binance* and *Meta Mask* [33].

MALWARE

Users communicate with the blockchain network through computers vulnerable to various malware. The presence of malware on a user's machine can compromise the blockchain's protective measures in several ways. The malware could take control of the user's account if the user's private key is saved on the device. Malware infecting a user's machine could gain access to other blockchain users' IP addresses and could use this information to target those users. Infected computers may also be used for crypto-jacking – where malware exploits a computer's computing resources to perform proof of work calculations for the attacker's gain [26].

The following sections describe the attributes of some popular blockchain-related malware.

ElectroRAT

Cybercriminals developed fake cryptocurrency apps in order to trick users into downloading a new strain of malware on users' computers with the end goal of stealing victims' funds, according to security firm *Intezer Labs*. The phony apps were called *Jamm, eTrade/Kintum* and *DaoPoker*. These apps were featured on dedicated websites called *jamm.to, kintum.io* and *daopker.com*, respectively. The first two apps seemed to have a simple cryptocurrency trading platform, whereas the third was a cryptocurrency gaming app. According to Intezer researchers, the apps also included a surprise in the form of a new malware strain called *ElectroRAT* that was hidden within the app and possessed the abilities to log keystrokes, take screenshots, upload files from storage, import files and execute commands on the victim's console, among other things. The malware was designed to capture cryptocurrency wallet keys and then exhaust victims' funds. User who have installed the trojanized applications should uninstall processes and remove all related files as soon as possible, transfer remaining funds to a new crypto wallet, and update passwords [8].

PCASTLE

Trend Micro first noticed a wave of attacks using *PCASTLE*, an obscured PowerShell script to target primarily China-based systems. XMRig, a crypto mining malware, was used in a number of attacks in 2018. Following that, the campaign peaked before levelling off, according to the security company [5]. In its most basic form, the PCASTLE malware is a cryptocurrency miner Trojan. When the PCASTLE Trojan arrives on a server, it launches the XMRig mining tool and begins mining *Monero*, its preferred cryptocurrency. All of the obtained coins are forwarded to the attacker's cryptocurrency wallet. Machines with Chinese IP addresses account for 92 percent of the PCASTLE attacks [15]. When one of the propagation methods succeeds, the first-layer PowerShell script is downloaded using a scheduled task or the RunOnce registry key. The PowerShell script in the first layer will attempt to obtain a list of URLs contained inside the script. The PowerShell command will be downloaded, executed and saved as a new scheduled mission. A PowerShell script will be run as part of the scheduled mission, which will download and run the second-layer PowerShell script. Once downloading and executing the third-layer PowerShell script, the malware will send device information to its command-and-control (C&C) server. Based on the aforementioned system information, the third-layer PowerShell script will download the cryptocurrency mining package, which will then be loaded into its own PowerShell process. It will also download the PCASTLE script part, which is in charge of other tasks, such as propagation. The campaign employs propagation techniques similar to those used in the previous campaign, but in a single PowerShell script (PCASTLE), also employs the EternalBlue hack, as well as brute force and the pass-the-hash methods. As long as there are systems to infect, the infection cycle will continue. This campaign, for example, makes use of an exploit for a flaw that has already been patched. As such, it is a good idea to use virtual patching in legacy or embedded systems. Furthermore, an effective multi-layered defensive strategy should be implemented. For example, additional protection protocols, such as activity detection, should be implemented to detect and prevent the execution of anomalous routines, unauthorized programs or scripts.

Sandboxes aid in the prevention of malware-related traffic by containing untrusted data and shellcode, while firewalls and intrusion prevention systems can block malware-related traffic. Authentication and encryption mechanisms help avoid unauthorized alterations of targeted applications. More robust account credentials can effectively reduce brute-force and dictionary attacks. The proper patching and updating of systems are important. Trend Micro endpoint security solutions with behavior monitoring features, such as the Smart Protection Suites and Worry-Free Business Security solutions, are examples of tools that can help protect users and businesses from these challenges by detecting malicious files, scripts and messages, as well as banning all associated malicious URLs [3].

Lemon Duck

The Lemon Duck cryptocurrency-mining malware has recently increased its operation, according to *Cisco Talos*, and is employing many techniques that are likely to be detected by malware detection software, but not immediately apparent to end-users. The Lemon Duck vulnerability has a cryptocurrency mining payload that exploits computer resources in order to mine the Monero virtual currency. To spread across the network, the malware uses a variety of methods, including emailing infected RTF files, psexec, WMI and SMB exploits, including the notorious *Eternal Blue* and *SMBGhost* threats that damage Windows 10 devices. Remote desk protocol (RDP) brute-forcing is also supported by some versions. The malware also employs tools like *Mimikatz*, which aid the botnet in increasing the number of systems in its mining pool. A PowerShell loading script is used to spread the virus, which is replicated from other infected systems via SMB, email or external USB drives. The code that exploits the BlueKeep vulnerability is also typically present. The malware has executable modules that are downloaded and controlled by the main module, which interacts with the C2 servers via HTTP. Malware that mine cryptocurrency can be expensive in terms of stolen computing cycles and power consumption. Although organizations should concentrate on safeguarding their most important properties, they should not overlook risks that are not specifically aimed at their infrastructure. To identify new resource-stealing threats such as crypto miners, defenders must be continuously cautious and track the actions of systems within their networks [36].

Ransomware

Ransom malware, often known as ransomware, is a type of malware that prevents individuals from using the network or accessing files, demanding payment in order for the user to regain access to databases and systems. In other words, ransomware normally works by encrypting and freezing the victim's computer system, then demanding a ransom to unlock it. Failing to comply with the requirements results in data being permanently lost. Phishing email attachments, corrupted software apps, corrupted external media and compromised websites are all ways for ransomware to spread. In certain circumstances, hackers have exploited the remote desktop protocol and other methods that do not require any user participation.

On May 7, 2019, the city of Baltimore (USA) found that its federal systems had been attacked with *RobbinHood*, a ransomware that had locked a huge number of crucial files. The city was compelled to shut down its systems to prevent the ransomware from expanding, but not before the ransomware infected voice mail, email and a database of parking charges. According to a printout of the computerized ransom note, the city could only retrieve its highjacked data for 3 Bitcoins per system or 13 Bitcoins in total. The initial attack aftermath was projected to cost the city USD 10 million, on top of the USD 8 million lost due to the city's inability to handle payments [38].

Mitigation

Organizations need to have an incident response plan in place for what to do in the case of a ransomware attack. Regular system and network scans should be performed, with antivirus products set to update signatures automatically. To prevent phishing attacks from reaching the network, anti-spam solutions need to be considered. The use of cautionary banners in all emails received from outside sources, reminding recipients of the risks of clicking on links and opening attachments is also highly recommended. If at all possible, a centralized patch management tool should also be considered. To avoid the execution of programs in frequent ransomware system areas, such as temporary files, application whitelisting and effective software restriction policies (SRP) are necessary. The use of virtual machines, logical and physical network separation and data isolation are other techniques that deserve consideration. Finally, the use of the least privilege principle should also be strictly enforced [27].

Implementing System Updates

Blockchain is created as a piece of software that runs on user computers. This program must be updated on a regular basis, as blockchain programmers put out updates that users must download and install diligently. The consequences of failing to perform the upgrades vary depending on the type of update. Failure to upgrade should restrict or eliminate the user's option to access and interact with the blockchain if the patch is merely to improve functionality. However, if the update addresses a security vulnerability, failing to install it could jeopardize the users' blockchain accounts or the blockchain infrastructure as a whole [26].

NODE-LEVEL ATTACKS

Blockchain nodes are similar to blockchain users as they have the same impact on the conservation and upkeep of the blockchain. Therefore, nodes are prone to almost all possible attacks on users. To protect nodes, practices like antivirus installation, patching and updating of software, and verification of all software configurations should be implemented.

According to Poston (2020), there are a number of notable node-level threats as follows [25]:

Software misconfigurations refer to errors or bugs in the software which may be intentional or unintentional, leaving the node vulnerable to attacks.

Denial of service attacks (DOS) affect processing nodes as these have the power to perform consensus in blockchains. Hence, a DoS/DDoS attack on nodes could bring a substantial portion of consensus control into the hands of the attacker.

Malicious transactions negatively affect blockchain's ability to perform data transactions, which can sometimes come from untrusted sources, making the blockchain vulnerable to attacks like the injection attack.

Blockchain-focused malware allows attackers to steal private keys, monitor addresses, launch DDoS attacks and/or filter or inhibit communication or transactions processing.

Shared vulnerability is another threat that can affect effective blockchain operations. To further explain, nodes are an integral part of the blockchain's infrastructure. A blockchain platform is built on nodes as they perform functionalities like storage, spreading and preserving of the blockchain's data.

Every node is a user, which means that all the security concerns of users apply to nodes as well. Even so, because of the increased involvement of nodes in the blockchain, a security breach can have a greater impact on a node than the breach would on a user [28]. If a node's private key is stolen, the account can be used to target the consensus algorithm by acting in the attacker's best interests. If a large enough number of nodes are compromised (over 50%), an attacker will be able to take ownership of the blockchain [25]. In other words, if one of the nodes is compromised, this could lead to compromise of other nodes as well. For instance, if an attacker gains access to a considerable number of nodes, then the attacker can tamper with the consensus algorithm, and available transactions or blocks, in favor of the attacker. This type of attack on nodes is called a *51% attack*. The attack gives control to the attacker, thus weakening and disrupting the entire ecosystem [2].

Failure of nodes to keep their own copies of the blockchain program up to date may also have a significant effect on the security of the blockchain. The failure of nodes to install a function upgrade can cause a disruption of the blockchain network which will affect how consensus is done, resulting in two networks with lesser protection [25].

Mitigation

A blockchain network should obtain node consensus in its process of block creation. Divergent blockchains may emerge for both benevolent and harmful reasons in practice. The blockchain protocol is set up to accept the blockchain that has the greatest amount of 'work' behind it. An attacker who constructs a divergent blockchain and builds it up to have more work than the actual blockchain can take advantage of a 51% attack. Checkpointing is meant to counteract this sort of attack by requiring nodes to store specific blocks on a regular basis and refuse to accept any diverging chain that does not contain these blocks [26].

A novel approach that can help to mitigate 51% attacks is proof of adjourn (PoAj), a proposed consensus protocol providing strong protection, irrespective of an attacker's hashing power. The suggested method's main notion is that it takes the network nodes out of all operations by introducing an adjournment period (AP). Every broadcast block would be treated as an initial block (IB) and won't be verified until the AP is completed. The AP is divided into two phases in this new design to conduct two unique responsibilities in the event of a large PoW assault. The first step involves performing verification tests on a predetermined number of IBs for a predetermined amount of time. The next step is only triggered if the first phase is successful. When the AP is completed, the freshly validated and picked block is put to the blockchain, enabling miners to resume mining [31].

Since processing nodes are also actively involved in the maintenance and security of a blockchain, they may not only be affected by similar security issues as users, but can be attacked in a number of other ways, such as misconfigured MSP services.

Membership service providers (MSPs) are an essential component used to manage identities on the blockchain network. The job of a provider is to authenticate clients who want to join the blockchain network. MSPs decide who will use the blockchain and what privileges they have in a permissioned blockchain. Therefore, if there are any errors or if some part of the blockchain is not functioning properly, the situation may lead to an access control failure. This failure can make the blockchain environment vulnerable to attacks, such as a DOS/DDOS attack. If a DOS/DDoS attack disables a network's MSPs, the network can become unusable, because an intruder could gain control of the network's access controls by compromising the MSP. For this reason, MSPs should implement security measures against DDoS, ransomware and other forms of attacks [25].

To add a layer of security in a membership service, *Intel SGX Technology* can prove helpful. Intel Software Guard Extensions' (SGX) remote corroboration feature helps identify trustworthy nodes on the blockchain. The SGX-enabled membership service may also improve privacy protection and defensive capabilities against malicious assaults. Finally, adopting this hardware-assisted technique might also help with accountability. However, due to the high cost of hardware, this method is not suitable for all membership service applications [13].

Another method entails the use of blacklisting of suspected malicious nodes. Typically, nodes on the blacklist are utilized as part of a botnet army to carry out an attack. However, since the number of such malicious nodes is often quite large, whitelisting may be explored, although scaling the process can prove difficult due to the often large number of nodes involved. When there is implicit trust between IoT services and devices, a whitelisting system will scale. The trusted ones are added to the trusted list at the smart contract level using this approach. As a result, only the most trustworthy users are permitted on the network at any time. The suggested trust approach uses an implicit blockchain consensus process to build trust, which eliminates the need for a third party to create trust between two communicating parties. The trust paradigm is straightforward. It advertises DDoS attacks by putting white- or blacklisted IP addresses on the public Ethereum blockchain and then spreading them to others [34].

BLOCKCHAIN NETWORK ATTACKS

In a blockchain network, processing nodes are tasked with transaction processing and offer different services such as block creation, ledger management, etc. Miners add approved transactions to the blocks and the nodes send and receive these approved transactions. Attackers try to find a flaw in the network to exploit and use in launching various network-level attacks. The following sections discuss various network-level attacks including Majority (51%), selfish mining, DDoS, and other attacks, along with a discussion of appropriate attack mitigation techniques.

FLAWED NETWORK DESIGN

A blockchain network can be run on the internal network infrastructure of a company or even on the Internet. The blockchain network using the infrastructure should be designed in such a way that it meets the needs of the blockchain's users. There is a requirement for a larger amount of bandwidth to reduce or avoid latency for peer-to-peer communications. For example, blockchains are designed to allow nodes to talk in a peer-to-peer fashion. If nodes are located in network segments with different security levels, this may violate the organization's network security protocols. It may result in the duplication of communication, or the communication being blocked to some of its peers. Thus, the infrastructure of the network must be properly set up, thereby protecting the blockchain from certain attacks [17].

POOR OVERALL NETWORK SECURITY

The blockchain network interacts over a conventional network infrastructure, implying that the blockchain's protection is contingent on the underlying network's security. Thus, even physical attacks on network infrastructure can compromise the blockchain's protection. An attacker who has access to communication links or network components may disrupt the blockchain's operation, thereby degrading its security. The physical security of communication infrastructure of an organization should be protected effectively. Poor logical security of the network can also affect the blockchain. Private blockchains may rely on the security controls of the underlying network (firewalls, segmentation, etc.) and may become vulnerable to an attacker circumventing these controls. Control of firewalls and other components could also allow an attacker to segment the blockchain network, leaving it vulnerable to attack [17].

51% AND DOUBLE SPENDING ATTACKS

In this type of attack, there is a group of miners that attain the majority of the network's *hash rate*. This hash rate refers to the measuring unit of the processing power of a blockchain network which in turn may enable blockchain manipulations, such as the ability to stop transactions from being checked, thus making them invalid. Using this attack mode, a detached chain is introduced to the network, thereby making it appear as a legitimate chain. The *double-spending* attack is enabled at this stage. Double spending refers to an instance where the malicious user sends the copy of a transaction to make it look legitimate in order to enable a digital token to be spent more than once. As previously mentioned, in blockchain policy, the longest chain rule is applied. If the malicious users have more than 50% of the total hashing power in the network, the malicious users would be able to operate the longest chain in the blockchain network which is the required proportion for a 51% attack [30].

The 51% attack can also enable attackers to exercise control over the price of a cryptocurrency by also enabling such an attacker to restrict transactions and/or cancel blocks. There are several instances of this type of attack that have occurred over the past few years. In 2016, a Bitcoin mining pool, *GHash.IO*, had gained control over 51% of the total hash rates. In another incident, a group of attackers named *51 crew* hijacked two Ethereum blockchains using the same attack methodology. In May 2018, malicious miners gained control over 51% of the hash rate in *Bitcoin Gold*. Finally, in June 2018, some of the blockchain currencies were also attacked which included *Monacoin*, *Zencash*, *Verge* and *Litecoin Cash* [29].

There are various mitigation techniques used for mitigating this type of attack, some of which are penalties for delayed block submissions, PirlGuard and merged mining. The next sections provide a brief explanation for each one of these mitigation techniques.

A penalty system for delayed block submissions is a security concept proposed by Horizen (2018) which aims at modifying the Satoshi-proposed consensus in order to secure a network against 51% attacks. Satoshi consensus is a set of rules which verify the authenticity of a blockchain network using a combination of a proof-of-work consensus algorithm on a Byzantine Fault Tolerance peer-to-peer network. An extensive increase to the attacking cost in order to discourage deceitful behavior by potential attacks or nodes is proposed by the penalty system. As such, a penalty is applied considering the length of time for which a block remains hidden from the blockchain network. On the basis of the interval length among the blocks, the time is calculated. During the period of notification to the entire network regarding the continuous fork, the miners, participants and exchanges are constrained from executing fraudulent transactions until the delay is removed. The primary focus in this approach is on privately mined chains and is not concerned about a network that is suffering from a fork [30].

Delayed proof of work (dPoW) is another control technique against the double-spending problem; a security solution provided by *Komodo* that served as the catalyst for the development of the dPoW consensus approach. To defend against a 51% attack on the Komodo blockchain, the Komodo chain gives permission to the entire chain to regulate malicious activities. To further explain, an additional layer of security is added, thereby preventing the attackers from executing the 51% attack, as the key property of this security chain is that it does not identify the longest rule. Furthermore, notary nodes that substantiate whether the hash is secured or not for the network are also being utilized. To perform the required tasks, the 64 special nodes throughout the globe are selected as part of the dPoW implementation [18].

The PirlGuard System is a security protocol developed for the Pirl blockchain to address 51% attacks. Inspired by the Horizen penalty system, once an alternative longer chain has been created by an attacker through a process of private mining and then presented to the processing nodes, the PirlGuard system will detect the fraudulent attempt and drop the malicious node. The malicious peer will then be penalized by being required to mine a set number of penalty blocks. The number of penalty blocks is determined based on the amount of blocks that the attacker had created through private mining. The Pirl master nodes in charge of notarizing the blockchain also act as enforcers of such penalties in order to maintain a fair and objective consensus mechanism on the Pirl blockchain.

ChainLocks is a security technique develop for the DASH cryptocurrency designed to provide protection against double-spending attacks while also providing instant transaction confirmation. To extend the active chain, every participant is required to sign the noticed block. The distinct block is verified by the majority of the participants – 60% or more – and a generation of P2P message (*CLSIG*) is done in order to notify each and every other node in the network regarding the event. The generation of *CLSIG* is not possible until and unless a sufficient number of members confirm its validity. As such, a valid signature for authenticity is included in the message and all the nodes can verify the signature on the network. Once the confirmation is achieved, the transaction cannot be nullified or reversed. Apart from 51% attacks, the ChainLock technique is also helpful for the mitigation of various security problems such as selfish mining.

Merged mining is yet another technique beneficial to cryptocurrencies with low hash rates. The hashing power of such cryptocurrencies could be increased by bootstrapping these on the other currency or currencies possessing a higher hashing power. This merging process is helpful for the mitigation of 51% attacks. The process involves the classification of the blockchain into a parent and auxiliary blockchain where both networks are sequential. Additionally, another benefit is the capability for the miners to mine more than a single block concurrently along with the network security enhancement. To reiterate, additional security is built up using the merged mining technique as there is a contribution to the overall hash rate by all the miners of both cryptocurrencies. The assumption from the Satoshi principle is that the blockchain network would always encompass a majority of the honest nodes. Nevertheless, the assumption turns out be false as it is seen that apart from the mining pools, individual attackers may also be able to achieve over half of the network hashing power [30].

DoS Attacks

It is extremely difficulty to execute a DoS attack on a blockchain as a blockchain has no single point of failure. This type of attack can only happen when there is a bottleneck in the network that can facilitate the attack [1]. Generally, blockchains create blocks within a certain range in size and at a specific rate. A large number of spam transactions could be created by an attacker, thereby potentially weakening the network. The number of transactions capacity might be a point of failure, hence inviting a possible DoS attack. Transaction flooding, block forger DoS, permissioned blockchain MSP DoS and artificial difficulty increases are included in the categories of DoS attacks discussed below.

Transaction flooding revolves around flooding the network with transactions, thereby increasing the queue size for transactions waiting to be added to the blocks.

Artificial difficulty increases is based on an attacker's attempt to temporarily increase the difficulty level of a proof of work consensus on a blockchain by engaging additional computing resources in order to push up the PoW degree of difficulty for all the other nodes, then removing the additional computing resources once the PoW difficulty has increased in order to create traffic congestion on the blockchain.

A *block forger DoS* involves the performance of a traditional DoS attack against the next block creator on a proof of stake blockchain in order to prevent the block from being added to the chain.

A permissioned blockchain membership services provider DoS, as its name suggests, entails a DoS attack on a permissioned blockchain MSP in order to cause users access denials to the blockchain.

The mitigation for various types of DoS attack varies on the basis of their implementation. In transaction flooding, users should try to clear flooded transactions from the queue by creating additional blocks intentionally or waiting until the attack is finished. For the increase in artificial difficulty, the difficulty level should be continuously monitored and adjusted accordingly so as to minimize the attack's impact. In permissioned blockchain membership services provider DoS attacks, the traditional DoS protection techniques for the nodes can be applied. Such traditional practices include building a stronger network infrastructure, or increased emphasis on stronger network security [32, 24].

Eclipse Attacks

In an eclipse attack, a single node is isolated from the rest of the network. Hence an eclipse attack facilitates double spending [24]. Bogus connection requests are used by an attacker, hence forcing the connections to restart, thereby significantly increasing overhead. In this type of attack the attacker manages to fully control a node's view of the distributed ledger and network operations

[7]. The attacker often proceeds with the execution of a double-spending attack against the isolated node, or they may perform a denial-of-service attack. The attacker also may opt to use the computational resources of the node for its own benefit as it relates to the consensus algorithm of the blockchain. In order to perform an effective eclipse attack, the attacker might require proper location, added power, scale or malware at its disposal. Location is helpful in eclipse attacks because an attacker may have the ability to intercept the users' messages before the messages reach the rest of the blockchain network using WiFi networks or physical cables.

To mitigate an eclipse attack, the number of connections should be increased, thereby reducing the probability of an attacker who can control all the nodes that the user connects to. A whitelist strategy can be used so that connections can be limited to known or trusted nodes. Proper mitigation strategies should improve the odds that an eclipse attack will be more easily detectable and therefore shorter in duration. Another useful strategy, if possible, is the use of permissioned, private blockchains (depending on the use case) in order to decrease the possibility of malicious nodes finding their way into the network, thereby making such scale-based attacks much more difficult to execute [32].

Replay Attacks

In a replay attack, the attacker spoofs the exchange of messages between two valid parties, thereby gaining access. In other words, the attacker steals the hash key and uses it again to make him/herself appear as a valid user [12]. To help ensure the authenticity and authority of the legitimate users, digital signatures are used while transacting, as the forging of digital signature is not possible for an attacker, thereby nullifying the fake transactions. A replay attack is used by an attacker in order to take the current transaction and resubmitting it to the blockchain as the latest transaction. Since the initial transaction was legitimate, the validity of its digital signature would be agreeable to the blockchain. By replaying the transaction and causing funds to be sent to the attacker, the attacker gets paid twice [26].

There is involvement of the same transactions on two distinct blockchains in a replay attack. For example, users may have the same assets on two ledgers when the cryptocurrency is forked into two separate currencies. To carry out a transaction, a user can choose either of the chains. The sniffing of the transaction data is done on one ledger along with the replay of it on the other ledger by the attacker. For example, in Ethereum, the transaction that is signed on one blockchain is acceptable to all Ethereum blockchains. Hence, transactions made by a user on the test network can be duplicated by the attacker on the public network. As such, countermeasures have been taken by Ethereum for the prevention of replay attacks through its use of *chainID* in transactions [29].

One-time private–public key pairs are used by some blockchains while elliptic curve-based encryption is being used by other blockchains to help improve the detection of the replay attacks [12]. By using nonces or distinctive values in blockchains during each transaction, security could be vastly improved. The mechanism of adding the nonce value would work in such a way that the transaction which is replayed should have a new nonce, or it will not be accepted. Note that any attempts by the attacker to alter the nonce value will result in an invalid digital signature, thereby causing the blockchain network to reject that transaction as invalid [26].

Routing Attacks

A routing attack involves the splitting of two or more isolated groups. Once accomplished, the transfer of information can be delayed. This attack strategy is useful in executing 51%, DOS and double spending attacks. Decentralization can be used to prevent this type of attack. If one connection is being intruded, then all other nodes are affected, but if there is decentralization, one can remove the one affected node, thereby stopping the attack [24].

To mitigate routing attacks, multi-homed nodes can be created. This means that nodes maintain connections with two or more different segments as a means to create more difficulty for an attacker in splitting the network. The more nodes that connect to other nodes in different network segments, the more communications channels an attacker would have to control. If the user uses the trusted communication or known routes selected for communication with nodes in other network segments, then the attacker won't be able to use the Border Gateway Protocol to disrupt nodes communications. The use of encrypted communications channels can also help to ensure that the attacker doesn't get the chance to monitor the network in order to affect network communications. Furthermore, network statistics monitoring should be effective in order to detect attempts by the attacker to try to create a significant amount of network latency as a prerequisite for launching various attacks [26].

Sybil Attack

A Sybil attack refers to the manipulation of online systems whereby an attacker tries to run various nodes on a particular blockchain network. Here, all the nodes are controlled by one person (the attacker) who may choose to deny the transmission of blocks, thereby stopping users from adding data to the network [1]. In a Sybil attack, the malicious user establishes fake nodes in large numbers appearing to be genuine to the other nodes on the network. The bogus nodes on the network for validating unauthorized transactions serve as participants in harming the network. Various devices such as Internet protocol addresses, and virtual machines are used as fraudulent nodes for the attack on the network. Aiming for corruption of a P2P network by creating various false identities, the Sybil attacker makes the assumption that each participating node represents a single real identity. Hence, the attacker gains the capability for denial of the transmitted blocks in order to override the real nodes [30]. In scenarios where large-scale Sybil attacks take place, the attacker controls the majority of the network's computing power. The numbering of the transactions might be changed by the attackers, thereby preventing the transactions from being confirmed. As such, a double spending attack could be enabled.

The consensus algorithm can be helpful in the prevention of this type of attack, as it makes it impractical for an attacker to implement the Sybil attack due to its high cost. Moreover, validation and authorization of participants adding data to the network can be done in order to verify the honesty of the network participants. Finally, the use of a permissioned or private blockchain can prevent the attacker from creating bogus accounts on the blockchain [1].

BLOCKCHAIN SYSTEM-LEVEL ATTACKS

Blockchain system-level attacks occur at the system layer of the blockchain infrastructure. The root cause of this type of attack can often be traced to coding errors on the backend side. System-level blockchain attacks include integer overflow, time warp, buffer out of bound, and race condition attacks as further explained below [26].

Integer/Buffer Overflow Attacks

Buffer overflow attacks involve the unauthorized execution of code, as the hacker will use up memory space not included in the buffer, thereby causing the function pointer to execute a malicious script. Buffer overflows are one of the most common types of arbitrary code execution vulnerabilities. In the majority of instances, hackers can use buffer overflows to write over sensitive settings in the application's memory resulting in access privilege escalation [10].

A buffer overflow attack was used in the *Bitcoin Hack.* Since Bitcoin is a cryptocurrency, it must validate transactions if the quantity of value in the sender's wallet is less than the amount in the sender's wallet. An integer overflow issue allowed exceptionally large transaction amounts to bypass the set of internal controls designed to detect such situations. An intruder took advantage of this flaw by initiating a transfer that sent 184 billion Bitcoins to an attacker-controlled account. This move gave the attacker control over 98 percent of all the Bitcoin supply ever created, effectively destroying the currency's value. The Bitcoin network decided to roll back the blockchain in order to undo the attack. This was a crucial decision since the distributed ledger is intended to be unchangeable, but the roll back move was required in order to maintain the cryptocurrency's value intact. The exploit demonstrated the significance of thorough endpoint protection of blockchain programs prior to deployment. An integer overflow vulnerability is a well-known programming issue that could have been discovered during a security audit [26].

Mitigation

Buffer overflows are caused by programming bugs, thus, the ideal way to control them is to better educate developers on how to avoid them. Buffer overflows are addressed in secure coding textbooks and various related best practices. Employing memory-safe programming languages, frameworks and libraries that can provide secure versions of functions likely to cause buffer overflows are examples of techniques to mitigate buffer overflow vulnerabilities. Application designers should employ ASLR (Address Space Layout Randomization) and PIE (Position Independent Executables); both use executable and address placement to limit the threats of buffer overflow attacks. ASLR was created for use in return-oriented programming, and works by making structure offsets more difficult to determine by randomizing memory locations [35]. Compiler flags and extensions should also be used. However, none of these techniques provides total safety, hence a layered security approach that involves code reviews and application security audits by security teams should be done on a regular basis [10].

Time Stamp Attacks

Blockchains nodes have a built-in counter that allows them to keep track of network time. During the boot-strapping stage, the network time is calculated by getting a version message from adjacent peers and determining its median. If the median time of a neighboring node is more than 70 minutes for any reason, the network time counter is reset to system time, thereby creating an opportunity for cybercriminals. An attacker can manipulate the block's network time by modifying the timestamp on each block. As a result, the block becomes disconnected from the network [23].

The *Verge Hack* used the time stamp attack. To further explain, *Verge* is a cryptocurrency dedicated to protecting its users' privacy. Verge was hacked by combining some of its built-in features in a creative way, such as its difficulty updates, flexible timestamps and consensus algorithm. These three components were used by the attacker to gain control of the Verge blockchain. Due to the lack of a synchronized time server on the blockchain, nodes could accept blocks or transactions with a timestamp of 2 hours behind the current time. Leveraging this feature to their benefit, the hacker made it look like Verge blocks were being generated at a considerably slower rate than expected by changing the timestamp of every other block to an hour in the past. As a result, the complexity of the attacker's mining technique (scrypt) was greatly reduced. As such, *scrypt mining* became absurdly simple, thereby allowing the hacker to carry out a 51% attack with less than 10% of the network's processing resources. Verge discovered a solution to correct the problem by reducing the timestamp window from 2 hours to 15 minutes. As a result, the attacker's effects on the mining difficulty were reduced. The most important lesson to be learned from this exploit is that complexity can compromise security. The combination of three previously mentioned elements incorporated into the Verge cryptocurrency facilitated the attack. The hack also showed how successful time warp attacks may be, as well as the dangers in frequent mining difficulty adjustments [26].

Mitigation

Locking the upper limit of the time frames is a key component in limiting the timestamp attack, as is using system time rather than network time. Also, time frames for acceptance should be limited only to known and trusted users [22].

BUFFER OUT OF BOUNDS ATTACKS

Buffers are temporary storage areas used by software before processing or transmission. However, some programmers neglect to safeguard them adequately. As a result, buffer attacks are still one of the most common attacks. Out-of-bounds reads and writes occur when reading and writing operations are performed outside of the buffer memory region. Unauthorized security breaches and writing to other portions of the software can result in unforeseen consequences [37].

The *EOS vulnerability* used the buffer out of bounds attack. To further explain, EOS is an open-source smart contract system. The EOS software, as a smart contract platform, must be able to decode and execute smart contract files. Malicious smart contracts were able to abuse the EOS blockchain software due to a buffer out of bounds writing flaw in its parsing function. The flaw was discovered and reported to EOS developers by *Qihoo 360* researchers. Their proof of concept showed that its Address Space Layout Randomization (ASLR) could be bypassed using a remote shell. As a result, the blockchain network and all nodes executing it could be fully compromised. The EOS developers were able to provide a patch on their *GitHub* page without fear of the weakness being exploited, since EOS had not yet been released. The found vulnerability emphasized the importance of external code reviews, as well as the fact that insecure smart contract platforms can permit a hostile smart contract to abuse the blockchain software's node [26].

Mitigation

Using a language that either prevents this vulnerability from occurring or offers constructs that make it easy to avoid this vulnerability is a good first step. In addition, the use of tools or extensions that nullify or eliminate buffer overflows and out of bounds when running or compiling the program is also highly recommended. When using the buffer in a loop, the buffer bounds need to be checked in order to make sure writing is not allowed beyond the allocated space [11].

RACE CONDITION ATTACKS

Hackers have leveraged system race conditions to rob money from online banks, brokerage firms, cryptocurrency exchanges, and even free Starbucks coffee! In computing and blockchain language, a race condition is any condition in which any two parts of code, which were initially to be executed one after another, get executed out of order. To further explain, in this type of attack the scheduling algorithm is able to change between the execution of threads, so it becomes difficult to predict what the execution sequence will be like. Race conditions are considered a vulnerability when they affect a security component used by an attacker to create a situation in which a delicate event is executed before the proper completion of the controls. Race condition vulnerabilities are sometimes known as *time of check/time of use* vulnerabilities due to this fact. Explained differently, in order to get around access constraints, race conditions attacks can be employed. Often, hackers employ race conditions attacks on financial institution websites. If a race condition could be discovered on a crucial function like fund transfer, cash withdrawal or credit card payment, the hacker may get access to large amounts of money [19].

The *Lisk Vulnerability* entailed a race condition attack. Lisk is a coin prone to an attack that took advantage of two built-in features. To further expand, Lisk utilizes the final 64 bits of a user's public key's SHA-256 hash as their blockchain address. Until the user executes a transaction, this address isn't linked to a specific public key on the blockchain. Because delivering money to an account does not link an address to a public key, this system became vulnerable. Some accounts with only a positive balance remain untouched. Finding a public key that can be linked to a certain address requires at most 2 to the power of 64 operations, and 'errors' could allow hackers to claim a separate address.

Following the discovery of the issue, Lisk developers released a notice outlining the problem and encouraging individuals to take steps to link their account's address to the public key. However, Lisk is still prone to such exploitations, including race conditions in which an adversary tries to get an address between the first transaction financing it and the transaction sending value to claim it [26].

Mitigation

Resources should be locked in order to ensure that both the processes are happening at the same time, and that no issues are being raised due to an out of order execution. Furthermore, adhering to secure coding and security principles, such as the least privilege principle, and regularly inspecting code will reduce the chance of software compromise [19].

BLOCKCHAIN SECURITY BEST PRACTICES

Various attacks and their best practices are listed in Tables 13.1–13.4.

TABLE 13.1
User-Based Attacks and Best Practices

	Applicable Attack	Best Practices
User-based attacks	Loss of private key	• Wallets can be used to address the issue of private key security as wallets can be used to store the private key. • Key rotation policy should be adopted to address the issue of loss of private keys [33].
	Malware	• When malware is activated, it comes along with files that need to be deleted, and related processes should be terminated and killed. • Checking and assessment of password policy should be done regularly [8].
	Failure to update blockchain software	• Update policy should be revised regularly. • Needed software patches should be applied expediently in order to keep the software up to date. • The use of antivirus software should be implemented along with filtering of e-mails [38].

TABLE 13.2
Node-Level Attacks and Best Practices

	Applicable Attack	Best Practices
Node-level attacks	Shared vulnerability	• Checkpointing is meant to counteract this sort of attack by requiring nodes to store a block(s) on the accepted chain on a regular basis and refusing to accept any diverging chain that does not contain such block(s) [26]. The suggested method's main notion is that it takes the network nodes out of all operations by introducing an adjourn period (AP) [31].
	Misconfigured membership service providers vulnerability	• SGX's remote corroboration feature helps identify each node on the blockchain as trustworthy [13]. • The trust list method helps distinguish the black- and whitelisted nodes [34].

TABLE 13.3
Network-Level Attacks and Best Practices

	Applicable Attack	Best Practices
Network-level attacks	51% attack (majority attack)	• A penalty system for delayed block submissions will require any user to pay a penalty for delayed block submission. The amount of time a block is hidden is taken into consideration, and a fine is applied accordingly. • Delayed proof of work (dPoW): The Komodo blockchain incorporates the delayed proof of work (dPoW) security solution. The primary attribute of this solution is that it does not follow the longest chain rule; thereby adding an additional security layer as to prevent malicious users from executing 51% attacks. • PirlGuard: The consensus algorithm is modified to defend against this type of attack. • ChainLocks: A network-wide voting process is used in ChainLocks that involves a 'first-seen' policy. All participants are required to sign the noticed block in order to extend the active chain. • Merged mining: Multiple cryptocurrencies are allowed to merge their mining activities together [30].
	DOS and DDoS attack	• Traditional practices for the mitigation of DoS include proper network infrastructure and network security along with the required system redundancies. • Flooded transactions should be removed from the queue by creating blocks [26].
	Sybil attack	• Direct and indirect validation of current and new members. Direct validation verifies upcoming members of a blockchain, while indirect validation allow current members to give authorization rights [1].
	Replay attack	• ChainID should be incorporated in transactions. Users who do not enable this wallet feature will remain vulnerable [29].

TABLE 13.4
System-Level Attacks and Best Practices

	Applicable Attack	Best Practices
System-level attacks	Integer/buffer overflow attacks	• Address Space Layout Randomization (ASLR) and Position Independent Code (PIE) should be used by developers [10].
	Time stamp attack	• Use the system time of the node in place of using the network time to make sure the upper limit is kept for the block. • Only trusted peers should be used in the case of time stamping [22].
	Race condition attack	• Safe concurrency alongside locking resources which help mitigate this type of attack [19].
	Buffer out of bounds attack	• One of the easiest ways to mitigate this type of issue is to develop the code in a language that does not allow such issues. • Safe buffer handling functions need to be used to mitigate the effects of buffer out of bounds attacks [35].

Inherent Security Measures

Currently, Hyperledger, Ethereum and Corda are some of the most prominent blockchain platforms. Ethereum and Hyperledger each have their own specific use cases, but Corda R3 is mostly being used in the financial services sector [14]. Given their various applications, this brief review will attempt to explain the three blockchain applications that have also been discussed in several other chapters of this book.

ETHEREUM

After Bitcoin, Ethereum is the second-largest cryptocurrency by market capitalization. However, unlike Bitcoin, Ethereum was not designed to function as a digital currency. Instead, Ethereum's creators set out to create a new type of global, decentralized computing platform that extends blockchain's security and openness to a wide range of applications. On the Ethereum blockchain, anything from financial tools and games to complicated databases is already functioning. And only the developers' imaginations may restrict its future possibilities. '*Ethereum can be used to codify, decentralize, secure, and exchange just about anything,*' according to the *Ethereum Foundation*, a non-profit organization.

The Ethereum blockchain is now securing ETH in the same way that Bitcoin's blockchain is securing Bitcoin. Every transaction is verified and secured by a massive amount of processing power provided by all of the machines on the network, making it very difficult for any third party to intervene. The underlying concepts underpinning cryptocurrencies contribute to their security; the systems are permissionless, and the software is open source, allowing a large number of computer scientists and cryptographers to analyze all elements of its network's security.

Applications that operate on the Ethereum blockchain, on the other hand, are only as safe as their creators make them. As such, any decentralized app considered for use should be thoroughly tested.

In 2021, the Ethereum protocol was updated in order to make it even quicker and more secure. Ethereum 2.0 (also known as ETH2) is a significant update to the Ethereum network. It is intended to help the Ethereum network expand while also improving security, performance and effectiveness [9]. Ethereum is also the first smart contract platform, allowing developers to use a Turing-complete blockchain platform to create smart contracts [26].

Ethereum Smart Contracts

The Ethereum Virtual Machine is where Ethereum smart contracts are executed. Contracts are written in the Solidity programming language (created for Ethereum), and each Ethereum Virtual Machine command has a gas value associated with it. A transaction containing a quantity of gas (fractions of an Ether) is used to pay for the computational work required in running the code to execute a smart contract.

The Ethereum Virtual Machine presently lacks parallelization capabilities, which means that all transactions in a block are processed in order. This is likely to change in the future if Ethereum adds sharing, which allows transactions that interact with separate areas of the distributed ledger's data to run in parallel [26].

Ethereum Security Measures

Ethereum currently provides basic built-in privacy and security features. Its primary built-in feature is the use of public keys for identity management, which provide users with a level of pseudo-anonymity.

Advanced security features (such as confidential transactions) can, however, be implemented via smart contracts in Ethereum. Support for the mathematical operations used in zero-knowledge proofs like zkSNARKS (Zero-Knowledge Succinct Non-Interactive Argument of Knowledge) is also planned for future developments in Ethereum [26].

HYPERLEDGER FABRIC

Hyperledger Fabric is currently the most promising Hyperledger blockchain. It offers a smart contract platform created by IBM and managed by the Linux Foundation. Hyperledger Fabric is designed as a blockchain platform for businesses. The characteristics of the Hyperledger blockchain are as follows. Network security is private, it is permission based and uses X.509 certificates for identity management and offers pluggable consensus algorithms (currently offers transaction ordering with planned full BFT solution) [26].

Hyperledger Fabric enables different companies and network participants to choose their own certificate authority and, as a result, employ a variety of cryptographic techniques for signing, verifying and attestation of identity. This is accomplished via an MSP procedure that operates at either the ordering service or channel levels [32].

Hyperledger Smart Contracts

Smart contracts, also known as chain code, on the Hyperledger blockchain may be written in *Node.js* or *Go* and executed within Docker containers. They are controlled by a third-party program that interacts with the blockchain. When it comes to transaction validation and execution, Hyperledger takes a somewhat different approach than Ethereum and other smart contract platforms.

Its control flow is execute, order and validate. To further explain, nodes run the code, verify its validity, and endorse it if it is legitimate. Transactions are arranged into blocks using the consensus protocol. The validate phase ensures that transactions fit the endorsement policy. Each transaction can have a required set of sponsors and would only be processed if the conditions are satisfied.

Hyperledger's control flow provides a handful of important advantages. To begin, transactions can be verified in parallel rather than sequentially during the execute phase. Next, validation and ordering are two separate steps that nodes can specialize in if they so choose. Because code can be easily evaluated and arranged by specialized nodes, and endorsement rules may be established to match the demands of the company, Hyperledger smart contracts are meant to appeal to business users [26].

Hyperledger Security Measures

There are several characteristics in Hyperledger that are aimed to increase its security. Conventional methods such as LDAP and OpenID Connect are supported by pluggable identity management. Logically, different blockchains have an option for nodes to belong to many channels as required by job functions. Data may also be shared privately via its *gossip protocol*, which only sends data to nodes on a need-to-know basis and is stored off-chain in separate client-side databases [26].

CORDA

Corda is a smart contract compatible platform for documenting and processing financial agreements that uses a distributed ledger [6]. Corda is different from typical blockchains in that no-one on the blockchain network has full visibility of the ledger. Only parties who are engaged in transactions have access to information, granted on a need-to-know basis. From the nodes' perspective, each node in the blockchain network stores its own copy of the present state of the distributed ledger. The ledger of a node is made up of many chains of irreversible states. Transactions take and generate states, and each node maintains a full record of both current and historical versions of all of its states [26]. Corda smart contracts keep track of the progress of financial agreements and other exchanging data amongst two or more identifiable parties in compliance with current and emerging laws [6].

Corda Smart Contracts

Corda is a smart contract platform based on the Java programming language. Contracts can be written in *Java* or *Kotlin*. A smart contract governs each state in Corda. A smart contract takes a transaction involving its state as an input and uses the rules specified in the smart contract to evaluate if the transaction is valid. To be acknowledged in Corda, transactions must be found to be both genuine and distinctive. A transaction's validity is determined by the parties involved. The uniqueness constraint for transactions is intended to ensure that no states are spent twice. Corda nodes can only see payments and states to which they were a participant, so they can't validate a transaction's uniqueness. To address this procedure, Corda blockchains employ a network of notaries who must sign an activity in order for it to be valid. A certificate is assigned to each state in each node's copy of the distributed ledger. A notary will only authorize a transaction if it is the designated notary for all of its previous input states. If the appointed notary refuses to transmit or sign, assets and transactions in Corda might be held captive. A malevolent notary can allow a double spending attack, making the blockchain difficult to reconcile. Therefore, notary trust and security are critical while using the Corda blockchain

Corda Security Measures

The safety of Corda is based mainly on its need-to-know concept and notary network. The consequence of a data leakage is mitigated because users can only observe and engage with transactions in which they have a stake. In contrast to previous blockchains, Corda uses point-to-point TLS-encrypted connections rather than peer-to-peer broadcasts (Table 13.5) [26].

SUMMARY AND CONCLUSIONS

Various blockchain attack vectors are explored systematically in this chapter. Some of the potential vulnerabilities that could lead to blockchain attacks stem from the blockchain infrastructure, architectural design and application context. For the contributing factors, various attacks are outlined by types, which includes various node-level, system-level, network-level and user-level attacks. Node-level attacks include misconfigured membership and shared vulnerability. Network-level attacks include 51% attack, Sybil attack, replay attack, among others. System-level attacks include buffer

TABLE 13.5
Comparison between Ethereum, Hyperledger Fabric and Corda

Security Measures	Ethereum	Hyperledger Fabric	Corda
Confidentiality		X	X
Operating capacity	X		
Protection of data for the transfer	X		
Identity authentication		X	X
Information authentication	X	X	X
Integrity of the system	X		
Works on different platform		X	X
Access control	X	X	X
Testability		X	X
Logging		X	X
Customization	X	X	X

Source: [26].

out of bound, race condition and time stamp. User-level attacks include malware and stolen private keys.

Defensive techniques are also presented for the attacks that can help mitigate various underlying blockchain vulnerabilities. In addition, inherent security measures are highlighted for Hyperledger, Corda and Ethereum platforms.

A number of real-life attack scenarios are presented and discussed in this chapter. Applying greater focus on implementing stronger security measures for various blockchain technologies should be the new aim for organizations in order to leverage blockchain technology effectively and securely.

CORE CONCEPTS

- Although blockchain technology is an ever-evolving technology with superior security features, it is comprised of various core components on which attacks can be performed.
- User-level attacks are those that occur as a result of user engagements. Users' carelessness, the theft of the private key or the existence of malware can all lead to blockchain security breaches. Mitigation techniques like hardware, paper and cloud wallets help protect keys from theft or compromise, along with an organizational key rotation policy. Meanwhile, routine system and network scans with antivirus products can help reduce or prevent the likelihood of malware-based attacks.
- Nodes are just as vulnerable as users. But a node itself may also be affected by shared and/or misconfigured membership service provider vulnerability. Shared vulnerability is vulnerability where an attacker steals the private key and obtains the target node with the aim of also compromising other nodes on the blockchain. A misconfigured membership service provider vulnerability is caused by control weaknesses on the blockchain leading to DoS/DDoS attacks. For additional protection, one can regularly update and patch the software, thereby mitigating any underlying vulnerability.
- Attackers seek out a network weakness to exploit and employ a variety of network-level attacks. For example, DDoS attacks occur when an attacker tries to delay or stop a network by flooding it with a huge number of transactions. In a Sybil attack, the attacker takes control of many fictitious identities. In a 51% attack, a miner gains control of most of the network's computing power. Such attacks can be mitigated by implementing permissioned blockchain membership service provider services, PirlGuard and/or merged mining.
- Blockchain system-level attacks occur on the system layer of the blockchain infrastructure. These attacks are produced by coding errors and code manipulation in the system's backend. Due care should be taken, and tools must be utilized to avoid such mistakes. Furthermore, restricting the resources to user's need, locking the upper time of time frame and using memory safe programming are some of the techniques used to tackle this sort of attack. System-level attacks require a lot of troubleshooting, but with the help of experts and competent developers, they are a manageable task.
- Blockchain has progressed rapidly over the years. Some powerful technological advancements include Ethereum, Hyperledger Fabric and Corda platforms. Ethereum's confidential transactions in smart contracts, along with information authentication and protecting the data on transfer; Hyperledger's ability to privatize the data via a gossip protocol in addition to identity authentication and testability; and Corda's notary system, are examples of enhancements currently in place on these blockchain structures.

ACTIVITY FOR BETTER UNDERSTANDING

Readers might want to learn more about some of the attacks discussed in this chapter. Table 13.6 provides the titles and links to some related YouTube videos to be used as a supplementary learning resource.

TABLE 13.6
Supplementary Learning Sources

Attack Discussed	Video Title	Direct link
Replay attack	*Blockchain 101 Ep 67 – What is a Replay Attack?*	https://www.youtube.com/watch?v=KWJTvnal8j4&ab_channel=HuobiGlobal
Sybil attack	*What are Sybil Attacks/Explained for Beginners*	https://www.youtube.com/watch?v=-EKhIBUQjcA
Buffer overflow	*Ethical Hacking: Buffer Overflow Basics*	https://www.youtube.com/watch?v=SOoJcrR4Ijo&ab_channel=NationalConsortiumforMissionCriticalOperations
Race condition	*Race Conditions and How to Prevent Them – A Look at Dekker's Algorithm*	https://www.youtube.com/watch?v=MqnpIwN7dz0&ab_channel=SpanningTree

REFERENCES

[1] Are Blockchains That Safe? How to Attack and Prevent Attacks. (2020, September 24). https://www.seba.swiss/research/are-blockchains-safe-how-to-attack-them-and-prevent-attacks

[2] Academy, B. (2020, October 21). What Is a 51% Attack? Binance Academy. https://academy.binance.com/en/articles/what-is-a-51-percent-attack

[3] Agcaoili, J. (2019, June 5). Monero-Mining Malware PCASTLE Uses Fileless Techniques. https://www.trendmicro.com/en_ca/research/19/f/monero-mining-malware-pcastle-zeroes-back-in-on-china-now-uses-multilayered-fileless-arrival-techniques.html

[4] Batabyal, A. (2020, April 30). *Bitcoin Paper Wallet.* 3 Best BTC Paper Wallets in 2020. https://coinswitch.co/news/bitcoin-paper-wallet

[5] Bisson, D. (2021, February 19). PCASTLE Malware Attacks Target China-Based Systems with XMRig. The State of Security. https://www.tripwire.com/state-of-security/security-data-protection/pcastle-malware-attacks-targeting-china-based-systems-with-xmrig/

[6] Brown, R. G., Carlyle, J., Grigg, I., & Hearn, M. (2016, August). *Corda: An Introduction.* ResearchGate. https://www.researchgate.net/publication/308636477_Corda_An_Introduction

[7] Cao, T., Yu, J., Decouchant, J., & Esteves-Verissimo, P. (2018, June 25). *Revisiting Network-Level Attacks on Blockchain Network.* https://orbilu.uni.lu/bitstream/10993/38142/1/bcrb18-cao.pdf

[8] Cimpanu, C. (2021, January 5). Hackers Target Cryptocurrency Users with New ElectroRAT Malware. https://www.zdnet.com/article/hackers-target-cryptocurrency-users-with-new-electrorat-malware/.

[9] Coinbase. (2021). What Is Ethereum? https://www.coinbase.com/learn/crypto-basics/what-is-ethereum.

[10] Constantin, L. (2020, January 20). What Is a Buffer Overflow? And How Hackers Exploit These Vulnerabilities. https://www.csoonline.com/article/3513477/what-is-a-buffer-overflow-and-how-hackers-exploit-these-vulnerabilities.html.

[11] CWE Content Team. (2009, October 21). *Common Weakness Enumeration.* CWE. https://cwe.mitre.org/data/definitions/787.html.

[12] Dasgupta, D., Shrein, J. M., & Gupta, K. D. (2019). A Survey of Blockchain from Security Perspective. *Journal of Banking and Financial Technology*, *3*(1), 1–17. https://doi.org/10.1007/s42786-018-00002-6

[13] Davenport, A., Shetty, S., & Liang, X. (n.d.). *Attack Surface Analysis of Permissioned Blockchain Platforms for Smart Cities.* IEEE. https://ieeexplore.ieee.org/document/8656983

[14] Geroni, D. (2021, April 6). Hyperledger vs Corda vs Ethereum: The Ultimate Comparison. 101 Blockchains. https://101blockchains.com/hyperledger-vs-corda-r3-vs-ethereum/.

[15] GoldSparrow. (2020, June 4). PCASTLE. Remove Spyware & Malware with SpyHunter – EnigmaSoft. https://www.enigmasoftware.com/pcastle-removal/.

[16] Gupta, R. (2019, July 11). *Hacking into BLOCKCHAIN: Is Blockchain Security a Concern?* https://www.plugandplaytechcenter.com/resources/hacking-blockchain-blockchain-security-concern/.

[17] Katrenko, A., & Sotnichek, M. (2020, October 8). Blockchain Attack Vectors: Vulnerabilities of the Most Secure Technology. Retrieved from https://tinyurl.com/y6cuyjey.

[18] Komodo: Advanced Blockchain Technology, Focused on Freedom. (2018). Available online: https://komodoplatform.com/wp-content/uploads/2018/06/Komodo-Whitepaper-June-3.pdf

[19] Li, V. (2020, February 27). Hacking Banks with Race Conditions. https://medium.com/swlh/hacking-banks-with-race-conditions-2f8d55b45a4b.

[20] McNamara, R. (2021, March 26). *Best Cryptocurrency Wallets*. Benzinga. https://www.benzinga.com/money/best-crypto-wallet/.

[21] Mitra, R. (2020, April 24). *5 Best Hardware Wallets: [The Most Comprehensive List]*. Blockgeeks. https://blockgeeks.com/guides/best-hardware-wallets-comparative-list-blockgeeks/.

[22] Mosakheil, J. H. (2018, May). *Security Threats Classification in Blockchains*. https://repository.stcloudstate.edu/msia_etds/48

[23] Muhammad, S., Spaulding, J., Njilla, L., Kamhoua, C., Nyang, D., & Mohaisen, D. (2019, March). *(PDF) Overview of Attack Surfaces in Blockchain*. https://www.researchgate.net/publication/331806569_Overview_of_Attack_Surfaces_in_Blockchain

[24] Poston, H. *Introduction to Blockchain: Network*. Blockchain Tutorial: Part 3 – Networks. https://ghostvolt.com/articles/blockchain_networks.html

[25] Poston, H. (2020, October 7). Targeting the Node. Retrieved February 1, 2021, from https://resources.infosecinstitute.com/topic/targeting-the-node/

[26] Poston, H., & Bennett, K. (2020). *Certified Blockchain Security Professional (CBSP) Official Exam Study Guide*. Blockchain Training Alliance.

[27] Ransomware: Facts, Threats, and Countermeasures. CIS. (2019, September 27). https://www.cisecurity.org/blog/ransomware-facts-threats-and-countermeasures/

[28] S., J. (2020, October 14). Blockchain: What Are Nodes and Masternodes? Retrieved March 23, 2021, from https://medium.com/coinmonks/blockchain-what-is-a-node-or-masternode-and-what-does-it-do-4d9a4200938f#:~:text=Nodes%20form%20the%20infrastructure%20of%20a%20blockchain.&text=They%20store%2C%20spread%20and%20preserve,transaction%20history%20of%20the%20blockchain

[29] Saad, M., Spaulding, J., Njilla, L., Kamhoua, C., Shetty, S., Nyang, D. H., & Mohaisen, A. (2019, March). Exploring the Attack Surface of Blockchain: A Systematic Overview. https://arxiv.org/pdf/1904.03487.pdf.

[30] Sayeed, S., & Marco-Gisbert, H. (2019, April 21). Settings Open Access Article Assessing Blockchain Consensus and Security Mechanisms against the 51% Attack. https://www.mdpi.com/2076-3417/9/9/1788/htm.

[31] Sayeed, S., & Marco-Gisbert, H. (2020). Proof of Adjourn (PoAj): A Novel Approach to Mitigate Blockchain Attacks. *Applied Sciences*, *10*(18), 6607. https://doi.org/10.3390/app10186607

[32] Security Model. Hyperledger. (2021). https://hyperledger-fabric.readthedocs.io/en/v1.0.5/security_model.html.

[33] Sharma, T. K. (2020, July 3). *Types of Crypto Wallets Explained*. Blockchain Council. https://www.blockchain-council.org/blockchain/types-of-crypto-wallets-explained/.

[34] Singh, R., Tanwar, S., & Sharma, T. P. (2019, November 15). *Utilization of Blockchain for Mitigating the Distributed Denial of Service Attacks*. Wiley Online Library. https://onlinelibrary.wiley.com/doi/full/10.1002/spy2.96.

[35] Static Analysis. (2019, August 28). *How to Detect, Prevent, and Mitigate Buffer Overflow Attacks: Synopsys*. Software Integrity Blog. https://www.synopsys.com/blogs/software-security/detect-prevent-and-mitigate-buffer-overflow-attacks/.

[36] Svajcer, V., & Huey, C. (2020, October 13). Lemon Duck Brings Cryptocurrency Miners back into the Spotlight. Web log. https://blog.talosintelligence.com/2020/10/lemon-duck-brings-cryptocurrency-miners.html.

[37] Tripwire Guest Authors. (2019, July 21). Six System and Software Vulnerabilities to Watch out for in 2019. https://www.tripwire.com/state-of-security/vulnerability-management/six-system-and-software-vulnerabilities/.

[38] Walker, M. (2019, July 6). Blockchain, a Barrier against Ransomware. https://thefintechtimes.com/blockchain-barrier-ransomware/.

14 Smart Contract Vulnerabilities, Attacks and Auditing Considerations

Maheswar Sharma, Keerthana Kasthuri, Parvinder Singh and Nynisha Akula

ABSTRACT

As its name implies, this chapter focuses on one of the most vulnerable components of blockchain technology, namely, smart contracts. The chapter discusses a total of nine smart contract-related attacks by taking a look at the root causes of such security breaches. These include reentrancy, access control, arithmetic, unchecked return value, DoS, bad randomness, race conditions, short addresses and timestamp dependency attacks.

In addition to coding errors and attack discussions, this chapter also discusses seven different smart contract audit methodologies. While – depending on the anticipated use case – not all seven audit methodologies need to be used in the course of smart contracts development, auditors need to have a basic understanding of these audit approaches to be able to recommend a proper mix of smart contract testing and evaluation before the launch of smart contracts. As mentioned, in Chapters 2 and 9, extreme caution must be exercised during the coding and testing of a smart contract, as smart contract development is akin to firmware development, in the sense that once deployed, a defective smart contract can no longer be fixed by applying a software patch to it. As such, the only workable solution is to kill the smart contract – assuming that such kill switch has been added to its design – and to replace the problematic smart contract with a fully functioning one.

Information systems auditors will also find the smart contract audit template a useful tool in planning and conducting smart contract audits.

SMART CONTRACTS SECURITY CONSIDERATIONS

Smart contracts have emerged as a major programming breakthrough in blockchain technology. Smart contracts are computer programs that are executed on platforms such as Ethereum with *Solidity* as the programming language used in compiling the code used on the Ethereum Virtual Machine.

Smart contracts are considered 'public' on the blockchain where users can interact with smart contracts by creating a transaction. While dealing with smart contract programs, there is a need to be familiar with the security risks associated with detecting and avoiding the smart contract code's major flaws. In this section, we examine the known security vulnerabilities that commonly occur during code execution [1].

DOI: 10.1201/9781003211723-14

Reentrancy Attack

Reentrancy is regarded as one of the most catastrophic smart contract attack strategies. This attack method can destroy the contract or facilitate the theft of sensitive information. A reentrancy attack occurs when a user builds a feature that allows an external function to call for another untrusted smart contract before it addresses its consequences. The vulnerability allows an attacker to execute the main function's recursive callback, causing an unintended loop that is repeated several times. *Single* function and *cross* function are two different types of reentrancy attacks used by attackers to breach the contract or capture valuable information (Figure 14.1) [7].

Solidity's call build (which is converted to a CALL instruction in EVM bytecode) causes a fraction of the callee's code to be executed – as defined in the fallback function. As a result, when the call construct is used, the creator may assume an atomic value transfer in which the code of another contract can be executed. By invoking the 'X' function ping with the address of the 'Y' account, 2 *wei* (the smallest denomination of ether) are transferred to the 'Y' account, and additionally, the fallback function of 'Y' is invoked. As the fallback function again calls the ping function with Y's address, another 2 *wei* are transferred before the variable 'sent' of contract 'X' was set. This looping goes on until all *gas* – the fee required to conduct a transaction or execute a contract on the Ethereum blockchain platform – of the initial call is consumed or the call stack limit is reached [14].

Real-Life Case Scenario

The effects of a reentrancy attack can be seen by studying the DAO (Decentralized Autonomous Organization) case, an association founded by developers to automate decisions and promote transactions with cryptocurrencies. DAO was one of the primary victims during the initial development process of Ethereum. Reentrancy played a vital role, and the total value of the contract was $150 million [7].

Access Control

Generally, smart contracts have a protected functionality which means that the owners and the other permissioned users are identified by their addresses. Access control systems can be deployed on smart contracts which are managed by the owners or the trusted third parties [4]. In Ethereum smart contracts, the purpose of the access control is to restrict access to components of the contract [6].

The vulnerability of access control is that an attacker can prevent users from gaining access or granting permissions to others to perform operations like currency transfers and/or contract self-destruction [5] (Figure 14.2).

Smart contract with reentrancy bug:

```
1. contract X{
2. bool sent=false;
3. function ping (address c) {
4. if (!sent) { c.call.value(2) ();
5. sent =true;}}
}
```

Smart contract exploiting reentrancy bug:

```
6. contract Y{
7. function(){
8. abc(msg.sender).ping(this);}
9. }
```

FIGURE 14.1 A sample code explaining the flaws leading to the reentrancy attack [14].

```
1. contract HOTTO is ERC20 {
2. owner = msg.sender;
3. function ABC() public {
4. owner = msg.sender;
5. distr(owner, totalDistributed);
6. }
7. function withdraw() onlyOwner public {
8. address myAddress = this;
9. uint256 etherBalance = myAddress.
10. balance;
11. owner.transfer(etherBalance);
12. }
13. }
```

FIGURE 14.2 A sample code explaining the flaws leading to access control vulnerability [5].

In the above example on *line 3*, by executing public function ABC (), the original owner is modified to msg.sender (*line 4*) without verifying the code. An attacker can also omit the only owner in *line 7* to withdraw all the Ethereum in the contract [5].

Real-Life Case Scenario

A parity wallet was a common multi-signature cryptocurrency wallet for the Ethereum smart contract, which was used for setting withdrawal limits, transaction logging and creating rulesets for required signatures [10]. A theft of 150,000ETH was carried out by the attacker due to the vulnerabilities observed on the parity wallet. Two transactions were sent by the attacker to the contract where the first transaction was used to gain ownership of the multi-signature wallet and the second one was used to transfer all the funds. Later, the attacker called the initWallet function in the first transaction. The initWallet function had no checks to ensure that it was not called after the contract was initialized. The complete ownership of the wallet was gained by the attacker by exploiting the parity wallet with the entire value being transferred to the attacker's account [11].

Arithmetic Over/Underflows

The Ethereum Virtual Machine (EVM) specifies fixed-size data types for integers, which means a specific range of numbers can be represented by an integer variable. For example, A *uint8* can only store numbers in the range 0–255. Trying to store 256 into a *uint8* will lead to overflow as the maximum value it can include is 255 and hence result in 0.

In the Solidity programming language, an unsigned integer is denoted as *uint256*. Each uint256 is restricted to a size of 256 bits which represents integer values from 0 to $2^{256}-1$. In the contract, if the assigned integer value is larger than the given range, the integer resets itself to 0 and if the value is smaller than the range, it assigns a larger value in the range. For example, when a positive number is subtracted from 0 it will result in an integer of $2^{256}-1$ [8].

Smart contracts result in an *overflow* and *underflow* vulnerability when an assigned value is incremented above the given range that makes contracts vulnerable to arithmetic overflow or underflow. Solidity can handle up to 256-bit numbers and incrementing this number by 1 can result in an overflow (Figure 14.3) [7].

To exploit the flaw, an attacker will call the function (batch transfer) with parameters. The attacker can enter a negative value in _value (line 2), which can lead to a reverse transaction; i.e., where the amount will be debited from the acceptor's account and credited to the sender. An attacker

```
1. uint256 total = uint256(cnt) *_value;
2. function batchTransfer(address[] _acceptors, uint 256 _value) public whenNotPaused returns (bool) {
3. uint cnt = _acceptors.lenth;
4. uint256 total = uint256(cnt) * _value;
5. require(cnt>0 && cnt<=20);
6. require (balances[msg.sender] >=total);
7. balances[msg.sender] =balances[msg.sender] .sub(total);
8. for (uint i=0;i<cnt;i++) {
9. balances[_acceptors[i]] = balances [
10. _acceptors[i]].add(_value);
11. Transfer(msg.sender, _acceptors[i] ,_value);
12. }
13. return true;
    }
```

FIGURE 14.3 Sample coding flaw leading to arithmetic over/underflow attacks [7].

can input two addresses in the receiver's function for the token smart contract to transmit ether to both addresses. This vulnerability can be avoided by using scalar data types instead of compound data types [7].

Real-Life Case Scenario

4chan's Group created the *Proof of Weak Hands Coin* or POWH using smart contracts. Even though it was set up as a Ponzi scheme, people continued to invest in it. However, it turned out that the developers of the POWH coin failed to secure all operations and were unable to put in place adequate protection against overflow and underflow attacks. By exploiting this coding oversight, an unknown hacker was able to steal 2000 ETH worth at least $1,000,000 [8].

Unchecked Return Values

There are many ways of performing external calls in Solidity, in which transfer is one of the methods used frequently to send ethers to the external accounts. The send function is also used, which performs a message call, and the Ethers are transferred to the specified address. On the other hand, for more flexible external calls, the CALL opcode is employed in Solidity directly. If any of these operations are succeeded or failed, the call and send functions will return a Boolean value indicating the status. A transaction in which the call and send functions are executed will not return, if the external call fails and the execution of the calling function is continued naturally, whereas the function will return a Boolean false value. Here, the developer will fail to check for the return value of false by expecting a return to occur if the external call fails (Figure 14.4) [1].

The above example represents a gaming contract in which the winner receives a winPrice of ether and it has a little left for anyone to withdraw. On line 11, *send* is used without verifying the response. A winner whose transaction is failed will allow the *payedOut* to be set to 'true' in any event, whether the ether was sent or not. The transactions can fail if a contract gives a fallback function on purpose or by running out of gas. Here the vulnerability is that anyone can withdraw the WinnerPrice with the *withdrawLeftOver* function [1].

Real-Life Case Scenario

Etherpot was a smart contract lottery. It was a failure due to the incorrect use of block hashes. To decide a winner in Etherpot, the hashes of a block – specifically a decision block – were used.

```
1 contract Game
2 {
3   bool public payedOut = false;
4   address public winner;
5   uint public winPrice;
6
7   // ... extra functionality here
8
9   function sendToWinner() public {
10      require(!payedOut);
11      winner.send(winPrice);
12      payedOut = true;
13   }
14
15   function withdrawLeftOver() public {
16      require(payedOut);
17      msg.sender.send(this.balance);
18   }}
```

FIGURE 14.4 A sample code explaining the flaws leading to the vulnerability of unchecked return value [1].

Theoretically, it is said that the winner could be calculated and reported as the winner at any time. But the block-hash function returns a 0 until the winner quits within the 256 blocks of the decision block. This contract was affected by an unchecked call value [1].

DoS Attacks

Denial of service attacks can take several forms as follows:

DoS with block gas limit: Each block on the Ethereum blockchain has a gas limit. A block gas limit has the advantage of preventing attackers from building an infinite transaction loop. However, if a transaction's gas consumption reaches this limit, the transaction will fail. This can result in a DoS attack in several ways [15].

Gas limit DoS on the network via block stuffing: Regardless of whether the smart contract does not contain an unbounded loop, the malicious actor can use a series of transactions with a higher gas price to prevent other transactions from being processed in the blockchain for multiple blocks. Here, several transactions were issued by the attacker which consumes the entire gas limit. The attacker tries to fill the entire block with a large number of transactions and if the gas price is high among those transactions, it can prevent other transactions from being processed [15].

DoS with unexpected revert: A DoS (denial of service) attack occurs when a user tries to transfer funds to another user and the functionality relies on the transaction being successful. If the funds are transferred to a smart contract which is created by the attacker, this allows the attacker to create a fallback function to reverts all the payments (Figure 14.5) [15].

In this case, if an attacker bids from a smart contract with a fallback function that reverses all payments. They will never be refunded, and no one will ever be able to outbid them. In the Revert array function code, a user wants to pay an array of users by iterating through the array, and the user wants to make sure each one is paid fairly. The issue is that if one payment fails, the function is reverted, and no one is compensated [15].

```
1. contract Auction {
2. address currentLeader;
3. uint highestBid;

4. function bid() payable {
5. require(msg.value > highestBid);

6. require(currentLeader.send(highestBid)); // Refund the old leader, if it fails then revert
7. currentLeader = msg.sender;
8. highestBid = msg.value;
9. }
10. }

Revert array function:
1 address[] private refundAddresses;
2. mapping (address => uint) public refunds;
3. // bad
4. function refundAll() public {
5. for(uint x; x < refundAddresses.length; x++) { // arbitrary length iteration based on how many
   addresses participated
6. require(refundAddresses[x].send(refunds[refundAddresses[x]])) // doubly bad, now a single
   failure on send will hold up all funds
7. }}
```

FIGURE 14.5 A sample code explaining the flaws leading to DoS with unexpected revert vulnerability [15].

Real-Life Case Scenario

A pyramid scheme that collected a considerable amount of Ether was *GovernMental.* It had acquired 1100 ethers at one point, but was susceptible to DoS vulnerabilities. To extract the ether, the contract modification involved the elimination of a major mapping. The deletion of this mapping had a gas expense that met the block gas cap, and therefore the 1100 ether could not be withdrawn [1].

Bad Randomness

A random number generator is needed for some smart contracts. Random number generation can be accomplished in a number of ways, such as using a secret value, basing the random number off of mining statistics or using an external random number generator oracle. Since smart contract code is visible on the blockchain, using an external oracle is the only secure way for random number generation (Figure 14.6) [16].

The code above has random numbers as a part of smart contract-based gambling games and will only use the hash of a previous block on the blockchain as a source of randomness which is published to everyone after a block is created. Ethereum will only store the last 256 block hashes and will return 0 for anything before those hashes. Therefore, the attacker can create a smart contract for the previous 256 blocks which checks if the current block is a winner or not. If it is so, the attacker calls the function to claim the prize [17].

Real-Life Case Scenario

Smart Billions lotteries, and other online games and lotteries, have relied on unpredictable data such as the 'block creation time' for tickets to a sporting events. Here, the attacker or the block miner can adjust the publication time to manipulate the result [18].

```
1. function play() public payable {
2. require(msg.value >= 1 ether);
3. if (block.blockhash(blockNumber) % 2 == 0) {
4. msg.sender.transfer(this.balance);
5. }
6. }
```

FIGURE 14.6 A sample code with flaws leading to bad randomness vulnerability [17].

Race Condition

A race condition is an attack that occurs in a multiuser environment where the user's execution of race code leads to undesirable events [1]. A blockchain race condition may occur when miners compete with the smart contracts' participants to insert the miner's transaction before the original transactions in the block by altering the permissions in the contract [2].

On the Ethereum blockchain, blocks are formed by a pool of transactions processed by Ethereum nodes. The blocks are valid when the miner solves the 'proof of work' mechanism. Transaction fees on the Ethereum platform are referred to as 'gas,' with the price of these fees being set by miners. Usually, miners decide which transaction should be included in the block and select the transaction from a set of transactions that are ordered according to the 'gas price' of each transaction. As mentioned earlier, Ethereum is a public blockchain where users are allowed to view other users' pending transactions. An attacker can view the transactions that possess the solution to the proof of work mechanism in the contract, thereby gaining data from the transactions. The attacker then can create a transaction with the higher gas price and include it in the block in order to gain the ethers in the contract. In a race condition, the potential vector could modify the permissions in the contract and this modification could adversely affect the solver [1].

According to a 2021 Ethereum update (EIP-1559), the latest approach to transaction fees entails splitting the fee into a base fee in addition to 'tips' where the base fee is automatically computed based on a set of algorithm according to each transaction. Furthermore, transactions are now sent directly to the network instead of miners. To further explain, the term tips, refer to voluntary, additional fees paid by users to further expedite transaction processing. This latest approach to transaction fees also prevents bad actors from spamming the transactions (Figure 14.7) [10].

The above contract contains *2 ethers* with the preimage of its SHA-3 hash in *line 2*. For example, if any user discovers the solution to the hash value, they could call the solve function with the respected parameter to gain the ethers. On the other hand, the attacker may be watching the pool of transactions looking out for someone who submits a solution. Once the attacker knows the solution, the attacker submits a transaction with a higher gas price. The attacker then mines the transaction before the original one solves the proof of work mechanism and gets the 900 ethers. In this vulnerability, there is a chance of a miner getting bribed or a miner, itself, being an attacker [1].

```
1. contract FindHash{
2. bytes32 constant public hash = 0xb5b5b97fafd9855eec9b41f74dfb6c38f5951141f9a3ecd7f44d5479b630ee0a;
3. constructor() public pay{}
4. function solve(string solution) public {
   //if you can get a pre-image of hash, receive 900 ether
5. require( hash == sha3(solution));
6. msg.sender.transfer(2 ether) ;
7. } }
```

FIGURE 14.7 Sample coding flaw leading to a race condition vulnerability [1].

Real-Life Case Scenario

Bancor is a well-known example of a platform that facilitates the exchange of digital assets without the interference of *buyers* and *sellers* on the Ethereum blockchain [3]. Ivan Bogatty and his team members have recorded an attack on Bancor. As the nature of Ethereum is public, the transaction is visible to everyone when it gets broadcasted on the network [1]. These transactions are available in the pending transactions section and are waiting to be mined. If an attacker gets enough time to front-run the pending transactions and buy the order with a slightly higher price than the original one, that attacker can realize a profit [3].

Short-Address Attack

A short-address attack is an attack performed on a third-party application. A short-address or parameter attack shows how parameters can be manipulated by an attacker in order to access tokens in the contract. Even though there are no real-life examples for this attack, the conceptual scenario explained in this section helps explain how the parameters can be manipulated in this type of attack [1, 26].

A short-address attack allows an attacker to bypass parameters less than the designated length; for example, sending an address of 35 hexadecimal characters instead of 40 hexadecimal characters. In such an instance, an Ethereum Virtual Machine fills the zeros by default at the end of the address to match the expected length. This leads to an attack when third-party applications do not validate the input (Figure 14.8) [1].

As mentioned before, the main flaw in a short address is the acceptance of an address that is less than the expected length. By considering the above code, line 6 enables an attacker to steal bytes from the _amount argument and when the smart contract code is executed, it transfers 256 times more tokens than expected.

To further expand, the ERC20 standard is commonly used for creating tokens on Ethereum. According to *GOLEM*'s tech team, a short-address attack allows an attacker to pass a parameter less than the designated length. The Application Binary Interface (ABI) specification adds trailing zeros

```
1. contract MyToken {
2. mapping (address => uint) balances;
3. event Transfer(address indexed _from, address indexed _to, uint256 _value);
4. function MyToken() {
5. balances[tx.origin] = 100;
   // tx.origin is the address externally owned account transaction which is unsafe
   }
6. function transfer(address to, uint amount) returns(bool solution)
   // prepends the address to the expected length by adding zero bits to make 32 byte long.
   {
7. if (balances[msg.sender] &lt; amount) return false;
8. balances[msg.sender] -= amount;
9. balances[to] += amount;
10. transfer(msg.sender, to, amount);
11. Return true;
12. }
13. Function getbalance(address addr) constant returns(uint) {
14. return balances[addr];
   } }
```

FIGURE 14.8 Sample coding flaw leading to short address attack [13].

to make up for the required length of the address. In this attack, 1 byte of zeros is added to the short address. When a parser goes through the whole address, if it has an underflow, an extra zero is added to make it uint256, and the transfer function is called. Here, the amount in the transfer function is multiplied by 256 (or 2 to the power of 8), and tokens are transferred accordingly [1].

Timestamp Dependency

Timestamp dependency is another vulnerability that can be exploited by corrupt miners. This occurs due to an imperfect interpretation of timekeeping in smart contracts, which results in the Ethereum network being disengaged from a synchronized global clock. For example, the Ethereum smart contract uses the current timestamp to produce random numbers and awards a prize based on the result. It will work as follows: when the timestamp is used by a smart contract to produce random numbers, the corrupt miner will post a timestamp within 30 seconds of the block's validation, thus altering the contract's performance to his advantage. As a result, the outcome of the random number generator can be changed (Figure 14.9) [8].

The Run.sol contract's code snippet illustrated above utilizes the block's timestamp value to produce a random integer, which is then included in a crucial calculation process. In *line 2*, the timestamp of the block is allocated as a random number to a private variable salt. The salt variable is then used to measure the values of the random function's parameters *x*, *y* and *seed*. The code implements the condition where the random function is called in *line 4*. The block's timestamp is used to determine the random function's return value, which is then assigned to the variable roll. The variable roll is then tested for a state; and if it passes, the submit function is executed as a vital call. A malicious miner will take advantage of this call by adjusting the local system's timestamp [8].

Real-Life Case Scenario

GovernMental was a Ponzi scheme that accumulated significant amounts of ether. The GovernMental contract collected players' ethers in rounds and only paid out a single winner in each game. The contract was awarded to the player who finished last in each round (by at least a minute). The consistent

```
1. function random() // TheRun.sol --
2. uint256 constant private salt =block.timestamp;
3. function random(uint Max) constant private
4. returns (uint256 result){ //get the best seed for randomness
5.  uint256 x = salt * 100 /Max;
6. uint256 y = salt * block.number / (salt%5) ;
7. uint256 seed = block.number/3 + (salt %300) + Last_Payout + y;
8. uint256 h = uint256(block.blockhash(seed));
9. return uint256((h/x)) % Max + 1 // random number between 1 and Max
10. }
Attacker function: -
1. //TheRun.sol -- call random() function
2.  //winning condition with deposit > 2 and having luck
3.  if( (deposit > 1 ether ) && (deposit > players[Payout_id].payout) ){
4.  uint roll = random(100); // create a random number
5.  if( roll % 10 == 0 ) {
6.  msg.sender.send(WinningPot);
7. WinningPot=0;
8.  }
9.  }
```

FIGURE 14.9 Sample coding flaw leading to timestamp dependency vulnerability [8].

winner of this game was a miner who was able to adjust the timestamp as to make it look that the player was the last to join [9].

SMART CONTRACT AUDITING CONSIDERATIONS

Smart Contract Auditing

Smart contract auditing is all about validating the underlying code. This type of audit helps in identifying the vulnerabilities, and also ensures that the contract code is reviewed before the contract is deployed on the blockchain. Currently, smart contract auditing methodologies employ (i) samples of transactions to analyze the risk of material misreporting, (ii) an annual point-in-time audit approach and (iii) a backward audit approach.

Smart contracts are adopted by business organizations to improve the efficiency of their business. Auditors have an option to buy data analytical tools from various vendors. Auditors can also integrate multiple audit tools for transparent and timely audit reporting [19, 25].

Table 14.1 outlines the various smart contract testing/audit methods along with their respective strengths and challenges to help ensure that a new smart contract is functioning according to its intended purposes.

As presented in Table 14.1, all audit methodologies have their advantages and disadvantages. As such, it is important to note that to have a high degree of assurance that a smart contract will fulfill its intended purposes, some of the auditing methodologies presented in Table 14.1 ought to be considered by the development or smart contract audit team.

The following section provides basic information related to tools used in each type of testing/audit method presented in the previous section.

Control Flow Analysis Tools

McCabe IQ

A detailed review of each feature in the attack map can be done with the flowgraph, annotated source code listing. Path information is tested to ensure that the control flow paths displayed are accurate, stable and compliant with specifications [31].

McCabe IQ's Data Dictionary functionality is another strategy for a more thorough investigation into forbidden functions that are being used in the program code. Users can scan for data elements and function calls in the data dictionary and classify the paths within a function that includes those calls [31]. The tool is available on GitHub at https://github.com/PyCQA/mccabe.

Ethereum Virtual Machine

The EVM consists of a single virtual machine built on a stack with a custom instruction format. Each instruction is portrayed as a one-byte opcode. On the data stack, arguments are transferred. The push instructions that are used to push constants onto the stack are the only exception. These constants are directly translated into the bytes of the instruction. As the EVM has an explicit marker for simple block entries, retrieving the basic block boundaries is tractable [30]. The tool is available on GitHub at https://github.com/exthereum/evm.

Taint Analysis Tools

TAJ

Taint Analysis for Java (TAJ) is a method designed to be sufficiently reliable in generating a low false-positive rate and be versatile enough to allow the evaluation of large applications. TAJ provides a range of strategies for generating practical effects on incredibly large programs though they are

TABLE 14.1
A Summary of Smart Contract Auditing Techniques

Methodology	Definition	Strengths	Challenges
Control flow analysis	Conventional compiler optimization method for evaluating valuable compile-time knowledge [32].	Can be used in multiple compilers [32].	Large numbers of false positives. Analyzing code that cannot be compiled [35].
Expert code analysis	Code is reviewed by a Solidity coding expert for accuracy and potential flaws [17].	A 'fresh pair of eyes' look at the code to detect any issues [17].	Work done is performed by humans and thus not an automated tool; therefore a higher probability that certain flaws may not get detected [17].
Taint analysis	Seeks to classify variables with user-controllable input that have been 'tainted' and tracks them to potential weak functions often known as 'sink' [34].	Can easily recognize web application flaws [34].	Taint analysis tools do not address critical requirements. Spills knowledge about its setup, procedures and internal concerns [33].
Dynamic code analysis	Dynamic code analysis functions like an attacker who tries to find vulnerabilities by adding malicious code into the required functions in a program [8].	The vulnerabilities which cannot be identified by static analysis can be identified through the dynamic analysis method [8].	When the code is updated, the users of the smart contracts will not be guaranteed that the new code will implement the same logic as the original contract code [27].
Manual code analysis	The developer manually inspects the code line by line before the contract is deployed permanently [29].	Helps minimize the number of test configurations [29].	Limited pair of traces can be inspected manually [29].
Vulnerability-based scanning	Vulnerability-based scanning helps in identifying the security issues and vulnerabilities which led to security attacks in smart contracts [8].	Manages security issues and vulnerability management. Smart contracts enclose many real-world vulnerabilities such as reentrancy, unchecked external call, timestamp dependency [8].	Challenges in vulnerability-based scanning relate to codifying privacy and performance issues [24].
Symbolic execution	An extension of a normal execution containing a simple computation. Symbolic execution does not need any real-world objects as semantics as it uses the arbitrary symbols in the programming language [23].	Relatively simple when compared to other analysis methods [23].	A problem in constraint solving [20]. Issues in dealing with loops as it never terminates [20].

limited to a given time or budget for memory. TAJ advocates a prioritization approach that targets the research on parts of the web application that are likely to partake in taint dissemination when implementations are incredibly large and the end-user still wants the analysis to terminate in a limited period or remain within a specified memory usage level [27]. The tool is available on GitHub at https://github.com/TJAndHisStudents/TaintFlowAnalysis.

DYTAN

Dynamic Taint Analysis adequately manages knowledge flow within a program due to data and control flow and enables the analysis to be tailored in different dimensions. SQL injection attacks are avoided through dynamic tainting, in which attackers send maliciously constructed strings to a web server to enter the underlying database. Most dynamic-taint-based SQL injection approaches work by tainting and monitoring unsafe files which are inputs from the user. Once a query string is submitted to the database, it is reviewed to ensure that the string or sections of it have not been generated using any corrupted data [28]. This tool is available on GitHub at https://github.com/behzad-a/Dytan.

Dynamic Code Analysis Tools

MAIAN

The dynamic tool MAIAN considers the execution traces of smart contracts which are the sequences recorded on the blockchain while running a contract. MAIAN also considers the vulnerability categories along with the execution traces. A transaction will request a smart contract to run a function every time, leading to the exploitation of an execution path for a given input. So, the execution traces may have a chain of effects. Therefore, to find the exploits on the specified execution traces of a smart contract, the MAIAN tool uses systematic techniques. MAIAN consists of two major components, namely, symbolic analysis and concrete validation. The symbolic analysis component has the input of contracts bytecode and the analysis specifications with the vulnerability category. MAIAN continues its execution until it reaches an uncertain trace with a set of vulnerability properties. All the executions are done on EVM, which will simplify the symbolic execution of smart contract bytecode. Symbolic variables are taken as inputs for every execution trace such that the symbolic analysis component will return the concrete values for the given symbolic variables if the contract is exposed as vulnerable. The role of the concrete validation is to verify the results of symbolic analysis and confirm the accuracy of bugs found in the smart contract. The state of the contract on the main Ethereum blockchain is not affected by MAIAN during the analysis [8]. This tool has been available on GitHub under an MIT license since March 2018 [21].

ContractLarva

This tool is used for runtime verification in smart contracts. A user can demonstrate the properties of a given smart contract by using Dynamic Event Automata (DEA) – a specification language used by ContractLarva. The tool generates a new Solidity contract by taking the specifications of the given contract as input. The new contract functions the same as an original contract, but a code is added to the contract to check the runtime verification and react accordingly in case of any observed violations. Although ContractLarva is implemented on many smart contracts to deploy runtime verification and recovery, it faces many challenges that need to be addressed [27]. This tool is available on GitHub under an Apache-2.0 license [21].

Vulnerability-Based Scanning Tools

Mythril

An automated tool for analyzing smart contracts. The analysis is based on byte code. It uses taint analysis and control flow analysis to find the vulnerabilities. Mythril uses the command line as an interface [22]. Vulnerabilities found through this tool are detailed and documented. ConsenSys has developed and maintained the Mythril on Github under MIT license in September 2017 [21].

Securify

Another automated user interface-based security analysis tool used for Ethereum smart contracts based on byte code and detects vulnerabilities such as reentrancy, input validation, etc. [22]. This tool provides added assurance that if a pattern is detected, the detection implies that the contract code possesses the corresponding security property. The tool is in Java and has been available on Github under Apache 2.0 license from 2018 [21].

SmartCheck

An open-source, user interface-based, automated tool that runs the Solidity source code and checks for vulnerabilities and other security issues [22]. SmartCheck converts the code into Extensible Markup Language (XML) syntax and specifies vulnerabilities as XQuery path expressions which are used in search of patterns in an XML tree. The latest version of the tool is closed source, and it has twice as many patterns for checking vulnerabilities. The tool is written in Java and can be accessed via the SmartCheck company website [21].

Symbolic Execution Tools

DART

One of the earlier tools in symbolic execution capable of performing random testing through dynamic analysis in order to find the bugs which are caused by library functions. DART checks for standard errors such as violations, assertions, crashes and non-terminations, using input with a single path testing without employing path selection mechanisms [20]. The tool is openly available on GitHub. The DART software development kit includes VM, dart2js, core libraries, etc.

Manticore

A symbolic execution tool that can find combination paths. Manticore compiles Solidity code to bytecode to analyze the traces of vulnerabilities and report them in a source code. The *Trail of Bits* company developed this tool to use the Python API. The tool has been openly available on GitHub under an AGPL-3.0 license since 2017 [21].

Oyente

A tool used as a reference point and that is used for various projects. It executes the byte code symbolically and checks the traces of the contract which is executed multiple times. Using Oyente, a Solidity compiler is needed to obtain a byte code. Oyente is a command-line tool that is written in the Python programming language. The tool has been openly available on GitHub under a GPL-3.0 license since 2016 [21].

Table 14.2 provides a summary of the discussed tools along with their purposes and analysis methods.

TABLE 14.2
A Survey of Tools with Their Purposes and Analysis Methods

	Purpose		Analysis Methods					
Tools	Security Issues	Exploits	Dynamic Analysis	Manual Analysis	Symbolic Execution	Vulnerability Scanning	Control Flow Analysis	Taint Analysis
ContractLarva			X					
MAIAN	X	X	X		X			
DART	X				X			
Manticore	X				X			
Oyente	X				X			
Mythril	X	X				X		
Securify	X					X		
SmartCheck	X					X		
McCabe IQ	X						X	
EVM				X			X	
TAJ			X					X
Dytan	X					X	X	X

SUMMARY AND CONCLUSIONS

As the usage of digital money, blockchain technology and smart contracts continue to gain attention, a system to categorize the possible vulnerabilities in a form that the public can see and access is required. This chapter provides a plethora of empirical facts to assist smart-contract developers, security researchers and security auditors in better understanding the smart contract vulnerabilities, and the various methodologies to identify such flaws.

Ethereum smart contracts are written in Solidity – a Turing-complete programming language with a high degree of abstraction. Information was gathered, examined and used to classify current flaws in smart contracts on the Ethereum blockchain after conducting a literature analysis of past research and other relevant resources. As a result of this study, a comprehensive master list of well-documented vulnerabilities was presented in this chapter.

CORE CONCEPTS

- Ethereum is an extended framework for smart contracts – computer programs that are executed by a network of mutually distributed nodes, without a need for third-party authorization.
- Since smart contracts handle and transfer valuable assets, it is equally important to secure smart contract implementations against attacks that will aim at stealing and manipulating the assets.
- The security vulnerabilities of the smart contracts are analyzed and the drawbacks in the programming languages which lead to the vulnerabilities are provided. The attacks which exploit the vulnerabilities of the smart contracts to steal the money are also highlighted.
- Ethereum smart contracts use various software tools depending on their methodology. These tools are automated and check for the vulnerabilities, bugs and security measures in the contract before being deployed to the blockchain.
- Vulnerability-based scanning tools such as Mythril, Securify and SmartCheck are used to find the vulnerabilities in the code. Some of these tools are used for runtime verification in smart contracts such as ContractLarva, etc.

ACTIVITY FOR BETTER UNDERSTANDING

Smart contracts application testers and information systems auditors interested in developing a blockchain-related audit checklist are encouraged to study the smart contract audit checklist below with the following questions in mind:

S/No.	Smart Contract Attack/ Vulnerability	Audit Check	Yes	No	Auditor's Notes
1.	Reentrancy attack	Is the recursive call back functioning properly? [27]			
2.	DoS attack	Is the fallback function working properly? [6]			
3.	Timestamp dependency	Is the clock synchronized? [27] ---------------- Can the random number generator function be rigged? [24]			
4.	Arithmetic over/ underflow attack	Is the input being validated? [24]			
5.	Short address	Does the attacker alter the parameters? [1]			
6.	Race condition	Could the attacker modify the permissions in a contract? [1]			

S/No.	Smart Contract Attack/ Vulnerability	Audit Check	Yes	No	Auditor's Notes
7.	Access control	Will the attacker gain privilege to perform operations? [5]			
8.	Unchecked return value	Can the developer check the return value? [1]			
9.	Bad randomness	Does the attacker predict the hash values of the previous block? [17]			

1. Go through each one of the nine areas in the checklist and test your smart contract vulnerability knowledge using the checklist above, without looking at the chapter content. How many did you get right on your first try?
2. Once again, go through the provided audit checklist in detail to see if there are additional smart contract vulnerability considerations that were not covered in the audit checklist?
3. In what ways and at what specific stages of smart contract application development, testing or auditing can the checklist prove useful?
4. In your opinion, of the nine smart contract attacks discussed in the chapter and listed in Table 14.2, which vulnerability/attack is the hardest to mitigate?

REFERENCES

[1] Antonopoulos, A. M. (2018). *Mastering Ethereum: Building Smart Contracts and DApps.* O'Reilly Media. Retrieved from https://cypherpunks-core.github.io/ethereumbook/09smart-contracts-security.html

[2] Wang, X., He, J., Xie, Z., Zhao, G., & Cheung, S. C. (2019). ContractGuard: Defend Ethereum Smart Contracts with Embedded Intrusion Detection. *IEEE Transactions on Services Computing*, *13*(2), 314–328. Retrieved from https://arxiv.org/pdf/1911.10472

[3] Eskandari, S., Moosavi, S., & Clark, J. (2019, February). Sok: Transparent Dishonesty: Front-Running Attacks on Blockchain. In *International Conference on Financial Cryptography and Data Security* (pp. 170–189). Springer, Cham. Retrieved from https://arxiv.org/pdf/1902.05164

[4] Maesa, D. D., Mori, P., & Ricci, L. (2018). Blockchain Based Access Control Services. *IEEE International Conference on Internet of Things (iThings) and IEEE Green Computing and Communications (GreenCom) and IEEE Cyber, Physical and Social Computing (CPSCom) and IEEE Smart Data (SmartData).* doi:10.1109/Cybermatics_2018.2018.00237

[5] Zhang, Q., Wang, Y., Li, J., & Ma, S. (2020). Ethploit: From Fuzzing to Efficient Exploit Generation against Smart Contracts. *International Conference on Software Analysis, Evolution and Reengineering.* doi:10.1109/SANER48275.2020.9054822.

[6] Liu, B., Sun, S., & Szalachowski, P. (2020). Smacs: Smart Contract Access Control Service. *International Conference on Dependable Systems and Networks (DSN).* IEEE. doi:10.1109/DSN48063.2020.00039.

[7] Sayeed, S., Marco-Gisbert, H., & Caira, T. (2020). Smart Contract: Attacks and Protections. 6–7.Retrieved from https://hackingdistributed.com/2016/06/18/analysis-of-the-dao-exploit/

[8] Praitheeshan, P., Pan, L., Yu, J., Liu, J., & Doss, R. (2019). Security Analysis Methods on Ethereum Smart Contract Vulnerabilities: A Survey. Retrieved from https://arxiv.org/pdf/1908.08605

[9] Atzei, N., Bartoletti, M., & Cimoli, T. (2017). A survey of attacks on Ethereum smart contracts. Retrieved from https://link.springer.com/content/pdf/10.1007%2F978-3-662-54455-6_8.pdf

[10] Andrew, M., Zhicheng, C., & Somesh, J. (2018). Smart Contracts and Opportunities for Formal Methods. Springer, Cham. doi:https://par.nsf.gov/servlets/purl/10098553

[11] Palladino, S. (2017, July). The Parity Wallet Hack Explained. *OpenZeppelin blog*. Retrieved from https://blog.openzeppelin. com/on-the-parity-wallet-multisig-hack-405a8c12e8f7

[12] EIP-1559: A Proposal to Update Transaction Fees on Ethereum. Retrieved from https://www.gemini.com/cryptopedia/ethereum-improvement-proposal-ETH-gas-fee#section-ethereum-transaction-fees

[13] Programmer Sought. (2018). ERC20 short address attack. Retrieved from https://www.programmersought.com/article/44231449349/

[14] Grishchenko, I., Maffei, M., & Schneidewind, C. (2018). Foundations and Tools for the Static Analysis of Ethereum Smart Contracts. *International Conference on Computer Aided Verification.* Retrieved from https://link.springer.com/chapter/10.1007/978-3-319-96145-3_4

[15] Kaden, Z. (2019, October 24). Smart Contract Attack Vectors. Retrieved May 9, 2021, from https://github.com/KadenZipfel/smart-contract-attack-vectors/blob/master/attacks/dos-gas-limit.md

[16] Kim, K. B., & Lee, J. (2020, November). Automated Generation of Test Cases for Smart Contract Security Analyzers. doi: 10.1109/ACCESS.2020.3039990

[17] Poston, H., & Bennett, K. (2019). Certified Blockchain Security Professional official Exam Study Guide. Retrieved from https://blockchaintrainingalliance.com/products/cbsp-official-exam-study-guide

[18] Metcalfe, W. (2020). Ethereum, Smart Contracts, DApps. *In Blockchain and Crypt Currency.* doi:https://doi.org/10.1007/978-981-15-3376-1_5

[19] Rozario, A. M., & Vasarhelyi, M. A. (2018). Auditing with Smart Contracts. *International Journal of Digital Accounting Research.* Retrieved from https://pdfs.semanticscholar.org/3122/b35e03fb9758108550327d493e2e4748ad2d.pdf

[20] Duraibi, S., Alashjaee, A. M., & Song, J. (2019). A Survey of Symbolic Execution Tools. *International Journal of Computer Science and Security.* Retrieved from https://www.cscjournals.org/manuscript/Journals/IJCSS/Volume13/Issue6/IJCSS-1519.pdf

[21] Di Angelo, M., & Salzer, G. (2019). A Survey of Tools for Analyzing Ethereum Smart Contracts. *IEEE International Conference on Decentralized Applications and Infrastructures (DAPPCON).* Retrieved from https://publik.tuwien.ac.at/files/publik_278277.pdf

[22] Parizi, R. M., Dehghantanha, A., Choo, K. K., & Singh, A. (2018). *Empirical Vulnerability Analysis of Automated Smart Contracts Security Testing on Blockchains.* arXiv. Retrieved from https://arxiv.org/pdf/1809.02702

[23] King, J. C. (1976). Symbolic Execution and Program Testing. *Communications of the ACM.* Retrieved from http://www.cs.umd.edu/class/fall2014/cmsc631/papers/king-symbolic-execution.pdf

[24] Dika, A., & Nowostawski, M. (2018). Security vulnerabilities in Ethereum smart contracts. International Conference on Internet of Things (iThings) and IEEE Green Computing and Communications (GreenCom) and IEEE Cyber, Physical and Social Computing (CPSCom) and IEEE Smart Data (SmartData). Retrieved from https://www.researchgate.net/publication/333590995_Security_Vulnerabilities_in_Ethereum_Smart_Contracts

[25] Androulaki, E., Barger, A., Bortnikov, V., Cachin, C., Christidis, K., De Caro, A., & Yellick, J. (n.d.). Hyperledger Fabric: A Distributed Operating System for Permissioned Blockchains. Retrieved from https://dl.acm.org/doi/abs/10.1145/3190508.3190538

[26] Manning, D. A. (2018). *Solidity Security: Comprehensive List of Known Attack Vectors and Common Anti-Patterns.* Retrieved from https://blog.sigmaprime.io/solidity-security.html

[27] Colombo, C., Ellul, J., & Pace, G. J. (2018). Contracts over Smart Contracts: Recovering from Violations Dynamically. *International Symposium on Leveraging Applications of Formal Methods.* Retrieved from http://www.cs.um.edu.mt/gordon.pace/Research/Papers/isola2018a.pdf

[28] Clause, J., Li, W., & A. O. (2007). Dytan: A Generic Dynamic Taint Analysis Framework. *Proceedings of the 2007 International Symposium on Software Testing and Analysis.* Retrieved from https://citeseerx.ist.psu.edu/viewdoc/download?doi=10.1.1.83.1353&rep=rep1&type=pdf

[29] Kolluri, A., Nikolic, I., Sergey, I., Hobor, A., & Saxena, P. (2019). Exploiting the Laws of Order in Smart Contracts. *Proceedings of the 28th ACM SIGSOFT International Symposium on Software Testing and Analysis.* Retrieved from https://arxiv.org/pdf/1810.11605.pdf

[30] Rodler, M., Li, W., Karame, G. O., & Davi, L. (2020). EVMPatch: Timely and Automated Patching of Ethereum Smart Contracts. Retrieved from https://www.usenix.org/system/files/sec21summer_rodler.pdf

[31] McCabe. (n.d.). *Software Security Analysis: Control Flow Security Analysis with McCabe IQ.* Retrieved from http://www.mccabe.com/pdf/Appnote-ControlFlowSecurityAnalysis-BannedFunctions.pdf

[32] Allen Frances. (1970). *Control Flow Analysis.* Retrieved from https://www.cs.columbia.edu/~suman/secure_sw_devel/p1-allen.pdf

[33] Tripp, O., Pistoia, M., Fink, S. J., Sridharan, M., & Weisman, O. (2009). *TAJ: Effective Taint Analysis of Web Applications*. Retrieved from https://www.academia.edu/download/45166189/TAJ_effective_taint_analysis_of_web_appl20160428-2695-14z7zlh.pdf

[34] Avancini, A. & Ceccato, M. (2010). Towards Security Testing with Taint Analysis and Genetic Algorithms. *Proceedings – International Conference on Software Engineering*, 65–71. Retrieved from https://citeseerx.ist.psu.edu/viewdoc/download?doi=10.1.1.455.3178&rep=rep1&type=pdf

[35] Dewhurst, R. (2020). *Static Code Analysis Control.* OWASP. https://owasp.org/www-community/controls/Static_Code_Analysis

15 Blockchain-as-a-Service

Ramya Bomidi, Srija Guntupalli, Sanober Mohammed and Bhargav Putturu Theja

ABSTRACT

This final chapter focuses on Blockchain-as-a-Service (BaaS) applications. As a newer technology, BaaS applications are cloud-based applications that enable organizations to leverage cloud-based resources to quickly build, implement, host and manage various blockchain projects. In the BaaS service delivery model – as is the case with other cloud solutions – the responsibility for hardware, software purchase and maintenance is passed on to the cloud service provider for greater ease and heightened efficiency; thereby, enabling entities to solely focus on the logic and workflow of the blockchain network.

The chapter discussion begins with a general overview of cloud attributes, such as on-demand service, broad network access, resource pooling, rapid elasticity and measured service. Next, various cost-saving characteristics, cloud deployment models (public, private and hybrid), computing roles and service models (IaaS, PaaS and SaaS) are explained as a way to familiarize readers with the full attributes and capabilities of cloud computing.

The second part of this chapter explains what a BaaS platform is, what its advantages and challenges are, as well as how BaaS differs from other computing approaches, such as serverless computing and Platform as a Service (PaaS). Readers are also provided with a basic understanding of how BaaS applications function, as well as what factors to consider when selecting a BaaS provider. Several popular BaaS platforms provided by entities such as IBM and Oracle, in addition to several BaaS applications under development, are also briefly discussed in this chapter.

A comparison of nine BaaS offerings is presented at the end of the chapter based on factors such as framework, pricing and special features. Furthermore, studies introduced by researchers regarding non-commercial BaaS platforms and governance within the BaaS context are also briefly discussed. The final activity in this section aims at providing interested readers in opting for a cloud-based, blockchain solution with some preliminary information to start a business case aimed at implementing BaaS within their own organizational setting.

INTRODUCTION

The development of blockchain technology and its uses in business have prompted many enterprises to experiment with the technology. Leading businesses and technology providers have begun offering Blockchain-as-a-Service (BaaS) to encourage and facilitate organizational adoption of the technology. Before diving into the details of BaaS, a preliminary understanding of cloud concepts, including its characteristics, types, deployment models and service models are discussed in the next section. The following sections provide additional technical insights about Blockchain-as-a-Service (BaaS) concepts including working models, BaaS service providers comparisons, as well as some business use cases.

DOI: 10.1201/9781003211723-15

A History of Cloud Computing

The popularity of cloud computing continues to increase every year. The American National Institute for Standards and Technology (NIST) definition for cloud computing states that the technology is '*a model for enabling ubiquitous, convenient, on-demand network access to a shared pool of configurable computing resources (e.g., networks, servers, storage, applications, and services) that can be rapidly provisioned and released with minimal management effort or service provider interaction*' [17]. In a nutshell, cloud computing may be loosely described as a more efficient use of computing that involves the Internet and its infrastructure [3].

During the 1960s, John McCarthy first introduced the concept of cloud computing to the public. However, the underlying cloud computing idea dates back as far as the 1920s with the invention of the earliest mainframe computers. With mainframe computers, users could access data and processing power from anywhere through a centralized platform. In the 1990s, the exploding popularity of the Internet led to the widespread availability of virtual computers resulting in the modern cloud computing infrastructure. As more businesses became familiar with cloud services, the popularity of cloud computing increased. Many organizations, such as Amazon, Microsoft, IBM and Google, have become prominent competitors in the marketplace for cloud computing services [3].

CLOUD COMPUTING AT A GLANCE

The cloud model is comprised of five essential attributes, four deployment models and three service models described as follows [16, 17].

Cloud Computing Attributes

All cloud service providers are expected to offer the following five essential attributes of cloud computing for better quality service and flexibility:

On-demand service: Cloud services are set to be on-demand. Customers can therefore take advantage of cloud services as needed, without having to rely on their service provider directly.

Broad network access: The broad network access attribute enables cloud customers to use cloud services without experiencing bandwidth bottlenecks. Modern technologies, such as advanced routing techniques, load balancing and multisite hosting are used to combat latency.

Resource pooling: The resource pooling characteristics of the cloud enable a cloud to remain financially viable by making major capital investments in equipment and then using cloud assets to generate income from a large base of cloud customers. The arrangement also provides an opportunity for various clients to forego major capital expenditures and instead opt for cloud-based subscriptions. Consequently, the use of cloud services leads to a significant reduction in capital expenditures, while increasing an organization's subscription expenses.

Rapid elasticity: Cloud computing offers scalability, which enables cloud customers to scale up or down quickly based on demand.

Measured service: Cloud systems are pay-as-you-go. Customers are only charged for the cloud resources used and nothing more; a similar concept to major utilities, such as water, gas or electricity.

Cloud Deployment Types

Each deployment type discussed below offer users distinct advantages and disadvantages, based on location considerations, customization capabilities and shared cloud services. As such, there are four basic types of cloud deployment models [36, 37]:

Public cloud: Public cloud platforms are the most common forms of cloud computing services. As mentioned before, all resources, such as software, hardware, personnel and facilities, are managed and owned by the cloud service vendor and made available to anyone willing to pay for these services; hence, the name 'public cloud'. Among the public cloud vendors, Amazon Web Services, Google and IBM are a few examples.

Private cloud: Similar to traditional IT environments, private cloud services are owned and operated by various organizations for the exclusive use of the organization's customers and/or employees. Remote access web connections enable such platforms to also be accessible by approved external users. Compared to public cloud services, private cloud platforms often have higher security measures because such platforms are only accessible to trusted users within the organization. The main benefits of using the private model include greater data privacy and security. However, an organization will need to spend money on purchasing the needed software and equipment for implementing a private cloud. A common example of a private cloud application is an internally hosted SharePoint site.

Hybrid cloud: Hybrid clouds are comprised of a mix of public and private cloud computing models. The main benefits of this model include cost-effectiveness, higher security and increased deployment flexibility. The North American healthcare industry is a good example of an effective hybrid cloud where private cloud solutions offer security. By contrast, a public cloud that is connected to healthcare facilities, insurance providers and other interested parties can provide timely access to potentially life-saving medical data [37].

Community cloud: Also referred to as 'affinity cloud'. With the community model, a cloud platform is shared by an affinity group with similar business goals. The services are often mostly managed by personnel within the affinity group or by a third party. The main advantage of using a community cloud model is the smaller capital investments. The main drawback is its shared resources. A common instance of a community cloud is gaming communities, such as the PlayStation network [23, 37].

CLOUD COMPUTING SERVICE MODELS

In addition to the deployment models, cloud computing is offered through three different service models – Infrastructure as a Service (IaaS), Platform as a Service (PaaS), and Software as a Service (SaaS) – to satisfy specific business challenges within an organization. All three models are deployed according to a pay-per-use policy [35].

Infrastructure as a Service (IaaS): The IaaS represents the digital server instances hosted without the need for physical access to the hardware. As a pay-as-you-go system, IaaS platforms provide hardware-related resources including storage and virtual servers. A high level of administrative authority is granted to the customers within their server instances, as well as the ability to set their network security level. IaaS offers the primary benefit of using the most recent technology, which provides users with faster access to services.

Platform as a Service (PaaS): The PaaS model provides the infrastructure and services needed to develop applications over the web [35]. Since the developers do not have to identify the scalable requirements in advance, the PaaS model offers a lot of flexibility and productivity to application developers. On the Internet, PaaS provides for low-cost, reliable and large-scale application growth. Apart from its advantages, PaaS cloud technology's main drawback is the 'lock-in' to a specific PaaS platform experienced by users, making platform transitions and changes very challenging at times [38].

Software as a Service (SaaS): The SaaS type offers all the IaaS and PaaS features along with the specific software created entirely or in part. SaaS is mostly provided in the form of web applications and includes some web storage or other network connections with non-remote

TABLE 15.1
Shared Responsibility Matrix

Responsibility	Infrastructure as a Service (IaaS)	Platform as a Service (PaaS)	Software as a Service (SaaS)
Governance risk and compliance	Customer	Customer	Customer
Data security	Customer	Customer	Customer
Application security	Customer	Customer	Customer/service provider
Platform security	Customer	Customer/service provider	Service provider
Infrastructure security	Customer/service provider	Service provider	Service provider
Physical security	Service provider	Service provider	Service provider

Source: [23].

applications. As the software and other components are hosted at the service provider location, the users can access the software services through a web browser [35].

Shared responsibilities: In cloud computing, the customer and cloud service provider share responsibilities in terms of physical security, data security and controls. The shared responsibility is often referred to as 'segregation of duties' between the customer and provider to secure the services based on service type or deployment model [23].

Table 15.1 outlines the shared responsibilities between customers and cloud service providers based on various cloud deployment schemes.

Enterprise/customer responsibility: The cloud customer is always ultimately responsible for data protection and account access control [23]. Meeting the demands of compliance standards is often a challenge for cloud service providers, as they must comply with multiple IT process management requirements. It is important to reiterate that the cloud customer is always ultimately liable for applicable compliance mandates [24].

Cloud service responsibility (CSR): The cloud provider is always in charge of protecting the physical facility where the data center is located, irrespective of the service type [23].

Responsibility by service type: Monitoring and auditing are important for ensuring the security of the platform. A service provider of IaaS only manages facilities; as such, the customer is responsible for implementing the necessary protections to safeguard applications, virtual machines and data.

Even though PaaS models are built on top of IaaS models, the cloud service providers are responsible for operating systems (OS) maintenance, while the customers are responsible for data and application security. In both the PaaS and SaaS models, when a customer updates their installed software, the service provider shares the responsibility of updating the hardware and ensuring applicable security controls [23, 24].

CLOUD COMPUTING ROLES AND RESPONSIBILITIES

Furthermore, understanding cloud computing's various attributes and models is necessary; but, so is understanding its major roles and responsibilities, as described below [21].

Cloud service providers (CSPs): Cloud providers are vendors offering cloud services. As mentioned, such entities own the cloud computing data center, and manage its various hardware, software and human resources, in addition to monitoring service provision and platform security. Cloud data centers enable cloud providers to build, configure, deploy and operate cloud services [21].

Cloud customer: A cloud customer can be a person or an organization that purchases or leases various cloud services [23].

Cloud user: Cloud users generally refer to the users or corporations who utilize the computing services online from anywhere and anytime. It is important to note that the term cloud users refers mainly to the application end users; whereas cloud customers purchase and pay for the cloud services. For example, when an organization purchases an Office 365 subscription to use, the organization acts as a customer, and the users who utilize the services are the cloud users [21, 23].

Cloud access security broker (CASB): CASB is a security enforcement point, sitting between the end-user and the cloud that acts as an independent, third-party identity and access management enforcer. CASB act as a central point for monitoring and managing access to cloud resources [10].

Regulators: In addition to making sure organizations comply with their regulatory mandates, regulators maintain the integrity of the organization. Organizations are examined for security and performance by regulators [23].

BENEFITS OF CLOUD COMPUTING

Cloud computing provides many benefits to an organization. Of these, cost reduction is the main benefit derived by implementing cloud storage solutions within an organization, as these organizations will only pay for the services being utilized. In addition, cloud computing also reduces upfront IT costs and eliminates the need for technical personnel responsible for maintaining servers and local infrastructure.

The concept of multitenancy is another aspect of cloud computing that involves sharing servers and applications among multiple users. Using a cloud in a distributed and shared manner, different users and applications can work together with greater efficiency and reduce costs by sharing common infrastructure.

Since the data and applications are no longer stored locally, data will always be available, even if access to a desktop or laptop computer is lost. Furthermore, the data will not be lost even in the case of a failure of the service since the cloud service provider will have backups [26].

WHAT IS BAAS?

Due to blockchain's distributed, immutable and transparent properties, the trend of developing blockchain applications has gained popularity in recent years. As part of the various innovations introduced into the IT industry, Blockchain-as-a-Service (BaaS) was developed to support the blockchain features in the cloud platforms. The BaaS concept refers to an 'as-a-service' cloud-based solution designed to facilitate the implementation of a blockchain platform that allows users to build, host and manage blockchain applications from a central cloud [40].

Blockchain-as-a-Service provides operational support capabilities just like IaaS, PaaS and SaaS cloud models. BaaS is considered a major milestone in the blockchain market, presenting a strong opportunity to increase the adoption of distributed ledger technology across the business world. In addition, the conceptual model of BaaS is derived from that of SaaS cloud computing model [32].

By acquiring a BaaS model, the customer can focus on the logic and workflow of the business network, relieving itself of the responsibility for hardware, software and expertise behind the scenes, and passes this responsibility to the service provider [20]. Several BaaS service providers, such as IBM, Oracle and Microsoft, are currently available on the market and provide BaaS services as ready-to-go or platform-based solutions. BaaS solutions that are ready-to-go resemble customizable

SaaS solutions, whereas platform-based solutions allow customers to manage and build applications themselves [11]. In general, most BaaS implementations offered by larger companies use open-source projects, such as Ethereum, Hyperledger Fabric, Quorum and R3 Corda [20].

How Does the BaaS Model Work?

To set up a BaaS platform, an organization subscribes to a blockchain service with a service provider. The blockchain infrastructure will be designed following the customer's requirements and will be backed by a service level agreement. The agreement defines the terms for bandwidth, monitoring, security controls and maintenance of the infrastructure. According to the customer's business requirements, the service provider may configure blockchain networks based on any distributed ledger, such as Ethereum, Hyperledger Fabric, R3 Corda, Chain Core or Quorum. The service provider then implements the essential resources and supports the required technology on behalf of the customer. As a result, the customer may focus on business rather than performance and network issues [32].

In many ways, web hosting service providers and BaaS are similar. Because the website is visited by many users every day, maintenance and infrastructure issues can be handled by a web hosting service provider – such as Amazon Web Services (AWS), Microsoft Azure or Google. Similarly, BaaS relieves the customer from having to manage the framework of a blockchain application [4].

Advantages of BaaS

With the BaaS model, companies can avail themselves of several advantages without investing in developing the blockchain platform on-premise. These advantages are discussed below:

Cost and time savings: The implementation of blockchain technology on-premise may involve a substantial capital investment upfront. Even after investing, if the outcome does not meet expectations, an organization may suffer a capital loss [29]. However, in the case of BaaS, since the infrastructure is outsourced, companies do not have to make an investment in further expanding the organizational IT infrastructure. Consequently, the customers can save both time and money by not having to build, manage and maintain the platform on-premise [20].

Reduction in personnel costs: Due to the limited number of IT services required on-premise for implementation of BaaS architecture, the customer's personnel will not be involved in the management of the equipment, power and cooling needed to implement the blockchain. Consequently, the costs associated with hiring and training employees are greatly reduced, allowing the organization's existing IT staff to focus on business needs.

Ease of use: BaaS provides quick access to a customizable template that can be easily integrated into existing applications, eliminating the need to build the architecture from scratch [32]. Thus, management and operation of the platform can be made easier without the need for any technical expertise [27].

Pricing and customer support: Generally, BaaS service providers provide subscription-based services [32]. As a result, customers can scale up or down the level of service based on demand, reducing the risk associated with a pay-per-use model. Moreover, the BaaS provider offers the customer a complete range of support services, including patching, monitoring and upgrading of the systems [20].

BaaS Challenges

Although BaaS has several benefits in terms of cost, personnel and ease of use, there are also challenges that BaaS adoption may bring to an organization.

Privacy and data compliance: BaaS implementation could generate privacy and data compliance issues. Considering that third parties provide hosting services for BaaS, data may be stored across multiple networks, leading to concerns regarding compliance. While blockchain regulations have increased in the past few years, regulators still need to fill important blockchain-specific gaps [29, 32].

Limited options: Currently, customers have a relatively limited scope of options when choosing a provider for BaaS services. As an example, BaaS is predominantly dominated by large companies such as Amazon, Microsoft and Oracle, as BaaS is a cloud-based service [29].

Limited resource availability: At this time, there are limited human resources available on the market with in-depth knowledge of blockchain technology. As a result, resources without proper planning and expertise might increase the probability of an organization failing [32]. Additionally, blockchain is still an emerging technology; as such, there may be a risk of an organization failing to properly implement needed BaaS requirements within its existing network environments, which may increase the time and cost associated with the integration of a BaaS initiative [29].

Current and Anticipated Future Interest in BaaS

BaaS has driven the blockchain economy's market growth among both big and small enterprises by raising investment in various blockchain technology-based projects. Furthermore, the increase in decentralized applications has led to a rise in blockchain technology's use across various sectors like retail, finance and Internet banking. For example, the government of the Republic of Korea invested approximately 800 million USD into expanding blockchain technology in 2019 [5].

Many large corporations have introduced their BaaS platforms recently, including IBM's Blockchain and Amazon's Managed Blockchain. In general, platforms offering Blockchain-as-a-Service typically adopt one specific blockchain technology, such as Hyperledger Fabric or Ethereum, as their underlying platform. As an example, Ethereum Blockchain-as-a-Service (EBaaS) by Microsoft Azure is built using the Ethereum framework; whereas, IBM's blockchain uses Hyperledger Fabric. The service provider manages the BaaS environment to ensure the infrastructure's flexibility, operationality and accessibility. In addition, service providers continue to develop BaaS tools in order to meet growing market demands [30].

The global BaaS market is poised to grow as more companies adopt the technology. The market extends from Asia-Pacific, Africa, the Middle East, Europe and Latin America. Among these regions, North America led with 2.51 USD billion in market size by investing in blockchain solutions across multiple industries, including financial services, defense and many others. A significant number of BaaS providers are also integrating blockchain solutions with advanced technologies such as artificial intelligence (AI) and machine learning (ML) to improve their security features. The *Fortune Business Insight* forecasts that the global BaaS market size will reach 24.94 billion USD by 2027, representing a compound annual growth rate (CAGR) of 39.5% during the forecast period. Nevertheless, the multiple benefits of BaaS for small and medium-sized businesses (SMEs) are expected to lead the market's development in the future [5].

Because of its large potential applications in numerous fields, BaaS is poised to experience a similar viral acceptance as web applications. A blockchain serves as a verification mechanism for information that is shared over a network. Reshma Murthy, President of Frontier Solutions, stated that '*BaaS will play a very important role. Software-as-a-Service and Infrastructure-as-a-Service are just emerging as ways to save money on resources that would otherwise be too expensive. BaaS can be used in the same way and provides an opportunity for organizations to try blockchain while they are still determining how well it will be accepted across their industry*' [20].

HOW IS BAAS DIFFERENT FROM SERVERLESS COMPUTING?

The term Function-as-a-Service (FaaS) often refers to serverless computing. The FaaS server provides essential services for hosting a business in a serverless environment. The main idea behind serverless computing involves resource allocation being handled by a third party, along with computing and storage decoupling, and scaling instances.

Serverless computing and BaaS share little in common. As opposed to serverless computing, BaaS relies on stateful transactions. Unlike BaaS, in which smart contracts enable cryptographically signed messages that are immutable, FaaS lacks this feature. However, the two technologies are both distributed and event-driven [39].

BAAS SERVER-SIDE CAPABILITIES

Blockchain technology ensures that digital transactions are highly secure and reliable because of its server-side functionality. Without proper server-side security, there will be a higher chance of any malicious external party gaining complete access to the system.

In general, a decentralized application is a serverless application based on Ethereum which utilizes an HTML/JavaScript front end and can interact with a backend. There are two components to a decentralized application (DApp): a smart contract that acts as the server on the Ethereum network and an interface for users to interact with the smart contract [12].

A smart contract has four characteristics, including observability, verifiability, privity and enforceability. These properties help smart contracts to record the transaction by creating high confidentiality, resilience and scalability. As a reminder, examples of smart contract platforms include Hyperledger Fabric, Corda Enterprise and Ethereum, which has experienced a significant increase in its use over the past few years. Each smart contract has different settings, consensus mechanisms and supporting languages to meet the business's requirements. Smart contracts offer tokenization written in supported languages and executed in particular environments, such as virtual machines and peer-to-peer networks [22]. Blockchain databases are a decentralized network of nodes that verifies new additions to the database and can enter new data into the database. By reaching consensus, a blockchain is protected from tampering. Developers can construct the business logic using the blockchain and have that logic written in a specific programming language running in the execution environment [20].

HOW IS BAAS DIFFERENT FROM PAAS?

While the concept of SaaS is at the core of the BaaS model, it can be explicitly rewritten within PaaS because of the availability of infrastructure support [27]. Comparing the BaaS and PaaS models becomes clearer when considering their similarities and differences. The PaaS model is for creating applications without having to manage its infrastructure, while the BaaS model uses a blockchain to build customized applications [38]. The advantages of BaaS are similar to those of PaaS in terms of cost reduction, scalability and subscription [33]. Like BaaS, PaaS also gives enterprises complete control over application development without worrying about infrastructure and storage [27]. The BaaS model also provides other benefits, including enhanced data security, transparency, privacy and immutability [2]. Generally, BaaS services distribute containers into the cloud and create a peer-to-peer network with them. The containers include software and data required to run blockchains as distributed nodes [20]. Furthermore, BaaS manages the forking and consensus mechanisms on its own [27].

HOW DO BAAS APPLICATIONS RUN?

Through its transparent nature, blockchain technology provides a trusted environment by delivering information to the public throughout its entire network while promoting trust and integrity. BaaS

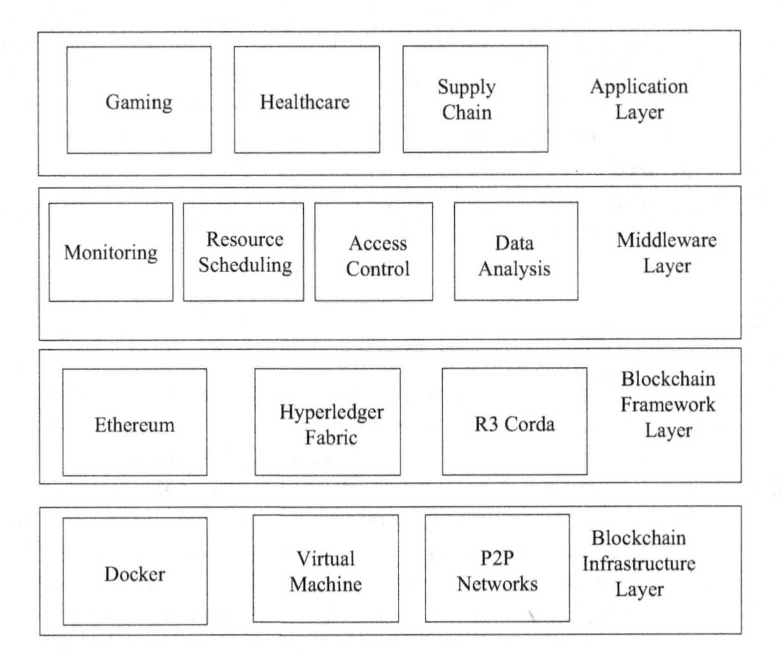

FIGURE 15.1 Blockchain-as-a-Service architecture.

Source: [22].

provides a majority of the applications and required expertise behind the scenes, making it easier to focus on the logic.

Four main layers currently constitute the current architecture of BaaS platforms, namely, the application, the middleware, the blockchain framework and the blockchain infrastructure layers. Generally, each layer can be seen as a single container with multiple functional components within it. Figure 15.1 illustrates the BaaS architectural overview.

Application layer: With the BaaS architecture, the application layer is the highest level through which users can interact with the BaaS services that support its back-end. As blockchain technologies have evolved from their traditional role of a cryptocurrency base, different types of tools support the network in performing the necessary functions. As with all platforms, blockchain applications are created on top of the blockchain infrastructure. Several blockchain providers, including *Amazon*, *Oracle* and *Samsung*, provide their customers with predefined or generic applications for direct use [19]. This approach is more efficient for developers and does not require a complete understanding of blockchain's internal mechanisms. As a result, BaaS providers require only the deployment of decentralized applications that run on the blockchain infrastructure.

Middleware layer: The middleware layer serves as an interface layer that allows the BaaS application layer to connect with the blockchain framework layer in the BaaS architecture. The resource scheduling, data analysis, monitoring and access control services are considered basic manipulation activities carried out within this layer. Resource scheduling algorithms are used to determine how to allocate limited resources to ensure maximum utilization. The monitoring service monitors and audits the performance of the network, notifying developers of any problems that may exist within the network. Additionally, blockchain data can also be recorded and analyzed to improve blockchain applications. In a BaaS implementation, access control can provide high assurance by avoiding unauthorized access to data by using the correct authorization and authentication measures. By providing these manipulation services, the middleware layer facilitates BaaS to be highly reliable and scalable.

Blockchain framework layer: The blockchain framework layer is a server for smart contracts that houses the smart contract platform in the BaaS architecture. Smart contracts run on a platform that supports tokenized programs written in a particular language and executes them automatically when certain conditions are met. The layer supports Ethereum, Hyperledger Fabric, R3 Corda and other platforms [22].

Hyperledger Fabric is a widely used distributed ledger technology. Many providers like *Huawei*, *Oracle* and *IBM* offering BaaS services utilize Hyperledger Fabric as an underlying blockchain platform. Hyperledger Fabric is a modular blockchain platform governed by the Linux Foundation and IBM which has numerous applications in various fields. Furthermore, Hyperledger Fabric is a permissioned public network in which new members must be selected and approved before joining. Therefore, only members of the network who have been approved have access. In the context of enterprise blockchain solutions, these capabilities are particularly valuable. Another distributed ledger technology is Ethereum, a public and permissionless protocol. Ethereum provides a generic platform for various types of applications and transactions. Among DLT providers, Quorum, a permissioned version of Ethereum, facilitates the privacy of transactions and smart contracts. R3 Corda, in addition to Ethereum and Hyperledger Fabric, is also a permissioned distributed ledger technology protocol used primarily for financial transactions [19].

Blockchain infrastructure layer: As far as the infrastructure goes, most providers run their blockchain platforms on hardware provided by the vendor. A typical setup includes hosting the blockchain in a cloud offering from the provider. Larger IT service providers usually offer such a setup, as they have a cloud infrastructure in place. There may be instances where the provider hosts the BaaS on third-party infrastructure. For example, IBM offers its customers the choice between hosting BaaS in their cloud or using third-party infrastructure like *AWS* [19]. In general, the BaaS architecture requires the development and implementation of the blockchain platform, which involves a distributed network requiring computing resources to implement smart contracts. The computational resources can either be a physical or virtual machine or a docker. For example, the Ethereum smart contract utilizes virtual machines, whereas the Hyperledger Fabric smart contracts make use of dockers. The smart contracts should be kept isolated in a sandbox to prevent the spread of viruses or bugs. There are two limitations to the growth of blockchain: the complications in deploying blockchain infrastructure and the high associated operating and maintenance costs. BaaS however discards the tedious task of setting up the infrastructure, enabling the developers to focus on the applications [22].

Consensus Mechanism

Most of the BaaS services employ Crash Fault Tolerance (CFT) or Byzantine Fault Tolerance (BFT). At present, the most used consensus mechanisms for BaaS provided by Hyperledger Fabric are *Kafka* and *Raft*, both of which are CFT ordering services. The CFT setup assigns a leader node that is supposed to be trustworthy. Since all nodes in the network will replicate the entries of the leader, the leader must be trusted. As CFT consensus or ordering services may only be accessed by those who have been granted permission to do so, leaders are presumed to be trustworthy. If the current leader crashes, a new leader will be appointed. By taking advantage of transaction finality, CFT also helps the consensus process move quickly. Kafka CFT can be used by *Alibaba*, *Amazon*, *Huawei*, *Oracle*, *SAP*, *IBM* and *Samsung* when built using Hyperledger Fabric, whereas Raft is only available from *Baidu*.

Quorum is compatible with Raft as well as Istanbul BFT – a variant of the BFT consensus. In contrast to CFT algorithms – which use a block proposer or leader to decide whether to approve a block – the Istanbul BFT algorithm requires multiple votes for every block. A network is capable of handling one-third of a node's defects when using BFT. Therefore, Istanbul BFT can be used in cases where competitors or other network participants cannot be trusted. Most BaaS providers offering Quorum on their platforms, such as *Alibaba*, and Microsoft *Azure* uses Istanbul BFT [19].

CRITERIA FOR SELECTING A BLOCKCHAIN AS A SERVICE PARTNER

There are several BaaS providers on the market today. However, choosing the best platform requires extensive research and analysis. Due to the lack of proper guidelines and best practices in the market, the selection process is a bit challenging. The following are a few important factors that should be considered when selecting a provider [32].

Previous track record: Organizations should select a provider with extensive experience in BaaS deployment when choosing a provider. Obtaining the provider's prior experience can be achieved by looking into their portfolio, major current clients, audit results and client testimonials.

Quality commitment: Quality commitment is a critical factor that an organization must consider for the successful implementation of projects, as a low-quality infrastructure can lead to business failure. When consumers place their trust in purchasing products and services, the service provider should provide the best quality service for its customers. Therefore, an organization must ensure that the vendor adheres to quality, established processes and standards.

Ease of use: Business results are impacted significantly by user-friendly processes and systems. Therefore, enterprises should ensure that blockchain-enabled systems are easily adopted by their employees.

Integration: In addition to several layers of business processes, workflows and many more, companies should adopt a BaaS that can be integrated with the system without disrupting the legacy network.

Process control: Consider whether the implementation plan covers any security gaps. Several key controls, such as the control of data flow, computing resources, active monitoring tools, and so on, are essential in maintaining an environment that is secured against hackers. When it comes to distributed ledgers, organizations should aim for the most robust outcome.

Back-end services: The back-end services – mainstream technologies and integration of popular services features – should be evaluated extensively. Enterprises should look out for features like data security, costing control, integration and process control.

Rapid provisioning: Rapid provisioning facilitates the quick deployment of blockchain networks while minimizing the risks involved in the process. Hurdles such as variations in web browsers, firewalls, databases, hardware and application servers complicate the quick deployment of networks. Therefore, enterprises should ensure that BaaS providers deploy and manage the blockchain environment effectively and efficiently with minimal errors or bugs.

Data security: The enterprise must ensure that the data are protected and remain within the platform's boundaries as established by the BaaS provider.

Cost control: Enterprises should also review pricing and post-deployment support options being offered by the provider. Choosing a vendor with a reasonable subscription price is equally as important as evaluating hidden charges and making a decision based on budget constraints [2].

BAAS PLATFORMS

As discussed in the previous sections, the BaaS model allows an organization to enjoy the benefits of a smart-contract platform while the service providers typically handle the infrastructure. Some popular commercial BaaS providers that can be integrated with an enterprise are discussed below.

IBM BaaS

IBM offered a BaaS solution based on Hyperledger Fabric on IBM Cloud in 2017 and since then has made available two versions of the platform, IBM Blockchain Platform 2.5.2 and IBM Blockchain Platform for IBM Cloud. IBM Blockchain Platform 2.5.2 allows the deployment of blockchain components on any Kubernetes platform, whereas IBM Blockchain Platform only allows deployment on the IBM Cloud Kubernetes cluster [18]. The Kubernetes serves as the infrastructure in IBM BaaS architecture [22]. The developers can create blockchain applications with the help of Bluemix. Using 'Bluemix', 'Hyperledger Fabric' and the IBM cloud, users can develop a DevOps and deploy Chaincode. Meanwhile, the business logic can be maintained using Chaincode, which is written in Go and Node.js. The IBM BaaS smart contract supports JavaScript, Go and Java languages [27]. The IBM blockchain platform supports various operational tools to ensure governance and flexibility over the platform [22]. The users have complete control over the operation, governance and implementation of blockchain components. The primary objective of the IBM BaaS solution is to offer high availability, enterprise-ready to go and disaster recovery options to the enterprise [27].

Oracle

Oracle offers Oracle Blockchain Cloud Services (OBCS) based on Hyperledger Fabric, which is an open-source platform.

The BaaS solutions offered by Oracle feature two key capabilities: a turn-key sandbox and independent software vendors. The turn-key sandbox environment is for developers. The Independent Software Vendors (ISVs) make deployment of blockchain technology easy in terms of onboarding and generating smart contracts and regardless of the vendor [27]. A composable-based architecture is used in the platform, which means all the systems and processes have been designed to be API-enabled. Oracle Blockchain Cloud Service runs on a blockchain network that validates the nodes and validates the ledger, responding to queries by executing smart contracts, which represent business logic. Through client SDKs or REST APIs, external applications can invoke transactions or run queries. After reaching the consensus, the transactions are added to the blocks. The transactions are transparent and shared but the access can be restricted based on their ranking. Previously, ordering nodes were managed by a cluster of Kafka and Zookeeper machines. However, this setup is increasingly resource-intensive when applied to the cloud. As a result, a consensus plugin RAFT was launched for managing the ordering service and to prepare it for enterprise-level production networks. With RAFT, a group of order nodes cooperate in order to create blocks of orders based on a dynamic leader-based model. Participation of member organizations is thus most fair and uniform, even when numbers are large. Additionally, Oracle provides a better encryption service for securing sensitive information [12, 16].

Microsoft Azure BaaS

As part of its partnership with Consensys – a leading provider of cloud computing services – Microsoft also offered Ethereum Blockchain-as-a-Service (EBaaS) on its Azure platform in late 2015. Azure BaaS supports Ethereum, Corda, Quorum and Hyperledger Fabric [27]. With Azure BaaS, customers can also deploy enterprise-grade templates with distributed ledgers without incurring costs. However, when an application is built on the blockchain, the customer is responsible for the development and costs. Microsoft Azure provides three offerings in BaaS: Azure Blockchain Service, Azure Blockchain Workbench and Azure Blockchain Development Kit [12].

Azure Blockchain Service: Azure Blockchain Service is a fully managed blockchain service that allows customers to create and operate a blockchain network in a matter of minutes. Azure offers complete end-to-end solutions for infrastructure management and network governance in a simplified form with a built-in consortium network. Ethereum Quorum ledgers are supported by the Azure Blockchain Service, which uses the Istanbul Byzantine Fault Tolerant consensus mechanism. The customer does not need to be concerned about the underlying infrastructure but can focus exclusively on application development and business logic. As part of its retirement, Azure's Blockchain Service will be replaced by Consensys' Quorum Blockchain Service in September 2021 [28].

Azure Blockchain Workbench (ABW): The Azure Blockchain Workbench is a set of Azure services that allow businesses to develop and manage blockchain applications in addition to transferring their business processes across organizations. Microsoft's Azure Blockchain Workbench facilitates the development of blockchain applications, allowing developers to concentrate their efforts on building business logic and smart contracts. Azure Blockchain Workbench is popular for its integration capabilities, which simplify the creation of blockchain applications using the workbench's web applications and REST APIs. With these APIs, customers can replace different distributed ledger technologies, storage products and databases. Furthermore, ABW can be integrated with all popular Microsoft products, including Excel, SharePoint, Office 365 and others.

Azure Blockchain Development Kit (ABDK): There are code samples included in the Blockchain Development Kit, which is a GitHub repository designed to integrate with Azure services. As a part of the Azure platform, Microsoft offers a blockchain development kit built upon Microsoft serverless technologies and integrated with SaaS from Microsoft and third parties. In addition to being able to connect to multiple data sources, the Azure Blockchain Development Kit can also integrate with legacy applications and protocols, such as Excel, Office 365, etc. With the RESTful APIs provided by Azure Blockchain Workbench and Azure Blockchain Development Kit, interacting with blockchain nodes has never been simpler or more secure [12].

Amazon AWS BaaS

With Amazon Managed Blockchain, AWS offers a fully managed platform as a service for building and deploying blockchain networks using open-source distributed ledger technologies such as Ethereum or Hyperledger Fabric. The platform provides customers with a choice of instance types with customizable computation and memory capacity to suit their blockchain application requirements. Launching the network is performed using AWS blockchain templates. Following the availability of the network, Managed Blockchain's voting API will enable members to vote on the network's participation. With the addition of a new member, Managed Blockchain offers the ability to launch and configure multiple peer blockchains that will process transaction requests and store a copy of the ledger. As part of the managed blockchain ecosystem, AWS Key Management Services serves as the certificate authority. As a result, new users who join a Managed Blockchain will not be required to set up hardware security modules. With network scalability, the network adapts to increased transaction volume and load. Furthermore, new members do not require special hardware to join the cloud. Amazon Managed Blockchain is used as part of the Hyperledger Fabric framework to enhance the 'ordering service' that enables trust across blockchain networks. The Quantum Ledger Database (QLDB) records all uncommitted transactions on the blockchain, ensuring a transparent and secure ordering service [12].

Alibaba

Alibaba Cloud is one of the most notable and patent-rich vendors in the market today. The BaaS platform is built using Alibaba Cloud Container Service for Kubernetes clusters, which supports Hyperledger Fabric, Ethereum and the financial-grade blockchain technology Ant Blockchain. In addition to supporting public deployments, the infrastructure facilitates private deployments as well. With Alibaba Cloud Blockchain-as-a-Service, users can create a production-level blockchain environment and can graphically manage the operations.

Alibaba BaaS offers services around the globe due to its distributed data centers. Since the consortium blockchain network is built upon the principle of multitenant isolation, each participant is free to manage its resources independently. In addition, the deployment is simple and easy to integrate with other cloud services and products, and also has a flexible ecosystem that can support a wide range of applications and plugins. Alibaba Cloud Blockchain-as-a-Service includes a default Anti-DDoS Basic Service. Additional services offered by the provider include product traceability, supply chain financing and data asset tracking to keep track of data ownership rights. Services provided by this provider may be customized to meet specific business needs while maintaining advanced security and privacy protection [1].

Accenture

In terms of *Accenture*'s BaaS offerings, the organization provides a different perspective compared to other providers in the market. The company's consulting services are aimed at providing organizations with a clear understanding and selection of the appropriate BaaS solution. Accenture's market strategies encompass consulting, strategy and implementation. Concerning the BaaS solution, Accenture contributes by providing a questionnaire and an extensive briefing about the services, followed by a design and implementation model of the platform based on business requirements, together with a deployment environment for the platform. Additionally, the vendor also provides advanced training to developers to meet their goals [2].

Baidu

Baidu, a popular web search engine in China, introduced the 'Xuperchain' blockchain platform and is now offering the 'Baidu Trust' BaaS platform for enterprises [2]. Unlike Microsoft, the Baidu Trust is a self-developed platform. The service takes advantage of distributed ledger technology to facilitate transactions and their tracking. Every component of the blockchain that will be built into a business's infrastructure can be customized and configured flexibly. A Baidu Trust asset-backed securities exchange product using blockchain technology was created in China and has seen success in asset securitization and asset exchange businesses [13].

Huawei

As of early 2018, *Huawei* introduced a Blockchain-as-a-Service platform, known as Blockchain Cloud Service (BCS). With Huawei BCS, blockchain connectivity can be managed and deployed much more easily, reducing the barrier to the adoption of blockchain technology. The BCS provides full-lifecycle management and GUI-based smart contract creation, deployment and commissioning. Data integrity and accuracy are ensured through the integration of industry-specific blockchain applications with the blockchain platform, ensuring data reliability and security. The blockchain network can be accessed via a software development kit (SDK) or a RESTful (Representational State Transfer) API, and monitoring platforms linked to Blockchain Cloud Service can be utilized to monitor data and resources in real time, and to send notifications as necessary. Due to Huawei

Cloud's high performance, blockchain ledgers can be stored efficiently and on time. The Huawei BCS is distinguished by its ability to reduce the risk of failure and no vendor lock-in [6].

SAP BaaS

The *SAP Blockchain Services* are offered in two main categories: *SAP Cloud Platform* and *SAP HANA Blockchain Services.*

SAP Cloud Platform Blockchain Service (SCP): Using the SAP Cloud Platform (SCP) Blockchain Service, blockchain-based applications can be developed quickly and efficiently on cloud platforms with little risk involved when integrated with the SAP landscape. The SCP Blockchain Service supports Hyperledger Fabric and MultiChain.

SAP HANA: *SAP HANA* does not run on the blockchain itself, but rather configures the properties of the connection. Unlike blockchain platforms, SAP HANA Blockchain Service is not a blockchain platform, but rather a means of connecting existing blockchain platforms to the database. SAP HANA is designed to enable analytics using the transaction data in the SAP system. The SAP HANA platform can connect to the distributed ledger technology by utilizing the SCP Blockchain Service. With the HANA database, blockchain data can be accessed using SQL queries. Analytics requires the consolidation of data. Consequently, SAP HANA Blockchain Service is required [15].

Table 15.2 provides a comparison of BaaS platforms. Refer to the access links for more information regarding pricing and features.

BAAS BUSINESS USE CASE

Businesses have found a more cost-efficient, and often more effective way to leverage technology by subscribing to services rather than building their own platforms. Similarly, BaaS enables projects to quickly scale up from a proof-of-concept stage to a full-scale production phase. The following are some notable use cases for BaaS solutions that demonstrate why the solution is both versatile and valuable:

Food Traceability with Amazon Managed Blockchain: Nestlé

Nestlé is one of the largest food and beverage companies in the world and is the world's largest coffee producer with a distribution network spanning more than 190 countries. Transparency in supply chains is very important to consumers as they are interested in knowing where their food is produced and processed. To view the entire process – from farm to packaging – as well as share the different attributes of coffee tasting such as location, plantation type, type of coffee, roasting level, etc., the use of real-time data collection becomes increasingly relevant when comparing consumer value and industry needs.

One of the main challenges associated with collaborating data is the time lag between retailers and customers. Nestle utilizes AWS to avoid time lags by implementing Amazon Managed Blockchain (AMB) as a means to invite vendors to collaborate on supply chain transparency. Nestlé prototyped quickly and co-funded with Amazon Web Services a proof-of-concept for chain of origin to efficiently accelerate its innovation. The chain of origin between the farm and the consumer consists of five levels of information flow. To begin, coffee is transported to the grading facility after being processed on the farm. During the second stage, the coffee is handed over to the broker before being roasted, at which point the roast level will be determined. The roasting process is complete once the product has been packaged in the packaging facility, where the packing suppliers and manufacturing

TABLE 15.2
Comparison of BaaS Platforms

Categories	Framework Support	Special Features	Pricing	Accessible Link
Azure	Ethereum, Corda, Hyperledger Fabric, Chain and Quorum	Built-in governance and encryption on hardware-level, easily integrated with Microsoft legacy applications.	Varies according to the plan, pay-per-use	https://azure.microsoft.com/en-ca/solutions/blockchain/#related-products
Alibaba	Hyperledger Fabric, Ethereum and Ant Blockchain	Facilitates rapid development and provides standard development kits and CA certificates.	Monthly or yearly subscription US$0.000472/GB/hour for storage	https://www.alibabacloud.com/product/baas
Accenture	Enterprise Ethereum Alliance, Hyperledger Fabric and Corda	Consulting services; partnered with IBM, Microsoft.	Based on consultation	https://www.accenture.com/ca-en/services/blockchain-index
AWS	Ethereum, Corda, Hyperledger Fabric and Quorum	Quick deployment with the help of AWS templates, backed by AWS key management service.	Pay-as-you-go	https://aws.amazon.com/managed-blockchain/
Baidu	Ethereum, Hyperledger Fabric and Xuperchain	Facilitates DApp development.	Pay-per-use	https://trust.baidu.com/?fl=1&castk=LTE%3D
Huawei	Hyperledger Fabric	Supports light nodes, monitoring and customization of BCS services.	Pay-per-use	https://www.huaweicloud.com/en-us/product/bcs.html
IBM	Hyperledger Fabric	Seamless connecting platform and reduces the risk of failure.	Deploy on IBM cloud 0.29 USD per hour	https://www.ibm.com/blockchain/platform
Oracle	Hyperledger Fabric	Integrates with Oracle and NetSuite and offers a fully comprehensive blockchain platform.	$0.75 Pay-as-you-go	https://www.oracle.com/ca-en/blockchain/
SAP	Hyperledger Fabric, and MultiChain integration	Integrated into SAP applications using the Blockchain as a Service (BaaS) abstraction layer.	Pay-per-use	https://www.sap.com/sea/products/hana.html

Source: [20].

information are recorded. Upon completion of packing, the product is transported to the carrier, which stores information about the distributors and retailers. At the end of the process, the customer supplement is completed, in which all information related to the sale of the product is recorded. Nestlé has been able to simplify the technological infrastructure used for the company to focus on adding value to the business, supporting farmers and roasters, and providing a superior taste experience for consumers with the help of Amazon Managed Blockchain Services [36]. The implementation of Amazon Managed Blockchain in Nestle helps in improving its coffee taste and accessibility, and helps consumers to know the provenience of their purchase. The main reasons behind Nestle choosing Amazon Managed Blockchain are its low cost, ease of use, integration with other AWS components and the fact that it supports Hyperledger Fabric. Real-time data collaboration enables Nestlé to deliver a premium coffee experience by connecting consumers with the artisanship of coffee. Through Amazon Managed Blockchain, clients can track their products from the farm to the consumer on the blockchain [14].

Using Amazon Managed Blockchain, all transactions are stored in a transparent, immutable and verifiable manner, allowing parties to interact in a more trusted and efficient way, with or without a centralized authority. One of the most important advantages of using this technology in the supply chain is the ability to detect errors, malpractices or any tampering immediately, and address the problem if necessary. Nestlé adopts blockchain validation as a solution across its various brands. Blockchain can be used to verify food provenance as well as the circumstances in which it was produced, due to its immutability [36].

Royalties Information for Publishers with Azure Blockchain Service

The video game industry has experienced rapid expansion in recent years. In addition to programming, video games also require music, special effects and graphics. To facilitate the provision of royalty information, a blockchain-based solution was implemented. There has been a reduction in customer access time statements access from 45 days to a few minutes. While video games have experienced rapid growth, the financial reporting systems used to compensate game publishers rely on an old-school accounting model in which spreadsheets are used to calculate and reconcile game royalties. Xbox Finance adopted blockchain technology in 2018 to eliminate the time-consuming process of processing royalties for game publishers and to make royalty information more accessible. Having emerged as a result of the advent of blockchain technology, digital ledgers can now be more easily connected.

Due to its tamper-proof nature, the digital ledger facilitates the recording of transactions by increasing transparency and making the system more user-friendly. Blockchain-based solutions enable smart digital contracts that execute themselves as transactions occur, codifying the logic of complex royalty agreements. The embedded logic enhances everyone's confidence in fair and accurate payments [25]. By leveraging the Azure blockchain service, the Xbox team was able to manage smart contracts and govern royalty relationships with different game publishers with ease. The service simplified publisher notes management and monitored network health. Azure blockchain service significantly reduced the effort involved in infrastructure security and maintenance, allowing for higher payment efficiency than an unmanaged ledger. Due to the implementation of this technology, getting access to the game sales data by publishers became effortless, as well as the wait-time in getting access being significantly shortened. The information retrieved from the Azure blockchain service relates to which games are selling well and which games are currently trending in the market.

Using the Microsoft Azure Blockchain technology, Blockchain-as-a-Service (BaaS) service is built on top of the Microsoft Azure platform. With these capabilities, setting permissions, monitoring and maintaining the network, as well as managing member credentials are made easier. Digital content can be obtained, and royalties can be calculated instantly with the aid of smart contracts by applying encoded logic [25]. Thus, the financial transactions can be tracked regularly, making forecasting and reporting more effective, and allowing consumers to focus on their core competencies. Enterprise resource planning accounting entries can now be generated directly from a blockchain. Additionally, artificial intelligence technology is integrated into the contract creation process and is hosted on Microsoft Azure [7].

Global Shipping Business Network Oracle Blockchain-as-a-Service: CargoSmart

CargoSmart – a global transportation management system – began experimenting with a blockchain solution to improve complex supply chain processes to create digital shipment documentation across the logistics industry. By utilizing the platform, the shippers and customs offices can work together more efficiently to construct an immutable audit trail with low latency. Several processes and procedures are involved in handling documents, including outdated paper processes, a wide range of technical capabilities, and the sharing of documents via different formats such as electronic data interchange (EDI) and emails.

The CargoSmart blockchain platform is designed to streamline the shipment documentation process and minimize conflicts, reduce document preparation time and avoid additional custom penalties. Thus, the time required to consolidate, validate and manage shipping information that is duplicated in various documents is eliminated, thereby making the logistics process more efficient and accurate. In collaboration with *Oracle Blockchain Cloud Services*, CargoSmart has created the *Global Shipping Business Network* (GSBN), which helps companies efficiently create blockchain networks to conduct more reliable and secure transactions [8]. According to CargoSmart's CEO Steve Siu, '*Blockchain promotes the concept of a consortium, so we invite shippers, forwarders, carriers, and terminals to participate*' [16]. With the decentralized governance, interoperability and enterprise-grade architecture, the platform is suitable for monitoring global shipments, organizing, and tracking across all the stakeholders [8].

The GSBN platform provides a single source of information for the shipping industry, with the following characteristics:

Open and extensible: The cooperative network may develop software and connect to consortium networks to enhance data interoperability and business efficiency.

Transparency: Peer-to-peer software allows data managers to share transparent records with other shipment investors, allowing them to act quickly on important milestones and maintain goods moving through the supply chain.

Digital baseline for standards: A widely accepted, trustworthy and comprehensive digital model provides a foundation for highly collaborative initiatives and business intelligence [16].

PROPOSED BAAS PLATFORMS

To address some of the flaws inherent in the commercial Blockchain-as-a-Service systems, the BaaS systems described below have been introduced by researchers. All these proposed BaaS systems have a common goal – to maintain the principal blockchain attributes of immutability, decentralization, persistence and auditability [31, 40].

FUNCTIONAL BLOCKCHAIN-AS-A-SERVICE

Function as a Service (FaaS), also known as serverless computing, is a way to deploy applications to the cloud rapidly and flexibly. The Functional Blockchain-as-a-Service model is based upon Function as a Service. The Functional BaaS model is a serverless architecture which, when compared to the traditional BaaS model, has several advantages. Furthermore, the robustness and hierarchical nature of the Function-as-a-Service (FaaS) model can be easily adapted to the FBaaS model. The architecture consists of an infrastructure layer composed of an operating system and a physical cluster built on AWS without using Amazon lambda functions; a component layer which implements basic functions such as authentication and authorization; a service layer which implements major functionalities such as storage and transactions; and a business logic layer which provides extensive and complex services [9].

NUTBAAS

The NutBaaS platform offers blockchain-related services through cloud computing environments. These services include monitoring and network deployment, smart contract testing and smart contract analysis. A NutBaaS system enables developers to concentrate on business logic. The name 'nut' refers to the provision of a 'hard' barrier (like that of a nutshell) to protect blockchain developers' applications. NutBaaS is composed of the resource, service application layer and business layer. Implementation of blockchain services typically requires the implementation of the resource layer; i.e., storage, databases, and networks. At the core of NutBaaS architecture is the NutBaaS service layer, which implements the majority of all blockchain-based fundamental and complex services. While the application layer brings the possibility of building individual applications to the technology, the business layer focuses on exploring more scenarios suitable for leveraging the technology to create additional business outcomes [40].

Full-Spectrum Blockchain-as-a-Service

FSBaaS integrates private and consortium blockchains, and is built on *Blockchain Lite*, a centralized private cloud-based blockchain platform, and Hyperledger Fabric, a distributed, open and secure blockchain infrastructure. Instead of restricting users to only one type of blockchain system, FSBaaS allows users to combine the best features from both a private blockchain system and a large consortium network. The Full-Spectrum BaaS architecture is comprised of blockchain tenants, blockchain nodes, business networks and business network solutions. The platform provides a flexible way for tenants to access both blockchain runtimes with a unified interface for programming principles and RESTful APIs [31].

Novel Blockchain-as-a-Service

Novel BaaS (NBaaS) is a proposed model which assists in improving the limitations related to PaaS-based BaaS. There are several different blockchain components in NBaaS, including peers, network members, transaction ordering by orders and the client part of the blockchain, the place where users operate on the network. Defining deployable components is essential to connecting a blockchain instance [31]. As part of NBaaS, deployable components can be deployed in a way that facilitates auditability and decentralization, however, NBaaS cannot prevent collusion between tenants and providers [40].

Unified Blockchain-as-a-Service

The uBaaS solutions are vendor-neutral, meaning that they are not tied to any specific cloud infrastructure provider. By using a front-end user interface that communicates with the back-end services through an application programming interface (API) gateway, a user can build and configure blockchain-based applications. Additionally, uBaaS supports deployment-as-a-service, which eliminates the need to lock into specific cloud platforms; design patterns, which help design smart contracts; and, data management to address the scalability and security challenges of blockchains [31].

Public Blockchain-as-a-Service

As a platform designed to provide a secure environment focusing on privacy and distributed equality, *Versus* has developed Public Blockchain-as-a-Service (PBaaS). By using PBaaS, any project can start its blockchain component by paying the quoted amount instead of notarization expenses to have hash power and cross-chain transactions. Each blockchain is independent and can be customized. Cross-chain transactions are an important component of growing the whole network, as they allow activity to flow freely between chains to reduce congestion. To create a chain, one must have a *VersusID*. Several blockchains of Versus will allow the exchange of popular currencies and tokens so that the capabilities are not limited to a single blockchain [34].

Although the above-proposed BaaS systems have similar goals, there is no overlap in the challenges they address. The primary objective of these studies was to make BaaS systems universal and general, rather than for their performance [31].

BAAS GOVERNANCE

Governance problems should be easier to resolve when there are fewer parties involved. A key aspect of governance in a BaaS context relates to the participants, as outlined in the following section.

PERMISSIONED CHAINS

BaaS applications typically involve private blockchains, dedicated to a particular use or application, and/or permissioned blockchains, restricted to a set of participants. Distributed ledger technology is not always necessary, especially where the activity is more centralized or if certain entities have more administrative authority. A database may be sufficient as auditing the database logs can provide acceptable assurance. Further assurance can be provided by immutability and integrity constraints – for instance, *secure audit34* and *forward integrity35* have been developed for append-only data stores, sufficient for auditing and tracking applications.

The use of distributed ledger technology (DLT) makes more sense in situations involving many participants, such as a larger business networks. Distributed ledgers (DLs) can be useful when parties plan to supervise each other's activities, or when parties wish to have some degree of autonomy, such as defining individual interests and functionality. DLTs may be appropriate or inappropriate depending on the risk appetite, trust concerns and balance of power of the participants. The major cloud service providers offer BaaS components in addition to more general cloud services.

OFF-CHAIN CONTROL

A key aspect of the governance of DL is who determines its functioning. BaaS consists of several parties – which include the service provider and tenants – each of whom has a different interest and incentive. SaaS services that resemble BaaS services involve the BaaS provider significantly more in establishing the DLT infrastructure. Tenants of BaaS may wish to consider whether the provider can unilaterally change the software; the client's ability to fork the provider's software for reuse; how the tenants, provider or other participants alter the ledger; and finally, whether the provider provides a distributed ledger running on another blockchain run by another community. Generally, tenants of BaaS will have greater control over their DLT infrastructure when their situation resembles that of PaaS [30].

SUMMARY AND CONCLUSIONS

This chapter has presented cloud computing concepts in order to enable readers to get a deeper understanding of the Blockchain-as-a-Service (BaaS) cloud infrastructure. Many large entities such as IBM, Microsoft, Amazon, SAP and Oracle provide organizations the capability to develop blockchain applications on the cloud. As such, deploying BaaS does not require significant technical knowledge and infrastructure, and scalability is achieved as the business grows. A brief discussion of several business use cases in the field of finance and supply chains were introduced to help readers better understand the underlying architectures of some commercial BaaS platforms.

CORE CONCEPTS

- Cloud computing describes the use of the Internet and its infrastructure. Cloud computing comprises five key attributes, four deployment models and three service models.
- The five essential characteristics of cloud computing are – on-demand service, broad network access, resource pooling, rapid elasticity and measured service. A cloud system is a pay-as-you-go system. Cloud computing enables customers to scale upward or downward automatically and rapidly based on demand. Broad network access enables cloud customers to utilize cloud services without experiencing bandwidth bottlenecks. Resource pooling makes it possible for a cloud to remain financially viable through significant capital investments and revenue from a large base of customers.
- A cloud deployment model can be classified into four categories – public, private, hybrid and community models. The public cloud is a cloud where all of the major resources are available to a public that is willing to pay for the service, while a private cloud is for the exclusive use of an organization. A hybrid cloud is a mix of both private and public. A community or affinity cloud platform is often shared across community groups with similar business objectives.
- Cloud computing offers Infrastructure-as-a-Service (IaaS), Platform-as-a-Service (PaaS) and Software-as-a-Service (SaaS) services depending on the needs of the customer. Cloud service providers providing IaaS offer hardware-related resources where the customer is ultimately responsible for the logical resources. The PaaS model provides a platform for a customer to build applications. The SaaS model offers all the IaaS and PaaS features along with the specific software created entirely or in part.
- Cloud service providers, customers, users, access security brokers and regulators play a crucial role in cloud computing. The benefits of the cloud computing solution include cost savings, reducing upfront IT expenses, providing backup services that prevent data loss and its support of multitenancy makes it easy to share servers and applications among multiple users.
- The BaaS concept refers to an 'as-a-service' cloud-based solution designed to facilitate the implementation of a blockchain platform that allows users to host and manage blockchain applications from a central cloud.
- An organization can set up a BaaS platform by subscribing to a blockchain service provided by a service provider based on any distributed ledger technologies, such as Ethereum, Hyperledger Fabric, R3 Corda, Chain Core or Quorum. A service-level agreement should be in place to assure that the blockchain infrastructure is developed according to a customer's specifications.
- BaaS offers several benefits, including cost and time savings, a reduction of personnel costs, ease of use, pricing and customer support. Privacy and data compliance, limited options and a lack of resources are some of the challenges that BaaS is currently facing.
- The Blockchain-as-a-Service architecture is divided into four layers – the application layer, middleware, framework and infrastructure layer. BaaS services typically implement Crash Fault Tolerance (CFT) or Byzantine Fault Tolerance (BFT) consensus mechanisms.
- Currently, there are several BaaS providers in the marketplace. Some of the most popular commercial providers of BaaS that can be integrated with an enterprise include IBM BaaS, Oracle, Microsoft Azure BaaS, Amazon AWS BaaS, Alibaba, Accenture, Baidu and Huawei. When choosing a BaaS service provider, the following factors should be considered: previous experience, ease of use, quality commitment, integration and process control.
- Blockchain technology researchers are proposing several non-commercial BaaS platforms, namely, Functional BaaS, NutBaaS, Full-Spectrum BaaS, Novel BaaS, Unified BaaS and

Public BaaS, in order to maintain the key blockchain attributes of immutability, decentralization, persistence and auditability.

- BaaS governance directs the participants who are involved in the BaaS platform. There are two main types of parties involved – permissioned or private chain and off-chain control.

ACTIVITY FOR BETTER UNDERSTANDING

Readers interested in learning more about the use of a BaaS platform are encouraged to visit the following website: https://which-50.com/nestle-pilots-blockchain-for-coffee-provenance/ and watch the video for a better understanding of the Nestle chain of origin.

After exploring the topics related to Blockchain-as-a-Service (BaaS) in this chapter, the following questions are designed to provide the readers with more insight into this technology.

1. Assuming that you were interested in adopting a blockchain solution in your organization, would your organization be better off with designing a blockchain use case or is it more cost-efficient and effective to opt for a BaaS offering?
2. What criteria would you use in the evaluation of a BaaS solution for your organization?
3. What are the security risks associated with Blockchain-as-a-Service, including geographical location, transactions, etc. that need to be considered when implementing a BaaS platform?
4. What factors contribute to the scalability of the BaaS?
5. What are some possible migration challenges when moving from one BaaS platform to another?

REFERENCES

[1] Alibaba. (n.d.). *What Is BaaS? Product Introduction.* Alibaba Cloud Documentation Center. Retrieved from https://www.alibabacloud.com/help/doc-detail/85263.htm?spm=a2c63.l28256.a3.1.764114a4XX73Xa

[2] Anwar, H. (2019, April 22). *Blockchain as a SERVICE: Enterprise-Grade Baas Solutions.* 101 Blockchains. https://101blockchains.com/blockchain-as-a-service/#5.

[3] Bairagi, S. I., & Bang, A. O. (2015). Cloud Computing: History, Architecture, Security Issues. *International Journal of Advent Research in Computer and Electronics (IJARCE) (E-ISSN: 2348-5523) Special Issue National Conference "CONVERGENCE 2015."* https://www.researchgate.net/publication/323967455_Cloud_Computing_History_Architecture_Security_Issues.

[4] Bhagat, V. (2020, September 8). *What Is Blockchain-as-a-Service and Its Business Benefits?* Techiexpert. Com. https://www.techiexpert.com/what-is-blockchain-as-a-service-its-business-benefits/

[5] *Blockchain-as-a-Service (BaaS) Market Size, Share and Trends, 2027.* (n.d.). Fortune Business Insights. Retrieved from https://www.fortunebusinessinsights.com/blockchain-as-a-service-baas-market-102721

[6] *Blockchain Service (BCS)_Blockchain Platform-HUAWEI CLOUD.* (n.d.). Huawei. https://www.huaweicloud.com/en-us/product/bcs.html

[7] Bourne, J. (2020b, December 16). *Xbox to Use Blockchain for Gaming Royalties with Microsoft and EY Collaboration.* The Block. https://blockchaintechnology-news.com/2020/12/xbox-to-use-blockchain-for-gaming-royalties-with-microsoft-and-ey-collaboration/

[8] CargoSmart. (n.d.). *CargoSmart Launches Blockchain Initiative to Simplify Shipment Documentation Processes – CargoSmart.* ArgoSmart Launches Blockchain Initiative to Simplify Shipment Documentation Processes. Retrieved from https://www.cargosmart.com/en/news/cargosmart-launches-blockchain-initiative-to-simplify-shipment-documentation-processes.htm

[9] Chen, H., & Zhang, L. J. (2018). FBaaS: Functional Blockchain as a Service. *Lecture Notes in Computer Science*, 243–250. https://doi.org/10.1007/978-3-319-94478-4_17

[10] Chuanyi Liu, Guofeng Wang, Peiyi Han, Hezhong Pan, & Binxing Fang. (2017). A Cloud Access Security Broker Based Approach for Encrypted Data Search and Sharing. *2017 International Conference on Computing, Networking and Communications*, 2–7. https://doi.org/10.1109/iccnc.2017.7876165

[11] Crepax, T., & Rao, S. P. (2020). Blockchain in the Cloud: A Primer on Data Security for Blockchain as a Service (BaaS). *SSRN Electronic Journal*. https://doi.org/10.2139/ssrn.3766900
[12] Dhillon, V., Metcalf, D., & Hooper, M. (2021). *Blockchain Enabled Applications: Understand the Blockchain Ecosystem and How to Make It Work for You* (2nd ed.). Apress. https://learning.oreilly.com/library/view/blockchain-enabled-applications/9781484265345/html/430562_2_En_14_Chapter.xhtml
[13] D'Mello, Y. (2018, January 19). *Baidu Has Launched Its Own Blockchain-as-a-Service (BaaS) Platform*. AiThority. https://aithority.com/technology/blockchain/baidu-has-launched-its-own-blockchain-as-a-service-baas-platform/
[14] Fritz, J., & Hamel, B. (2019). AWS. Enterprise Solutions with Blockchain: Use Cases from Nestlé, Sony Music, and Workday, 1–71. https://d1.awsstatic.com/events/reinvent/2019/Enterprise_solutions_with_blockchain_Use_cases_from_Nestle_Sony_Music_and_Workday_BLC204.pdf
[15] Gupta, R. P. (2018, December 25). *SAP HANA Blockchain Service – Explained in Simple Words*. SAP Blogs. https://blogs.sap.com/2018/11/15/sap-hana-blockchain-service-explained-in-simple-words/
[16] Hall, M. (2018, November 18). Oracle Blogs. https://blogs.oracle.com/blockchain/post/cargosmart-leads-global-shipping-consortium-formation-built-on-blockchain-technology
[17] Hogan, M., Hogan, M., Liu, F., Sokol, A., & Tong, J. (2013, May 24). *NIST Cloud Computing Standards Roadmap*. US, NIST, Department of Commerce. Retrieved January 26, 2021, from https://www.nist.gov/system/files/documents/itl/cloud/NIST_SP-500-291_Version-2_2013_June18_FINAL.pdf
[18] IBM. (n.d.). *IBM Cloud Docs*. Which IBM Blockchain Platform Offering Is Right for Your Business? Retrieved 2021, from https://cloud.ibm.com/docs/blockchain/index.html
[19] Kernahan, A., Bernskov, U., & Beck, R. (2021). Blockchain Out of the Box: Where Is the Blockchain in Blockchain-as-a-Service? *Proceedings of the 54th Hawaii International Conference on System Sciences*. https://doi.org/10.24251/HICSS.2021.520
[20] Kilroy, K. (2019). Blockchain as a Service. Retrieved February 2021 from https://www.oreilly.com/library/view/blockchain-as-a/9781492073475/
[21] Kushida, K. E., Murray, J. & Zysman, J. (2011). Diffusing the Cloud: Cloud Computing and Implications for Public Policy. *Journal of Industry, Competition and Trade*, 209–237. Retrieved from https://doi.org/10.1007/s10842-011-0106-5
[22] Li, X., Zheng, Z., & Dai, H. N. (2021). When Services Computing Meets Blockchain: Challenges and Opportunities. *Journal of Parallel and Distributed Computing*, *150*, 1–14. https://doi.org/10.1016/j.jpdc.2020.12.003
[23] Malisow, B. (2017). *CCSP (ISC)2 Certified Cloud Security Professional Official Study Guide* (2nd ed.). Hoboken, NJ: Sybex. Retrieved January 26, 2021, from https://learning.oreilly.com/library/view/isc2-ccsp-certified/9781119603375/
[24] Microsoft. (2019). *Shared Responsibility for Cloud Computing*. White Paper. https://azure.microsoft.com/mediahandler/files/resourcefiles/shared-responsibility-for-cloud-computing/Shared%20Responsibility%20for%20Cloud%20Computing-2019-10-25.pdf
[25] Microsoft. (n.d.-b). *Xbox Game Publishers Access Royalties Statements Even Faster Now That Microsoft Uses Azure Blockchain Service*. Microsoft Customers Stories. Retrieved 2021 from https://customers.microsoft.com/en-us/story/microsoft-financial-operations-professional-services-azure
[26] Müller, S. D., Holm, S. R., & Søndergaard, J. (2015). Benefits of Cloud Computing: Literature Review in a Maturity Model Perspective. *Communications of the Association for Information Systems*, *37*, 1–15. https://doi.org/10.17705/1cais.03742
[27] Onik, M. M., & Miraz, M. H. (2019). Performance Analytical Comparison of Blockchain-as-a-Service (BaaS) Platforms. *Lecture Notes of the Institute for Computer Sciences, Social Informatics and Telecommunications Engineering*, 3–18. https://doi.org/10.1007/978-3-030-23943-5_1
[28] P. (2021, March 15). *Azure Blockchain Service Overview – Azure Blockchain*. Microsoft Docs. https://docs.microsoft.com/en-us/azure/blockchain/service/overview
[29] Shi, J. (2019). The Application of Blockchain-as-a-Service (BaaS) and Its Providers in China. *Proceedings of the Fourth International Conference on Economic and Business Management*, 127–130. https://doi.org/10.2991/febm-19.2019.28
[30] Singh, J., & Michels, J. D. (2018). Blockchain as a Service (BaaS): Providers and Trust. *2018 IEEE European Symposium on Security and Privacy Workshops*, 67–74. https://doi.org/10.1109/eurospw.2018.00015

[31] Song, J., Zhang, P., Alkubati, M., Bao, Y., & Yu, G. (2021). Research Advances on Blockchain-as-a-Service: Architectures, Applications and Challenges. *Digital Communications and Networks*. https://doi.org/10.1016/j.dcan.2021.02.001

[32] Takyar, A. (2020, November 5). *Blockchain as a Service (BaaS) Guide for Beginners*. LeewayHertz – Software Development Company. https://www.leewayhertz.com/guide-to-blockchain-as-a-service/

[33] Technology, S. (2018, June 28). *Positioning SupplyBloc Technology as a Blockchain (Platform) as a Service (BaaS/PaaS)*. Medium. https://medium.com/supplybloc/positioning-supplybloc-technology-as-a-blockchain-platform-as-a-service-baas-paas-b5fd1453b148

[34] Verus. (n.d.). *Truth and Privacy for All. Community Driven Open Source Cryptocurrency*. Retrieved 2021 from https://verus.io/technology/PbaaS

[35] Wang, L., Ranjan, R., Chen, J., & Benatallah, B. (2017b). *Cloud Computing*. Amsterdam University Press. https://learning.oreilly.com/library/view/cloud-computing/9781439856420/xhtml/13_Chapter01.xhtml#ch1-5

[36] Wheeler, M. (2019, December 10). *Nestlé Using Blockchain to Authenticate Coffee Origin*. Food & Beverage Industry News. https://www.foodmag.com.au/nestle-using-blockchain-to-authenticate-coffee-origin/

[37] Xue, C. T. S., & Xin, F. T. W. (2016). Benefits and Challenges of the Adoption of Cloud Computing in Business. *International Journal on Cloud Computing: Services and Architecture*, *6*(6), 1–15. https://doi.org/10.5121/ijccsa.2016.6601

[38] Yasrab, R. (2018). Platform as a Service(PaaS). *The Next Hype of Cloud Computing*, 1–21. https://arxiv.org/ftp/arxiv/papers/1804/1804.10811.pdf

[39] Yussupov, V., Falazi, G., Breitenbücher, U., & Leymann, F. (2020). On the Serverless Nature of Blockchains and Smart Contracts. https://arxiv.org/abs/2011.12729v1

[40] Zheng, W., Zheng, Z., Chen, X., Dai, K., Li, P., & Chen, R. (2019). NutBaaS: A Blockchain-as-a-Service Platform. *IEEE Access*, *7*, 134422–134433. https://doi.org/10.1109/access.2019.2941905

Index

Note: Page numbers in *italic* denote figures and those in **bold** denote tables.

A

Accenture 276, **278**
Access control 18, 19, 35–36, 43–44, 246–247, *247*
Accounting
  double-entry 64–65
  single-entry 63
  triple-entry 64–66, 72, **73**
Address derivation 30–31, *31*
Address Space Layout Randomization (ASLR) 234, 235, **237**
Adjournment periods 228
Advanced Encryption Standard (AES) 23, 24, **26**, 87
Agile methodology 151
Alibaba Cloud 276, **278**
Altcoins 56
Alternative trading system (ATS) 80
Amazon Web Services (AWS) 275, 277–279, **278**
Ancile 133–134
Anonymity 11, 18
Anti-Money Laundering (AML) regulations 81–82, 181
API testing 157–158
Application auditing 165, **166–167**
Application configuration changes 163
Application design 127–137
  approaches 133–134
  considerations 127–131, 136, **136**
  design process 134–136, *135*
  fictional case study 131–133
  functional requirements 132
  personas 131–132
  technical requirements and tasks 132–133
  user stories 132
Application development 141–152
  architectural layers 144–146
  firmware versus software 141–142, **142**
  frameworks 142–144
  good practices 149–151, *150*
  Integrated Development Environment (IDE) 149
  tools 146–148, **148**
Application development risks **206**, 208–209
Application layer 145, 271, *271*
Application testing 151, 155–164
  API testing 157–158
  automation testing 163–164
  bug management 158–161, **161**
  challenges 156
  functional testing 157–158
  integration testing 157
  load testing 158
  mutation testing 158
  node testing 158
  opportunities to enhance strategies 163
  performance testing 158
  phases 157
  planning 161–162
  regression testing 157, 163
  smart contract testing 158
  test management 164
  tools 156, 161, **161**, **162**
  unit testing 157
  user acceptance testing 164
Architectural layers 144–146, 271–272, *271*
Arithmetic over/underflow attacks 247–248, *248*
Artificial difficulty increases 231
Artwork tokenization 55
Asset management 179–180
Asset misappropriation fraud 68, 96
Association of Certified Fraud Examiners (ACFE) 66–68, 175
Assurance 181
Asymmetric cryptography 25, **27**
Attacks 205, 207, 223–241
  51% attacks 227, 228, 229–231, 234, **237**
  buffer out of bounds attacks 235, **237**
  buffer overflow attacks 233–234, **237**
  denial of service (DoS) attacks 40, 174, 227, 228, 231, **237**, 249–250, *250*
  double-spending attacks 68–69, 229, 230
  eclipse attacks 231–232
  exchange wallet attacks 69
  on hash functions 31, 32
  on identity management 36
  malware 69, 224–227, **236**
  network-level 223, 229–233, **237**
  node-level 223, 227–228, **236**
  race condition attacks 235–236, **237**, 251–252, *251*
  ransomware 226–227
  replay attacks 232, **237**
  routing attacks 232–233
  social engineering attacks 69
  Sybil attacks 233, **237**
  system-level 223, 233–236, **237**
  time stamp attacks 234–235, **237**
  Turing completeness and 42
  user-level 223, 224–227, **236**; *see also* Smart contract attacks
Attesting 110
Automation testing 163–164
Automobile insurance 86–87
Azure BaaS platform 274–275, **278**, 279

B

Backups 173
Bad randomness attacks 250–251, *250*
Baidu 276, **278**
Bancor 252

Banking payments systems 78–79
Banking sector 78–79, 115
Baran, Paul 7
Batavia platform 78
Benevolent dictator for life 50
Billing fraud 68, 95–96
Binance 59
Birthday paradox 31
Bitcoin 2, 6, 42, 50, 56–57
  attacks on 229–230, 233–234
  cryptography **26**, **29**
  exchanges 58
  performance bugs 160
  triple-entry accounting 64
  wallets 39
Bitcoin address generation 32
BitcoinJ 156, **162**
Bitcoin mining 32
Bitfinex 59
Bittrex 58
Blacklisting 228
Blob file type 40
Blockchain-as-a-Service (BaaS) 263, 267–284
  advantages 268
  architecture 270–272, *271*
  challenges 268–269
  consensus 272–273
  defined 267–268
  governance 282
  interest in 269
  versus Platform as a Service (PaaS) 270
  platforms 274–277, **278**, 280–282
  versus serverless computing 270
  server-side capabilities 270
  service partner selection 273
  use cases 277–280
Blockchain teams 50
Blockchain technology
  attributes 10
  benefits 11
  defined 1–2
  history 2–3
  versus traditional databases 3, **3**, *4*, 127–128
  types 5–6, **6**, 18–19, **19**
Block ciphers 24
Block data 9, *10*, 30
Block forger DoS attack 231
Block gas limit 249
Block headers 9, *10*, 30
Block metadata 9, *10*
Block structure 9, *10*
Block stuffing 249
Blowfish algorithm 24, **26**
Bogatty, Ivan 252
Broad network access 264
Brute-force attacks 31
Buffer out of bounds attacks 235, **237**
Buffer overflow attacks 233–234, **237**
Bug management 158–161, **161**
Business continuity 171–174, 207
Buterin, Vitalik 2, 50, 105
Byzantine failure **208**
Byzantine Fault Tolerance (BFT) 5, 8, 57, 58, 230, 272
Byzantine Generals Dilemma (BZD) 3–5, *5*

C

CargoSmart 279–280
Cargo tokenization 55
Car insurance 86–87
Car Insurance Policy Framework (CAIPY) 87
Cash larceny fraud 67
Castro, Miguel 22
Centralized networks 7, *7*, 130
Certificate authorities 18, 34, 37
Chaining blocks 10
ChainLocks 230, **237**
Change enablement 183; *see also* COBIT 2019 implementation
Checkpointing 228, **236**
Check tampering fraud 67
Ciphertext 23, 24
Claimed signatories 34
Closed blockchains 19, **19**
Cloud access security brokers (CASBs) 267
Cloud computing 264–267
  attributes 264
  benefits 267
  deployment types 264–265
  roles and responsibilities 266–267
  service models 265–266, **266**
  shared responsibilities 266, **266**
Cloud customers 267
Cloud service providers (CSPs) 266–267
Cloud storage 174
Cloud users 267
Cloud wallets 224
COBIT 2019 implementation 171, 182–193, *184*, *185*, **215–219**
  phase 1 - What Are the Drivers? 183–184, **186**, **187**
  phase 2 - Where Are We Now? 184, **188**, **189**
  phase 3 - Where Do We Want to Be? 184–185, **190**, **191**
  phase 4 - What Needs to Get Done? 185, 192, **192–193**, **194**
  phase 5 - How Do We Get There? 193, **195**, **196**
  phase 6 - Did We Get There? 193, **197**, **198**
  phase 7 - How Do We Keep the Momentum Going? 193, **199**, **200**
Coinbase 58
Cold storage 38
Collaboration 129, 179
Collins, Reeve 58
Collision resistance 31
Commit file type 40
Communication plans 173
Community clouds 265
Completeness 28
Concurrency bugs 159
Confidentiality 18, 23
Configuration bugs 159

Consensus 6, 10, 20–23, *20*, 42, 180
attack mitigation 230, 233
auditing 165, **166**
Blockchain-as-a-Service (BaaS) 272–273
Corda 84, 119, 120, **122**, **145**
Ethereum 109–110, **122**, 143, **145**
Hyperledger 115, **122**, 143, **145**, 272
versus permissions 127–128
Quorum 144, **145**
risks 208, **208**
Consistency 129–130
Consortium-type blockchains 6
Construction industry 120–121
Consumer credit sector 82–84
Continual improvement 183; *see also* COBIT 2019 implementation
ContractLarva 256, **257**
Contract management 174–177, *176*, *177*
Control flow analysis 254, **255**, **257**
Corda 116–121, **122**, 144, **145**
components 117–119, *118*
consensus 84, 119, 120, **122**, **145**
governance 50
limitations 121
security measures 239–240, **240**
smart contracts 119–120, **122**, 240
use cases 84–85, 120–121
user privacy 121
when to choose *123*
CorDApps 117
Corda Testing Tool 156, **162**
Core developers 49, 50
Core dev team governance 50
COSO framework 213–214, **215–219**
Counterfeit goods 95, 178
Counterparty 58
COVID-19 pandemic 77, 85, 178
Crash failure **208**
Crash Fault Tolerance (CFT) 272–273
Credit cards 82–83
Cryptoanalysis 23
Cryptocurrencies 11, 49, 56–59, 144
in banking sector 79
characteristics 56
creation by forks 50–52
exchanges 58–59
risks **206**, 209
types 56–58
volatility 209 *see also* Tokens
Cryptography 23–28, 43
asymmetric 25, **27**
elliptic curve 28, **29**, 232
homomorphic 28, **29**
hybrid 25–28
private-key 224
public-key 25, **27**, 210, 224
symmetric 24–25, **26**
zero-knowledge proof 28, **29** *see also* Hash functions
Cryptosystem 23
Crypto wallets 37–39, *38*, *39*, 44, 69, 247

D

D'Agosta, Tristan 58
DART software 257, **257**
Dash cryptocurrency 57, 230
Data authentication 23, 31
Data control 180
Data Encryption Standard (DES) 24
Data integrity **3**, 23, 173–174
Data migration 182
Data protection 133–134, 206, 211–212
DeBeers 100
Decentralized applications (DApps) 54, 57, 147, 270
application layer 145, 271
design 128, 129–131, 132–133, 134, 136
Decentralized Autonomous Organization (DAO) attack 42, 50–51, 52, 155, 246
Decentralized finance (DeFi) 57
Decentralized networks 7–8, *8*, 130–131
Decryption 24 *see also* Cryptography
Defensive programming 41
Delaware Blockchain Initiative (DBI) 81
Delayed proof of work (dPoW) 230, **237**
Delegated Byzantine Fault Tolerance (DBFT) 23
Delegated proof of stake (DPoS) algorithm 21
Denial of service (DoS) attacks 40, 174, 227, 228, 231, **237**, 249–250, *250*
Design *see* Application design
Desktop wallets 224
Deterministic wallets 39
Developers 18
core 49, 50
scarcity of 128, 209, 269
Development *see* Application development
Development frameworks 142–144
choosing between 122–123, *123*
Quorum 55, 144, **145**
Ripple 53, 57, 144, **145** *see also* Corda; Ethereum; Hyperledger
Differential attacks 31
Diffie, Whitfield 25
Diffie-Hellman key exchange 25, **27**
Digital identities 81, 112, 129
Digital signatures 34–35, *34*, *35*, 43
Disaster recovery 171–174, 207
Distributed hash tables (DHT) 40
Distributed ledger technology (DLT) 85, 282
Distributed networks 8–9, *9*
Dogecoin 58
Domain name system (DNS) 44
Double-dipping fraud 86
Double-entry accounting 64–65
Double-spending attacks 68–69, 229, 230
Dynamic code analysis **255**, 256, **257**
DYTAN (Dynamic Taint Analysis) 255–256, **257**

E

Eclipse attacks 231–232
E. coli contamination 97
ElectroRAT malware 225

Elgamal, Taher 25
Elgamal algorithm 25, **27**
Elliptic curve cryptography 28, **29**, 232
Elliptic Curve Digital Signature Algorithm (ECDA) **29**, 35
Elliptic Curve Key Exchange (ECKE) 28
Embedded supervision principles 79
Employee security risks 207
Emulation testbeds 157
Encryption 24 *see also* Cryptography
Environmental perturbations **208**
Environment bugs 159
EOS 58, 235
Equity tokens 53
ERC-20 standard 53–54
ERC-223 standard 54
ERC-721 standard 54
ERC-777 standard 54
Error correcting code (ECC) 174
Error handling 149
Estonia 2
Ether cryptocurrency 51, 57, 106
Ethereum 2, 105–111, **122**, 143, **145**, 272
  bug issues 160
  components 106–108, *107*
  consensus 109–110, **122**, 143, **145**
  cryptography **29**
  currency 51, 57, 106
  Decentralized Autonomous Organization (DAO) attack 42, 50–51, 52, 155, 246
  gas fees 54–55
  governance 50
  integration issues 208
  limitations 110
  security measures 238, **240**
  smart contracts 41–42, 52, 106, 108–109, 110, **122**, 160, 238
  token standards 53–54
  Turing completeness 41–42, 106
  use cases 55, 87, 110–111, 176–177, *176*, *177*
  when to choose *123*
Ethereum Improvement Protocol 1559 (EIP 1559) 54–55, 251
Ethereum Tester 156, **162**
Ethereum Virtual Machine 41–42, 106, 108–109, 143, 146, 238, 247, 252, 254, **257**
Etherpot lottery 248–249
Ethers.js library 146–147
ETradeconnect platform 85
Exchange wallet attacks 69
Exonum Test Kit 156, **162**
Expense reimbursement fraud 68
Expert code analysis **255**
External consistency 130

F

Faulty coding 207
Federated Byzantine Fault Tolerance 23
51% attacks 227, 228, 229–231, 234, **237**
File System in User Space (FUSE) 40
Finality 10, 68, 79, 80, 208
Finance and accounting 95
Financial reporting 66
Financial sectors 77–89
  banking sector 78–79, 115
  consumer credit sector 82–84
  financial services sector 77–78, 116
  insurance sector 86–88
  investment management sector 80–81
  Know Your Customer (KYC) regulations 81–82, 116, 180, 181
  money laundering prevention 81–82, 116, 181
  stock exchange marketplaces 80
  trade financing sector 85–86
  use cases 81, 84–88, 115–116, 121
Financial statement fraud 66–67
Firmware, versus software 141–142, **142**
Flexible tag data-loggers (FTDs) 98
Food fraud 96–97
Food traceability 99–100, 277–279
Forced forks 52
Forecasting 94
Forks 50–52, *51*, 208
Fourth Industrial Revolution 178
Framework layer *271*, 272
Frameworks *see* Development frameworks
Fraud schemes 66–68
  asset misappropriation fraud 68, 96
  billing fraud 68, 95–96
  cash larceny fraud 67
  check tampering fraud 67
  in contract management 174–175
  counterfeit goods 95, 178
  expense reimbursement fraud 68
  financial statement fraud 66–67
  food fraud 96–97
  healthcare fraud 96
  insurance fraud 86–88
  inventory misuse and theft 68, 96
  payroll fraud 68
  purchase cycle fraud 69–71, **70**, **71–72**
  register disbursement schemes 67
  skimming fraud 67
  supply chain fraud 95–97
Fraud triangle theory 96
Front running 206
Full node wallets 37
Full-Spectrum Blockchain-as-a-Service (FSBaaS) 281
Functional Blockchain-as-a-Service 280
Functional consistency 130
Functional requirements 132
Functional testing 157–158
Function-as-a-Service (FaaS) 270, 280

G

Ganache 146, 151
Gas fees 54–55
Gas limit DoS attack 249
General Data Protection Regulation (GDPR) 211–212
General platform tokens 53
Geth 147

Ghosh, Arun 209
GHOST protocol 59, 110
Global Shipping Business Network (GSBN) 280
Gold assets tokenization 55
Good practices
  application design 136, **136**
  application development 149–151, *150*
  COSO framework 213–214, **215–219**
  purchase cycle fraud schemes 69–71, **71–72**
Gossip protocol 239
Governance 49–50, 282
GovernMental pyramid scheme 250, 253
Grigg, Ian 64
Group insurance 88

H

Hanyecz, Laszol 2
Hard fork bugs 160
Hard forks 52
Hardware wallets 38, *38*, 69, 224
Hash functions 30–32, *30*, *31*, 43
Hash rates 229–231
Healthcare
  fraud 96
  use cases 110, 120
Health insurance 87, 88
Health Insurance Portability and Accountability Act (HIPAA) 210–211
Hellman, Martin 25
Hierarchical deterministic wallets 39
Homomorphic cryptography 28, **29**
Huawei 276–277, **278**
Hybrid blockchains 6
Hybrid clouds 265
Hybrid cryptography 25–28
Hyperledger 111–116, **122**, 143–144, **145**
  benefits 112–113
  components 114–115, 143
  consensus 115, **122**, 143, **145**, 272
  design philosophy 113
  frameworks 111–112
  governance 50
  limitations 115
  security measures 238–239, **240**
  smart contracts 114, **122**, 239
  tools 113–114
  use cases 99–100, 115–116
  when to choose *123*
Hyperledger Burrow 111–112, 114
Hyperledger Caliper 113
Hyperledger Cello 114
Hyperledger Composer 113–114, 156, **162**
Hyperledger Explorer 114
Hyperledger Fabric 2, 112, 114, 115, 131, 144, **145**, 208, 238–239, 272
Hyperledger Indy 112, 115, 116
Hyperledger Iroha 112, 114, 115, 144, **145**
Hyperledger Quilt 114
Hyperledger Sawtooth 112, 114, 115, 143, **145**

I

IBM 2, 11, 78, 144, 179, 274, **278**
Identity management 36, 98, 231
Identity theft 206
Ijiri, Yuji 64
Immutability 10, 68, 127, 180, 181
Implementation
  challenges 181–182 *see also* COBIT 2019 implementation
India Trade Connect Platform 85–86
Infrastructure as a Service (IaaS) 265, 266, **266**
Infrastructure layer 146, *271*, 272
Infura 147
Initial coin offerings (ICOs) 52–53
Insider trading 206
Insurance sector 86–88
Integer overflow attacks 233–234, **237**
Integrated Development Environment (IDE) 149
Integration testing 157
Intel SGX Technology 228
Intended signatories 34
Internal consistency 130
Internal supply chains 94
Internet of Things (IoT) **29**, 111
Interoperability issues 182
Inter-planetary file systems (IPFS) 39–40, 44
Inter-planetary Name System (IPNS) 40, 44
Inventory management 94
Inventory misuse and theft 68, 96
Investment management sector 80–81
IoTA 57
IPFS 147–148
Irreversibility 56

J

JP Morgan 55, 144
Just bunch of keys (JBOK) 39

K

Key pair owners 34
Key person risks 207
'Kill switches' 41
Kim, Joyce 57
Knapsack algorithm 25, **27**
Know Your Customer (KYC) regulations 81–82, 116, 180, 181
Koblit, Neal 28
Komodo 230
Kotlin programming language 120
KPMG 165, 209
Kraken 58

L

Large state space 31
Ledger transparency risks 206, **206**
Lee, Charlie 57

Legal compliance *see* Regulatory compliance
Lemon Duck malware 226
Length extension attacks 31
Life insurance 88
Linq platform 81
Linux Foundation 2, 111, 143
Lisk cryptocurrency 235–236
Liskov, Barbara 22
List file type 40
Litecoin 57
Load testing 158
Logistics 94

M

McCabe IQ 254, **257**
McCaleb, Jed 57
McCarthy, John 264
Maersk 100, 179
MAIAN 256, **257**
Malicious transactions 227
Malware 69, 224–227, **236**
Manticore **162**, 257, **257**
Manual code analysis **255**
Marco Polo Network 86
Marketing 94
Market manipulation 206
MD4 algorithm 32
MD5 algorithm 32
Medicaid Integrity Program (MIP) 211
MedRec 110
Membership service providers (MSPs) 37, 44, 228, 231, **236**
Memory correction 174
Merged mining 230–231, **237**
Merkle, Ralph 25, 33
Merkle-directed acrylic graph (DAG) data structures 40
Merkle trees 32, 33–34, *33*, 43
Message Authentication Code (MAC) 23
Message integrity 31
Metacoins 58
MetaMask 110, 147
Microloans 83–84
Microsoft Azure BaaS 274–275, **278**, 279
Middleware layer 271–272, *271*
Miller, Victor S. 28
Misappropriation of assets 68, 96
Mislabeling fraud 97
Mist 147
Model-driven approach 134
Modular architecture 131
Monero 57
Money laundering prevention 81–82, 116, 181
Monolithic applications 131
Mucci Farms 97
Multitenancy 267
Musk, Elon 58
Mutation testing 158
Mutual funds 80–81
Mythril **161**, 256, **257**

N

Nakamoto, Satoshi 2, 64
National Association of Securities and Dealers Automated Quotations (NASDAQ) 81
National Institute of Standards and Technology (NIST) 24
National Security Agency (NSA) 32
Neo cryptocurrency 23, 58
Nestlé 277–279
Network administrators 18
Network layer 145–146
Network-level attacks 223, 229–233, **237**
Node additions 163
Node-level attacks 223, 227–228, **236**
Node operators 49–50
Nodes unreachable event 163
Node testing 158
Nonce 9, *10*, 30, 43
Nondeterministic wallets 39
Non-locality 31
Non-repudiation 24
Notary services 119
Novel Blockchain-as-a-Service (NBaaS) 281
NutBaaS platform 280–281

O

Object-Oriented Programming (OOP) 134
Off-chain governance 50, 282
Off-site storage procedures 173
Omission failure **208**
On-chain governance 50
On-chain public data 150–151
On-demand service 264
Online wallets 38
Ontology-driven approach 134
Open access 18, 130–131
Open blockchains 19, **19**
Open collaboration 179
Open governance 50
Operational risks **206**, 207–208, **208**
Oracle 274, **278**
Oracles 119, 148
Ordering-based consensus (OBC) algorithm 22
Oyente **161**, 257, **257**

P

Pacioli, Luca 64
Paper wallets 38, **39**, 69, 224
Parity Multisig bug 160
Parity wallets 247
Participants 17–18, 42
Password verification 31
Payment Card Industry Security Standard Council (PCI DSS) 209–210
Payments
  push versus pull 149–150, *150*
  risks **206**, 209
Payment service providers (PSPs) 79
Payment tokens 53
Payroll fraud 68

PCASTLE malware 225–226
Peanut Corporation of America (PCA) 97
Penalty system for delayed block submissions 230, **237**
Performance bugs 160
Performance testing 158
Permissioned blockchain membership services provider DoS attack 231
Permissioned chains 282
Permissionlessness 56
Permissions 127–128
Persistence analysis 52
Personal Information Protection and Electronic Documents Act (PIPEDA) 213
Personally identifiable information (PII) 115, 180
Personas 131–132
Phishing 69, 172, 226
Pierce, Brock 58
Pigeonhole Theorem 31
PirlGuard system 230, **237**
Plaintext 23, 24
Platform as a Service (PaaS) 265, 266, **266**
Poloniex 58
Populus **162**
Position Independent Executables (PIE) 234, **237**
Practical Byzantine Fault Tolerance (PBFT) 10, 22
Pre-image resistance 31
Privacy 18, 129, 182, 206
Private blockchains 6, **6**, 18–19, **19**
Private clouds 265
Private cryptocurrencies 56
Private-key cryptography 224
Processing nodes 18
Product distribution 178, 277–280
Production 94
Program management 183 *see also* COBIT 2019 implementation
Progressive web apps (PWAs) 128–129
Proof of activity (PoACT) algorithm 22
Proof of adjourn (PoAj) 228
Proof of authority (PoA) algorithm 22
Proof of burn (PoB) algorithm 21
Proof of capacity (PoC) algorithm 22
Proof of deposit (PoD) algorithm 22
Proof of elapsed time (PoET) algorithm 21, 115
Proof of importance (PoI) algorithm 21, 22
Proof of knowledge 28
Proof of stake (PoS) algorithm 6, 10, 21, 110
Proof of storage (PoS) algorithm 22
Proof of Weak Hands (POWH) coin 248
Proof of work (PoW) algorithm 6, 10, 20–21, *20*, 30, 43, 110, 230
Provenance 10
Pseudonymity 56
Public Blockchain-as-a-Service (PBaaS) 281
Public blockchains 6, **6**, 18, 19, **19**
Public clouds 265
Public cryptocurrencies 56
Public-key cryptography 25, **27**, 210, 224
Pull payment transactions 149–150, *150*
Purchase cycle fraud schemes 69–71, **70**, **71–72**
Push payment transactions 149–150, *150*

Q

Quorum 55, 144, **145**, 272, 273

R

Race condition attacks 235–236, **237**, 251–252, *251*
RAID (Risk, Action, Issues and Decision) model 163
Ransomware 226–227
RC4 algorithm 23, 25, **26**
Real estate tokenization 55
Reentrancy attacks 246, *246*
Register disbursement schemes 67
Registered investment advisors (RIAs) 81
Regression bugs 160
Regression testing 157, 163
Regulators 18, 267
Regulatory compliance 181, 209–213
  General Data Protection Regulation (GDPR) 211–212
  Health Insurance Portability and Accountability Act (HIPAA) 210–211
  Payment Card Industry Security Standard Council (PCI DSS) 209–210
  Personal Information Protection and Electronic Documents Act (PIPEDA) 213
Regulatory compliance risks **206**, 209
Remix 148, 161
Replay attacks 232, **237**
Research and development (R&D) 94
Resistance to change 181
Resource pooling 264
Reverse supply chains 94, 98
Ripple 53, 57, 144, **145**
Risk assessment 173, 214
Risks 205–209
  application development risks **206**, 208–209
  cryptocurrency and payment risks **206**, 209
  ledger transparency risks 206, **206**
  operational risks **206**, 207–208, **208**
  regulatory compliance risks **206**, 209
  security risks 206–207, **206**
Rivest, Ronald 25, 28, 32
Rivest Adi Adleman (RSA) algorithm 25, **27**, 28
RLPX 145
RobbinHood ransomware 226
Roles 17–18, 42
Roll Your Own Mechanism protocol 146
Routing attacks 232–233
Royal Mint Gold 53
Royalties information 279
Rule of code 50

S

Sales 94–95
Salesforce 174
Salmonella contamination 97
SAP Blockchain Services 277, **278**

Satoshi consensus 230, 231
Scalability issues 18, **122**
Schneider, Bruce 24
Scrypt mining 234
Second pre-image resistance 31
Secure Sockets Layer (SSL) certificates 32
Securify **161**, 256, **257**
Securities and Exchange Commission (SEC) 57, 80, 81
Security
  design for 129
  enhanced 11, 99
  use cases 180–181
Security bugs 160
Security failure **208**
Security risks 206–207, **206**
Security tokens 53
Selfish mining 51
Self-sovereign identity (SSI) model 36, 44, 116
Sellars, Craig 58
Semantic bugs 160
Semantic layer 145
Serverless computing 270, 280
Service layer 145
Service level agreements (SLAs) 173
Service start-ups 163
SHA-256 algorithm 32
Shared ledgers 9–10
Shared vulnerability 227, **236**
Shipping 279–280
ShoCard identity management 37
Short-address attacks 252–253, *252*
Signature authentication 31
Simplified Byzantine Fault Tolerance (SBFT) 22
Single-entry accounting 63
Skimming fraud 67
Skuchain EC3 network 85
Slither 161
Smart agreements 117
Smart Billions lotteries 251
Smart B/L 55
SmartCheck **161**, 257, **257**
Smart contract attacks 245–254
  access control attacks 246–247, *247*
  arithmetic over/underflow attacks 247–248, *248*
  bad randomness attacks 250–251, *250*
  Decentralized Autonomous Organization (DAO) attack 42, 50–51, 52, 155, 246
  denial of service (DoS) attacks 249–250, *250*
  race condition attacks 251–252, *251*
  reentrancy attacks 246, *246*
  short-address attacks 252–253, *252*
  timestamp dependency attacks 253–254, *253*
  unchecked return value attacks 248–249, *249*
Smart contracts 40–42, 44, 54, 129
  auditing 254–257, **255**, **257**
  bugs in 160
  in contract management 175–177, *176*, *177*
  Corda 119–120, **122**, 240
  Ethereum 41–42, 52, 106, 108–109, 110, **122**, 160, 238
  fraud mitigation 67, **71–72**
  Hyperledger 114, **122**, 239
  in insurance sector 86–87, 88
  for legal compliance 181
  risks 207
  testing 158
  triple-entry accounting 64–65 *see also* Smart contract attacks
Social engineering attacks 69
Soft forks 52
Software, versus firmware 141–142, **142**
Software as a Service (SaaS) 265–266, **266**
Software engineering strategies 134
Software failure **208**
Software misconfigurations 227
Software wallets 224
Solidity programming language 109, 110, 146, 238, 246, 247, 248, 257
Soundness 28
Sovrin identity management 37
Standardization, lack of 208
State objects 84
Stellar 57–58
Stock exchange marketplaces 80
Stolen private keys 224, **236**
Storage 127
Stream ciphers 24
Subject matter experts (SMEs) 163
Supply chains 93–102
  blockchain benefits 98–99
  components 94–95
  defined 93
  fraud risks 95–97
  security risks 207
  types 94
  use cases 99–100, 111, 178–179, 277–280
Sybil attacks 233, **237**
Symbolic execution **255**, 257, **257**
Symmetric cryptography 24–25, **26**
System-level attacks 223, 233–236, **237**
System updates 227, **236**
Szabo, Nick 40

T

Taint analysis 254–256, **255**, **257**
TAJ (Taint Analysis for Java) 254–255, **257**
Talent scarcity 128, 209, 269
Tamil Nadu, India 2
Technical requirements and tasks 132–133
Technical resource limitations 182
Temporal failure **208**
Testing *see* Application testing
Tether cryptocurrency 58
Thin node wallets 37
Time of check vulnerability 235
Time of use vulnerability 235
Time stamp attacks 234–235, **237**
Timestamp dependency attacks 253–254, *253*
Timestamps 9, *10*, 30

Tipping 206
Token holders 50
Tokenization 11, 55
Tokens 11, 49, 52–54, **53**
 standards 53–54
 types 52–53
 use cases 55
Toycoins 58
Traceability
 cryptocurrencies 56, 59
 fraud mitigation 68, 98
 in supply chains 98, 99–100, 277–279
Trade financing sector 85–86
TradeLens platform 100
Training considerations 182
Transaction clearance and settlements 80
Transaction flooding 231, **237**
Transaction lifecycle 42, *43*
Transaction tear-off 119
Transaction uniqueness 84, 120
Transaction validity 84, 120
Transparency 11, 98, 128, 131, 180, 206
Tree file type 40
Triple-entry accounting 64–66, 72, **73**
Truffle 146, 156, **162**
Trusted execution environment 21, 145
Trusted third parties 129, 213
Turing, Alan 41
Turing completeness 41–42, 106
Tyra, Jason 64
TZero initiative 80

U

Unchecked return value attacks 248–249, *249*
Unexpected revert vulnerability 249, *250*
Unified Blockchain-as-a-Service (uBaaS) 281
United Arab Emirates (UAE) 2–3
Unit testing 157
Universal Modeling Language (UML) 134
Untrusted code 42
Uport identity management 36
Usability gaps 182
Use cases
 asset management 179–180
 assurance 181
 Blockchain-as-a-Service (BaaS) 277–280
 contract management 174–177, *176*, *177*
 Corda 84–85, 120–121
 data control 180
 disaster recovery planning 171–174
 Ethereum 55, 87, 110–111, 176–177, *176*, *177*
 financial sectors 81, 84–88, 115–116, 121
 healthcare 110, 120
 Hyperledger 99–100, 115–116
 legal compliance 181
 product distribution 178, 277–280
 security 180–181
 shipping 279–280
 supply chains 99–100, 111, 178–179, 277–280
 tokens 55
User acceptance testing (UAT) 164
User-centered design approach 134
User interface (UI) bugs 160
User-level attacks 223, 224–227, **236**
Users 18
User stories 132
Utility tokens 53

V

Van Saberhagen, Nicholas 57
Verge cryptocurrency 234
Verification
 cardholder identity 83
 data integrity 173–174
 digital signatures 34–35, *35*
 Know Your Customer (KYC) regulations 81
 passwords 31
 user identities 37
Visual consistency 130
Vulnerability-based scanning **255**, 256–257, **257**
Vyper programming language 109

W

Wallets 37–39, *38*, *39*, 44, 69, 224, 247
Walmart 2, 11, 99–100, 179
Whitelisting 228

X

XCP 58
XMRig mining tool 225
XRP 144

Y

Yul programming language 109

Z

Zero-knowledge proofs 28, **29**, 238
Zero-proof authorization 83
Zero-Trust Architecture (ZTA) 12, **13**, 14
Zhoa, Changpeng 59

Printed in the United States
by Baker & Taylor Publisher Services